PLANT REPRODUCTION

PLANT REPRODUCTION

By

Dr. Shubhrata R. Mishra
Department of Botany
Vikram University
Ujjain (M.P)

DISCOVERY PUBLISHING HOUSE
NEW DELHI-110002

First Published-2005

ISBN 81-7141-955-0

Published by

DISCOVERY PUBLISHING HOUSE
4831/24, Ansari Road, Prahlad Street,
Darya Ganj, New Delhi-110002 (India)
Phone: 23279245 • Fax: 91-11-23253475
E-mail:dphtemp@indiatimes.com

Printed at
Arora Offset Press
Laxmi Nagar, Delhi–92

Preface

The present title "Plant Reproduction" is the compilation of a comprehensive account of the fundamental principles of plant reproduction. It incorporates and organizes the information concerning plant reproduction from relevant and authentic sources. It presents a connected and precise account of the subject matter of this important branch of Botany which forms an integral part of the studies undertaken by the undergraduate and post graduate studies of the subject. The principles have been outlined in considerable detail with illustrations along with the fundamental facts and theories that explain the basics of plant reproduction.

This text book represents a group effort, rather than product of any single individual. Foremost on the list stand the reviewers whose advice, comments and collective wisdom helped to shape this text into a final form. Their interest in the subject, their concern for the accuracy and method of presentation, and their experience with students of widely varying abilities and backgrounds made the review process an educational experience. To these individuals, who carefully recorded their comments, opinions, and sources. The author would like to express her sincere thanks.

Author is thankful to all her colleagues, friends, botanists and educationists who have accorded their full co-operation and well wishes during the preparation of manuscript of the present title.

The author expresses her gratitute to Mr. Wasan and staff of M/s Discovery Publishing House for their whole hearted co-operation in the publication of this book.

The author would feel greatful if her attention is drawn to any errors found in the book and would welcome suggestions for improvement.

Author

Contents

1

ASEXUAL PROPAGATION

Asexual propagation involves reproduction from vegetative parts of plants and is possible because the vegetative organs of many plants have the capacity for regeneration. Stem cuttings have the ability to form adventitious roots. Root cuttings can regenerate a new shoot system. Leaves can regenerate new roots and new shoots. A stem and a root (or two stems) can be grafted together to form a continuous vascular connection when properly combined as a graft. New plants can start from a single cell. Cells of tobacco pith and carrot root in aseptic cultures have regenerated entire plants identical with the one from which the original cultures were made. Any living cell of the plant appears to have all the genetic information needed to regenerate the complete organism.

Reasons for Using Asexual Propagation

Asexual propagation reproduces *clones.* Such propagation involves mitotic cell division in which there is duplication (usually) of the complete chromosome system and associated cytoplasm from the parent cell to the two daughter cells. Consequently plants propagated vegetatively reproduce all the genetic information of the parent plant. This is why the unique characteristics of any single plant are perpetuated in the establishment of a *clone.*

The asexual process is, particularly important in horticulture because the genetic makeup (genotype) of most of the valuable fruit and ornamental cultivars is highly heterozygous, and the unique characteristics of such plants are immediately lost if they are propagated by seed. Asexual propagation is necessary to grow cultivars that produce no viable seeds, such as some bananas, figs, oranges, and grapes. Propagation of some species may be easier, more rapid, and more economical by vegetative methods than by seed. Cotoneaster seed has complex dormancy conditions; leafy stem cuttings, on the other hand, root rapidly and in high percentages. Seedlings of some species grow more slowly than rooted cuttings. Some plants grown from seed have an extended *juvenile* period and during this time the plant may not only fail to produce flowers and fruit, but may exhibit other undesirable morphological features (e.g. thorniness) that are not present when propagation is by material in the adult stage. On the other hand, it may be desirable to maintain this juvenile stage indefinitely so as to facilitate propagation of difficult-to-root cuttings.

The Clone

Many fruit and ornamental cultivars are groups of plants propagated vegetatively, starting with an individual plant, usually one grown from seed, or a part of a plant, e.g., a *bud sport.* Such a group of plants taken collectively has been given the name *clone.* A *clone* can be defined as "genetically uniform material derived from a single individual and propagated exclusively by vegetative means, such as cuttings, divisions, or grafts". A discovery with any individual member of the clone, e.g., a method of propagation, a certain cultural practice, a method of disease control, or a cross-pollination requirement, applies in the same manner to all other members of that clone.

When a member plant of a certain clone requires cross-pollination to set fruit, it must be pollinated by a plant of a different clone, rather than by a different plant of the same clone, which in reality is just another part of the same plant. Many clones of horticultural interest have been discovered and perpetuated by man. The 'Bartlett' pear clone originated from seed in England about 1770 and has been maintained asexually ever since. The 'Winesap' apple clone originated similarly some 200 years ago. The original seedling tree has probably been dead for over 100 years. Buds taken from it and budded to rootstocks produced new trees having tops of the same genetic constitution as the original seedling tree.

By continuing this process of asexual propagation, thousands upon thousands of trees have since been grown whose budded tops, collectively, make up the 'Winesap' apple clone. Clones exist in nature, reproducing naturally by such structures as bulbs, rhizomes, runners, stolons and tip layers. The species *Lilium tigrinum* is said to be a single clone. Apomixis in some species of Rosaceae, Gramineae, and Compositeae makes possible the reproduction of some clones naturally by seeds.

A clone can perpetuate itself successfully in nature, sometimes better than can seed-propagated plants, as long as the environment remains reasonably constant. If the environment changes drastically, however, a clonally reproduced species will be at a disadvantage because it has no opportunity to evolve forms better adapted to the new environment. Similarly, cultivated clones have a disadvantage in that an adverse situation, such as disease or insect attack, can affect all members of the clone equally and may even destroy it. The concept of the clone, however, does not mean that all individual members are necessarily identical in all characteristics.

The actual appearance and behaviour of a plant, i.e., its *phenotype*, results from the interaction of its genes (*genotype*) with the *environment* in which the plant is growing. Consequently, within a given clone the appearance of the plants, or of fruits and flowers on different plants, may vary some what because of climate, soil, disease, and the like. In many years 'Barlett' pear trees grown in California produce round, apple shaped fruit, whereas 'Barlett' pears grown in Washington and Oregon are relatively long and narrow; this difference is believed to be due to different climatic factors. Variation in fruit shape within any one area and even within an orchard or tree may also be related to the presence, the absence, or the number of seeds within the fruit. Some plants produce leaves of different appearance when growing in shade than when growing in the sun.

Some water plants produce leaves quite different in appearance on the part of the plant submerged than on the part in the air. Within a single orchard, fruit trees of the same clone often differ to some extent because of differences in soil, water availability, rootstock, or competition from surrounding plants. Although the environment can modify the growth and appearance of individual members of a clone such changes are not permanent, since the genotype of the plants is unaffected by the environmental modifications. Further in this chapter several causes of more or less permanent changes within clones are discussed. The life of the clone is theoretically unlimited. An early belief, that has long prevailed, is that a clone deteriorates with age and can only be rejuvenated by seed propagation. Evidence now shows that modification, sometimes leading to deterioration, can indeed occur in particular clones, and must be guarded against in propagation.

The most significant factor seems to be virus infection. In fact, virtually any clone that is grown for any period of time is likely to become infected sooner or later, its success being determined by its ability to tolerate the virus. Genetic changes (mutations) may occur in the clone which, although not always degenerative in a strict sense, can produce off-type individuals that will reduce the value of the clone. An unfavourable environment may lead to progressive deterioration of a clone. For example, lack of sufficiently long post-bloom vegetative period for some bulbous plants may lead to gradual decline in vigor and productivity.

If a clone is maintained in a proper environment, and procedures are carried out whereby viruses, other pathogens, and off-type mutants are eliminated, a clone could be perpetuated indefinitely. Such procedures, described later in this chapter, are an important aspect of plant propagation. Another belief, proposed by some people, is that permanent genetic changes in the clone can be induced by the environment or by the influence of one genetically different plant on another if the two are grafted together. Although some aspects of the latter theory have not perhaps been examined critically, the prevailing belief among most scientists is that such permanent genetic changes in clonal material do not occur.

Changes in Clones Associated with Age

The growth cycle of a seedling plant involves a change from a *juvenile* phase to a *adult* phase. The two phases may be distinguished by distinctive morphological and physiological differences. The Juvenile-to-adult change may be referred to as *ontogenetic* because more or less permanent changes occur within the apical meristem as it grows, but the basic genetic information in the cells is not altered. New seedlings revert to the juvenile phase and undergo the same transition to the adult phase as did the parent. Evidence that no permanent genetic change is involved is found with nucellar seedlings, which are asexually produced by apomixis and reproduce genetically the same parent plant but only after having gone through the transitional juvenile phase. The juvenile-to-adult change is not the same as the vegetative to-reproductive change, although the two kinds of changes may be correlated. Likewise, the development of senescence is different from either of these phenomena.

The juvenile and adult phases may differ distinctly in appearance or they may show transitional development from one phase to the other. Leaf shape is a common means of identifying phase changes such as illustrated in *Acacia* and *Eucalyptus.* The juvenile phase of many plants–for example, citrus, pear, apple and honey locust–is characterized also by non flowering, excessive vigor, and thorniness; the adult phase by flowering and fruiting, reduced vigor, and thorniness. The juvenile phase of a number of plants, of which English ivy (*Hedera helix*) is a classic example, is a trailing vine with alternate palmate leaves. The mature, flowering form is an erect or semierect shrub with entire, ovate leaves, produced oppositely around the stem.

In certain conifers, such as junipers, the juvenile forms produce needle-shaped leaves, but the adult produces scale-like leaves. Cuttings taken from juvenile parts of many perennial plants can reproduce adventitious shoots and roots easily, whereas cuttings taken from the mature phase of the same plant are much less, if at all, capable of forming adventitious roots or shoots. The ontogenetic change from juvenile to adult as the seedling grows older occurs in the somatic (vegetative)cells and results in differences in the apical meristem in separate parts of the plant at various stages of development. Prevailing evidence suggest that a plant must attain a certain size before the adult phase appears. Consequently the growing points produced in different parts of an individual plant where such ontogenetic changes are taking place may differ considerably in their ontogenetic age. This explains why a portion of an individual plant may be in the juvenile phase and another portion in the adult phase.

The phenomenon that different growing points at different parts of a plant may perpetuate particular phases (juvenile or mature) it used in propagation has been called *topophysis.* For example, buds taken from the lower, juvenile portion of seedling trees of pear, apple, citrus, or honey locust, if used in budding, will produce nursery trees that are vigorous, thorny, and slow to flower; buds from the upper portion of the same tree can be used to produce nursery trees that are less vigorous, smooth-barked, and thornless, and that flower quickly. By proper selection of propagation material from a seedling plant, it is possible to propagate leaf persistent (juvenile) or deciduous (adult) beech trees, or to propagate fungus (*Keithia thusima*)-resistant (juvenile) or fungus-susceptible (adult) forms of *Thuja plicata.*

By such selection of propagating material, the juvenile forms of many conifers can be maintained almost indefinitely; this can be done in certain vines, such as *Hedera.* Another type of topophysis occurring in some plants in that the direction of shoot growth produced by a cutting may differ, depending upon whether the cutting was taken from an upright or a lateral branch. For example, cuttings of *Araucaria,* or of coffee, taken from lateral shoots will continue to grow in a horizontal direction, whereas cuttings made from upright shoots of the same plant will develop into upright plants. Juvenile-to-adult phase changes are important in the initial establishment of a clone from a seedling plant. Various horticultural techniques are designed to speed this transition. To attain the adult phase in the shortest time, selection of propagation wood from the upper, onto-genetically older part of the plant is desirable.

In most well-established clones, however, considerable vegetative propagation has already taken place, a flowering and fruiting pattern has become set, and the adult phase has been attained. Consequently plants propagated from material taken from different parts of most grafted or rooted plants do not exhibit such differences. However, sometimes juvenile characteristics can persist for several vegetative generations. In propagation by rooting cuttings it may be desirable to preserve the juvenile phase of the clone. This can be done to some extent by selection of propagation material, since juvenility tends to persist in the lower part of the plant, particularly in the crown and roots. Likewise, adventitious buds from roots or from heavily cut-back stems, or from "sphaeroblasts" (wart-like growths containing meristematic and conductive tissue) also tend to produce juvenile growth. The success in propagation by stooling is believed to be due to the perpetuation of the juvenile form.

Budsports

A branch which shows changes in one or more inheritable characters that can be perpetuated by asexual means in termed a *bud sport.* Bud sports can originate by any of the somatic mutations or chromosomal changes mentioned earlier. Many of these sports are chimeras and are discovered when a lateral shoot develops from a mutated area. In addition, a bud sport may result when there is a rearrangement of tissues of a chimera such that a shoot may arise from a deeper layer, either from the displacement of epidermal cells with cells of the L-II layer, or by formation of adventitious buds.

Somatic variants such as these are starting points of new clones and, if introduced as cultivars, are given names. Numerous bud sports have been discovered, and a great many more have undoubtedly gone undetected. Some have become important new clones in both fruit and ornamental species. In Florida, pink-fleshed grapefruits were found on a single branch of a tree in a grove of thousands of trees producing whitefleshed grapefruits: an obvious mutation. The seedless 'Washington Navel' orange probably arose as bud mutation of the Brazilian orange, 'Laranja Selecta'. The first citrus fruit to be patented originated in 1929 as a mutation: the red-fleshed 'Ruby' grapefruit. Mutations in apples are common, resulting in changes in fruit colour, size, and shape.

Polyploidy may give rise to "grant" sports in certain plants. Individual tetraploid grapevines occur in vineyards, growers referring to such vines with prodigious growth habits and poor fruitfulness as "bulls," "males," or "giants". Chromosome counts have shown them to be often only partial mutations, (periclinal chimeras) consisting of a single outer layer of diploid cells surrounding the mutated tetraploid inner tissues. Such giant sport canes in the grape are observed to arise at pruning wounds or near areas where a bud or shoot has been injured or killed. These mutant shoots seem to arise from deeply embedded dormant bud initials.

Spontaneous polyploidy also occurs in apples, producing giants sports. Fruits of these sports are usually larger, flatter, and more irregular in shape than those from the parent plant. Trees bearing such fruits are vigorous, with thick twigs, often having wide-angled crotches, which gives them a low, flat shape. Certain of such giant apple sports are diploid-tetraploid periclinal chimeras, with a $2n$ layer (34 chromosomes per cell), one, two, or three cell layers in thickness, over a $4n$ tetraploid (68 chromosomes per cell) interior. A certain 'Delicious' apple giant sport was found to be tetraploid in all internal tissues, but diploid in the epidermis. Many (possibly most) mutations produce sports inferior to the parent plant in one or more characteristics. If detected by the propagator, such sports are discarded. Sometimes, as in citrus, these may be propagated unknowingly and cause serious economic loss.

Graft Chimeras

While most chimeras occur naturally, they can also be established artificially by grafting. If scions are cut back severely, almost to the stock, adventitious buds may sometimes arise from the callus in the region of the graft union. Occasionally, a bud may develop which contains tissues of both stock and scion. The resulting structure is a chimera, in which the cells of the two graft components remain distinct regardless of how intermingled they become. Winkler artificially produced in Germany many graft chimeras, using species in the Solanaceae family. He prepared wedge or saddle grafts or tomato (*Lycopersicon esculentum*) on black nightshade (*Solanum nigrum*) and black nightshade on tomato, both of which are easily grafted and produce callus readily.

Winkler's grafts were made, following union, the graft was cut transversely at the junction, exposing tissues of both components. The cut surface soon became covered with a pad of callus derived from both the tomato and nightshade. From this callus, adventitious buds formed which developed into shoots. Most of these shoots had the appearance of either pure tomato or pure nightshade. Occasionally, however, shoots arose and grew normally which were different from either component. Some showed mixed characters, with half the shoot resembling tomato and the other half the nightshade. Winkler gave it the name "chimera," after the fabulous mythological monster that was part lion and part dragon. Graft chimeras in the periclinal form also developed. One was found to consist mostly of a nightshade tissue covered with a single layer of tomato epidermal cells. It bore fruit like that of the true nightshade and its seeds produced nightshade plants.

Graft chimeras among woody perennials have long been known to exist, and some have been perpetuated indefinitely by asexual methods. Probably the earliest known "authentic" chimera is one called the "bizzarria" orange. It is described as bearing a fruit which is citron on one side and orange on the other. It originated supposedly in 1644 in a garden in Florence, Italy, from a grafting operation in which a scion of sour orange (*Citrus aurantium*) was grafted on a stock of citron (*C. medica*). According to the story, the original scion failed to grow, but in the swollen callus around the graft a shoot developed bearing the unusual "bizzarria" orange were still being published 250 years after its origin, agreeing that it probably was a periclinal chimera.

Grafting medlar (*Mespilus germanica*) on a stock of the hawthorn (*Crataegus monogyna*) has led to several recorded instances of chimeras. Near Metz, France, in 1899, two adventitious shoots were found arising at the graft union of an old medlar tree. They had a different appearance from either the medlar or the hawthorn and were different from each other. They were named *Crataegomespilus asnieresi* and *Crataegomespilus dardari* and were regarded for many years as true "*graft hybrids*" believed to have resulted from a fusion of the nuclei of vegetative cells.

Both of these types, sometimes called hawmedlars, have been maintained vegetatively, grafted on the hawthorn and widely distributed to botanical gardens. It is likely, however, that these two forms are periclinal chimeras. There have been several strong proponents of the graft hybrid hypothesis. Winkler made his Solanaceae graft chimeras in the hope of gaining evidence for the graft hybrid idea. He believed that some of these unusual plant forms were true graft hybrids involving fusion of vegetative cell nuclei. Daniel is probably the most conspicuous supporter of this hypothesis and has written extensively on the subject although many of his views are not generally accepted.

2

Propagation by Cuttings

In propagation by cuttings, a portion of a stem, root, or leaf is cut from the parent plant, after which this plant part is placed under certain favourable environmental conditions and induced to form roots and shoots, thus producing a new independent plant which, in most cases, is identical with the parent plant.

The Importance and Advantages of Propagation by Cuttings

This is the most important method of propagating ornamental shrubs–deciduous species as well as the broad-and narrow-leaved types of evergreens. Cuttings are also used widely in commercial greenhouse propagation of many florists crops and are commonly used in propagating several fruit species.

For species that can be easily propagated by cuttings, this method has numerous advantages. Many new plants can be started in a limited space from a few stock plants. It is inexpensive, rapid, and simple, and does not require the special techniques necessary in grafting or budding. There is no problem of compatibility with rootstocks or of poor graft unions. Greater uniformity is obtained by absence of the variation which sometimes appears owing to the variable seedling rootstocks of grafted plants. The parent plant is usually reproduced exactly with no genetic change.

It is not always desirable, however, to produce plants on their own roots by cuttings even if it is possible to do so. It is often advantageous or necessary to use a rootstock resistant to some adverse soil condition or soil borne organism, or to utilize available dwarfing or invigorating rootstocks.

Types of Cuttings

Cuttings are almost always made from the vegetative portions of the plant, such as stems, modified stems (rhizomes, tubers, corms and bulbs), leaves, or roots. The reproductive parts of the plant are not ordinarily used, although such parts as the ovary, pedicel and petals, have been reported to form roots.

Several types of cuttings can be made; these are classified according to the part of the plant from which they are obtained:

- Stem cuttings
  - Hardwood
    - Deciduous
    - Narrow-leaved evergreen
  - Semi-hardwood
  - Softwood
  - Herbaceous
- Leaf cuttings
- Leaf bud cuttings
- Root cuttings

Many plants can be propagated by several of these different types of cuttings with satisfactory results. The type used would depend upon the individual circumstances, the least expensive and easiest usually being selected.

If the particular plant being propagated will root well by hardwood stem cuttings in an outdoor nursery, this method ordinarily used, because of its simplicity and low cost. Root cuttings of some species are also satisfactory, but cutting material may be difficult to obtain in large quantities. For species more difficult to propagate, it is necessary to resort to the more expensive and elaborate facilities required for rooting the leafy types of cuttings.

In selecting cutting material it is important to use stock plants that are free from diseases, moderately vigorous and productive, and of known identity. Stock plants that are diseased or injured by frost or drought, that have been defoliated by insects or diseases, that have been stunted by excessive fruiting, or that have made rank, overly vigorous growth should be avoided.

A commendable practice for the propagator is the establishment of stock blocks as a source of propagating material, where uniform, true-to-type pathogen-free mother plants can be maintained and held under the proper nutritive condition for the best rooting of cuttings taken from them.

Stem Cuttings

This is the most important type of cutting and can be divided into four groups, according to the nature of the wood used in making the cuttings: *hardwood, semi-hardwood, softwood,* and *herbaceous.* In propagation by stem cuttings, segments of shoots containing lateral or terminal buds are obtained with the expectation that under the proper conditions adventitious roots will develop and thus produce independent plants. The type of wood, the stage of growth used in making the cuttings, the time of year in which the cuttings are taken, and several other factors can be very important in securing satisfactory rooting of some plants.

Hardwood Cuttings (Deciduous Species)

This is one of the least expansive and easiest methods of vegetative propagation. Hardwood cuttings are easy to prepare, are not readily perishable, may be shipped safely over long distances if necessary, and require no special equipment during rooting.

The cuttings are prepared during the dormant season-late fall, winter, or early spring–from wood of the previous reason's growth, although with a few species, such as the fig or olive, two-year-old or older wood is used. Hardwood cuttings are most often used in propagation of deciduous woody plants, although some broad-leaved evergreens, such as the olive can be propagated by leafless hardwood cuttings. Many deciduous ornamental shrubs are started readily by this type of cutting. Some common ones are privet, forsythia, wisteria, honeysuckle, and spiraea. Rose rootstocks, such as *Rosa multiflora* are propagated in great quantities by hardwood cuttings. A few fruit species are propagated commercially by this method–for example fig, quince, olive mulberry, grape, currant, gooseberry, pomegranate, and some plums.

The propagating material for hardwood cuttings should be taken from healthy, vigorous stock plants growing in full sunlight. The wood selected should not be from extremely rank growth with abnormally long internodes, or from small, weakly growing interior shoots. Wood of moderate size and vigor is the most desirable. The cuttings should have an ample supply of stored foods to nourish the developing roots and shoots until the new plant becomes self-sustaining.

Hardwood cuttings vary considerably in length–from 4 to 30 in. Long cuttings, when they are to be used as rootstocks for fruit trees, permit the insertion of the varietal bud into the original cutting following rooting, rather than into a smaller new shoot arising from the original cutting.

At least two nodes are included in the cutting, the basal cut is usually just below a node and the top cut ½ to 1 in. above a node. However, in preparing stem cuttings of plants with short internodes, little attention is ordinarily given to the position of the basal cut, especially when quantities of cuttings are prepared and cut to length many at a time, by a band saw or paper cutter.

The diameter of the cuttings may range form ¼ in. to 1 or even 2 in., depending upon the species. The "mallet." the "heel", and the straight cutting are three different cuttings. The mallet includes a short section of stem of the older wood, whereas the heel cutting includes only a small piece of the older wood. The straight cutting, not including any of the older wood, is most commonly used, giving satisfactory results in most instances.

Where it is difficult to distinguish between the top and base of the cuttings, it is advisable to make all the basal cuts at a slant rather than at right angles. In large-scale operations, bundles of cutting material are cut to the desired lengths by band saws or other types of mechanical cutters rather than individually by hand.

Cuttings of the first (*Abies* sp) and pines (*Pinus* sp). are very difficult to root. In addition, there is considerable variability among the different species in these genera in regard to the ease of rooting of cuttings. As with many other plants, cuttings taken from young seedling stock plants root much more readily than those taken from older trees. Treatments with root-promoting substances particularly indolebutyric acid at relatively high concentrations, are usually beneficial increasing the speed of rooting and the percentage of cuttings rooted, and in obtaining heavier root systems.

Narrow-leaved evergreen cuttings are best taken between late fall and late winter. Rapid handling of the cuttings after the material is taken from the stock plants is important. The cuttings are best rooted in a greenhouse with relatively highlight intensity and under conditions of high humidity or very light misting but without heavy wetting of the leaves. A bottom heat temperature of 75° to 80°F (24° to 26.5°C) has given good results. Dipping the cuttings into a fungicide helps prevent disease attacks. Sand alone has given

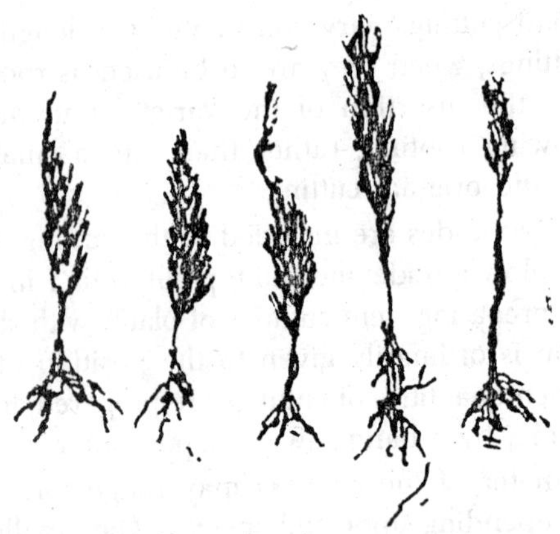

Fig. 2.1. Narrow-leaved evergreen cuttings.

good results as a rooting medium, as has a 1:1 mixture of perlite and peat moss. Some individual cuttings take longer to root than others. The slower-rooting ones can be restuck in the rooting medium, and often will root eventually.

The type of wood to use in making the cuttings varies considerably with the particular species being rooted. The cuttings are made 4 to 8 in. long with all the leaves removed from the lower half of the cutting. Mature terminal shoots of the previous season's growth are usually used. In some instances, such as with *Juniperus chinensis pfitzeriana*, older and heavier wood can also be used, thus resulting in a larger plant when it is rooted. On the other hand, some nurserymen use small tip cuttings, 2 to 3 in long, placed very close together in a flat for rooting. In some species, such as *Juniperus excelsa*, older growth taken from the sides and lower portion of the stock plant roots better than the more succulent tips. Cuttings of *Taxus* root best if they are taken with a piece of old wood at the base of the cutting. Such cuttings are also less subject to fungus attacks. In certain of the narrow-leaved evergreen species, some type of basal wounding is sometimes beneficial in inducing rooting.

Semi-Hardwood Cuttings

Cuttings of this type are usually made from woody, broad leaved evergreen species, but leafy summer cuttings taken from partially matured wood of deciduous plants could also be considered as semi-hard-wood. Cuttings of broad-leaved evergreen species are generally taken during the summer months from new shoots just after a flush of growth has taken place and the wood is partially matured. Many ornamental shrubs, such as camellia, pittosporum, euonymus, the evergreen azaleas, and holly are commonly propagated by semi-hardwood cuttings. A few fruit species, such as citrus and olive, can, also be started in this manner.

The cuttings are made 3 to 6 in. long with leaves retained at the upper end but removed from the lower end. If the leaves are very large, they should be reduced in size to lower the water loss and to allow closer spacing in the cutting bed. The terminal ends of the shoots are often used in making the cuttings, but the more basal parts of the stem will usually root also. The basal cut is usually just below a node. The cutting wood should be obtained in the cool, early morning hours when the stems are turgid, and kept wrapped in clean moist burlap and out of the sun at all times until the cutting are made.

It is necessary that leafy, semi-hardwood cuttings be rooted under conditions which will keep water loss from the leaves at a minimum; commercially they are usually rooted under intermittent mist sprays. Bottom heat and growth regulator treatments are also beneficial. Rooting media such as a 1:1 mixture of perlite and peat moss of perlite and vermiculite give satisfactory results.

Softwood (Greenwood) Cuttings

Cuttings prepared from the soft, succulent , new spring growth of deciduous or evergreen species may properly be classed as softwood cuttings. Many ornamental woody shrubs can be started by softwood cuttings. Typical examples are the hybrid French lilacs, forsythia, magnolia, weigela, and spiraea. Some deciduous ornamental trees such as the maples can also be started in this manner. Although fruit tree species are not commonly propagated by softwood cuttings, apple, peach, pear, plum, apricot, and cherry will root, especially under mist.

Softwood cuttings generally root easier and quicker than the other types but require more attention and equipment. This type of cutting is always made with leaves attached. They must, consequently, be handled carefully to prevent drying, and be rooted under conditions which will avoid excessive water loss from the leaves. Temperature should be maintained during rooting at 75° to 80°F (23° to 27°C) at the base and 70°F (21°C) at the leaves for most species. Softwood cuttings produce roots in 2 to 4 or 5 weeks in most cases. In general, they respond well to treatment with root-promoting substances. It is important in making softwood cuttings to obtain the proper type of cutting wood from the stock plant. This will vary greatly, however, with the species being propagated. Extremely fast-growing, soft, tender shoots generally are not desirable, as they are likely to deteriorate before rooting. At the other extreme, old, woody stems are slow to root in most cases.

The best cutting material has some degree of flexibility, but is mature enough to break when bent sharply. Weak, thin, interior shoots should be avoided as well as vigorous, abnormally thick, or heavy ones. Average growth from portions of the plant in full light is the most desirable to use. Some of the best cutting material is the lateral or side branches of the stock plant. Heading back the main shoots will usually force out numerous lateral shoots from which cuttings can be made. Softwood cuttings ordinarily are 3 to 5 in. long with two or more nodes. The basal cut is usually made just below a node. The leaves on the lower portion of the cutting are removed, but those at the upper portion are retained. If the upper leaves are very large, they should be reduced in size to lower the transpiration rate and to use less space in the propagating bed, although for the best rooting it is desirable to retain the maximum leaf area that is possible without wilting. All flower buds should be removed. In some large nurseries where quantities of cuttings are prepared, bundles of cutting material are rapidly chopped into uniform lengths by paper cutters.

The cutting material is best gathered in the early part of the day and should be kept moist, cool, and turgid at all times by wrapping in damp, clean burlap or placing in large polyethylene bags (kept out of the sun). Laying the cutting material or prepared cuttings in the sun for even a few minutes will cause serious damage. Soaking the cutting material or cuttings in water to keep them fresh is undesirable.

Herbaceous Cuttings

This type of leafy cutting is made from such succulent, herbaceous greenhouse plants as geraniums, chrysanthemums, coleus, and carnations. They are 3 to 5 in. long with leaves retained at the upper end. Most florists crops are propagated by herbaceous cuttings. They are rooted under the same conditions as softwood cuttings, requiring high humidity. Bottom heat is also helpful. Under proper conditions, rooting is rapid and in high percentages.

Although root-promoting substances are not required, they are often used to gain uniformity in rooting and development of heavier root systems. Herbaceous cuttings of some plants that exude a sticky sap, such as the geranium, pineapple, or cactus, do better if the basal ends are allowed to dry for a few hours in the air before they are inserted in the rooting medium. This allows the wounded tissues to dry, which tends to prevent the entrance of decay organisms.

Leaf Cuttings

In this type of cutting, the leaf blade or leaf blade and petiole, are utilized in starting a new plant. In most cases, adventitious roots and an adventitious shoot form at the base of the leaf. Rarely does the original leaf becomes apart of the new plants. One type of propagation by leaf cuttings is illustrated by *Sansevieria.* These leaf pieces are inserted three-fourths of their length into sand, and after a period of time a new plant forms at the base of the leaf piece. The original leaf cutting does not become a part of the new plant. The variegated form of *Sansevieria, S. trifasciata laurenti,* is an example of a periclinal chimera which will not reproduce true to type from leaf cuttings; to retain its characteristics, it must be propagated by division of the original plants.

In starting plants with thick, fleshy leaves, such as *Begoniarex,* by leaf cuttings, the large veins are cut on the undersurface of the mature leaf, which is then laid flat on the surface of the propagating medium. The leaf is pinned or held down in some manner, with the natural upper surface of the leaf exposed. After a period of time under humid conditions, new plants will form at the point where each vein was cut. The old leaf blade will gradually disintegrate.

Another method, sometimes used with the fibrous-rooted begonias, is to cut large, well-matured leaves into triangular sections each containing a piece of a large vein. The thin outer edge of the leaf is discarded. These leaf pieces are then inserted upright in sand with the pointed end down. The new plant develops from the large vein at the base of the leaf piece. The original leaf dies, not becoming a part of the new plant.

The African violet (*Saintpaulia*) is typical of leaf cuttings which can be made of an entire leaf (leaf blade plus petiole), the leaf blade only, or just a portion of the leaf blade. The new plant forms at the base of the petiole or midrib of the leaf blade.

Leaf cuttings should be rooted under the same conditions of high humidity that are used for softwood or herbaceous cuttings. Root-promoting chemicals are usually helpful.

Leaf-Bud Cuttings

This type of cutting consists of a leaf blade, petiole, and a short piece of the stem with the attached axillary bud. Such cuttings are of particular value for species that are able to initiate roots but not shoots from detached leaves. In such cases, the axillary bud at the base of the petiole provides for the essential shoot formation. A number of plant species such as the black raspberry (*Rubus occidentalis*), blackberry, boysenberry, lemon, camellia, and rhododendron can be readily started by leaf-bud cuttings, as well as many tropical shrubs and most herbaceous green-house plants usually started by stem cuttings. Red raspberries (*Rubus idaeus*) apparently do not reproduce by leaf-bud cuttings.

This method is particularly valuable when propagating material is scarce, because it will produce at least twice at many new plants from the same amount of stock material as can be started by stem cuttings. Each node can be used as a cutting. Leaf-bud cuttings should only be made from material having well-developed buds and healthy, actively growing leaves.

Treatment of the cut surfaces with one of the root-promoting substances may stimulate root production. The cuttings should be inserted in the rooting medium with the bud as soon as the plants become well established with good root development they can be transplanted to their permanent location.

Plants with Large Roots, Propagated Out-of-Doors

Root cuttings of this type are made 2 to 6 in. long. They are tied in bundles, care being used to keep the same ends together to avoid planting upside down later. The cuttings are then packed in boxes of damp sand, sawdust, or peat moss for about 3 weeks. During this storage period the cuttings should be kept cool, about 40°F (45°C). They should be planted 2 to 3 in apart in a well-prepared nursery soil with the tops of the cuttings level with, or just below the top of the soil.

Root Promoting Substances

Herbaceous Plants

Cuttings of most plants of this type have shown a definite increase in rooting when treated with growth regulators. The chief effect has been to speed the rate of rooting and to produce heavier clumps of roots. Plants which have shown good response include chrysanthemums, geraniums, carnations, begonias, poinsettias, English ivy, and African violets. Cuttings of most varieties of such herbaceous plants usually root so easily under proper care without growth-regulator treatments, however, that it is debatable whether their use is justified.

Broad-Leaved Evergreens

Propagation of this type of plant, by softwood or semi-hardwood cuttings, is often considerbly improved by growth regulator treatments, provided that the other necessary requirements of rooting are satisfied. The citrus species–orange, lemon, and grapefruit–as well as the olive are evergreen tree fruits that respond well to treatment. A great many ornamentals in this group have also given good response, including camellia, holly, rhododendron, box, daphne, euonymus, gardenia, oleander, escallonia, pittosporum, and pyracantha.

Narrow-Leaved Evergreens

For most species, if the cuttings are taken at the right time of year, and the proper material and concentration used, improved rooting can be expected to result from the use of root-promoting chemicals. Species found to respond to treatments include fir, juniper, spruce, pine, yew, aborvitae, and hemlock.

Deciduous Plants

Cuttings may be made of this type of plant either as hardwood cuttings during the dormant season or as softwood cuttings during the active growing period. Experience has shown a response to growth-regulator treatments with both types of cuttings. Plants of this type that have responded to treatments include the apple, pear, peach, plum, cherry, walnut, filbert, maple, birch, beech, and mulberry.

Environmental Conditions for Rooting Leafy Cuttings

For the successful rooting of leafy cuttings, the essential environmental requirements are proper temperature (65° to 75°F: 18° to 27°C), an atmosphere conductive to low water loss from the leaves; ample light; and a clean, moist, well-aerated, and well-drained rooting medium. There are many possible types of equipment that are satisfactory for providing these conditions-from a simple glass jar placed over a few cuttings stuck in sand to elaborate greenhouse benches with automatic mist control and automatically controlled electric soil cables below the cuttings. Probably one of the simplest devices, which is entirely satisfactory for rooting a limited number of cuttings, is an ordinary wooden box half-filled with the rooting medium, with a pane of glass or a sheet of polyethylene film over the top. If this is placed in a heated room in the winger near a window, cuttings of many species can be rooted in it. Such a box can also be used successfully in the summer for rooting cuttings of some plants, if placed out-of-doors in the shade. Cuttings of many plants can be rooted successfully with only artificial light, if placed under an ordinary large fluorescent lamp fixture.

In commercial operations, large-scale rooting of leafy cuttings is done in hotbeds, cold frames, or beds in greenhouses or plastic houses. In these structures, proper environmental conditions can be established and maintained.

Sanitation

Most commercial propagators recognize the value of maintaining strict sanitary procedures during all stages of making and rooting leafy cuttings. It is much easier to prevent attacks of disease organisms than to try to stop them. Losses could be considerable from a disease attack where hundreds of thousands of cuttings might become involved.

During the preparation of facilities for rooting the cuttings and while making and inserting them in the rooting medium, the following procedures, if observed closely, will aid considerably in preventing losses from attacks by noxious organisms.

Propagating benches in the greenhouse should be washed thoroughly with water and sprayed with a copper naphthenate solution (1 part to 5 parts paint thinner). (*Caution* : Creosote and certain phenolic materials used as wood preservatives give off fumes which are toxic to plants, and should not be used.) Flats for rooting the cuttings should be thoroughly washed, filled with the rooting medium, and then sterilized by steam or methyl bromide. The work benches where the cuttings are to be prepared should be washed thoroughly with water and sprayed with a Clorox (1 part to 4 parts water)solution or a formaldehyde (1 qt 38 per cent formaldehyde to 5 gal, of water)solution. Tools used in preparing the cuttings should be dipped in such solutions several times daily.

The cutting material itself should be free of insects and disease organisms. As the cutting material is gathered, it can be placed on and wrapped in large sheets of clean, washed black polyethylene. When full these are protected from the sun and brought into the propagating room as soon as possible. The cutting material can be placed on a large raised wire rack and washed thoroughly with water. Mist nozzles should be placed over this rack to keep the material fresh while the cuttings are being prepared. After the cuttings are made, they may be soaked for about 10 minutes in a mild fungicide, such as Morsodren (1tsp per 5 gal, of water). Following this, the cuttings are allowed to drain in a wire rack and are then treated with a root-promoting compound just before they are stuck into the propagating medium.

Efforts to conduct propagating operations under clean, sterilized conditions are of no avail, however, unless all components are included–the cuttings themselves, the flats, the rooting medium, the area where the cuttings are prepared, the tools used in making the cuttings, and the rooting bench. Keep the watering house nozzle off the floor where it could pick up and later distribute harmful organisms.

Preparing the Rooting Frame and inserting the Cuttings

The frames or benches should preferably be raised or, if on the ground, equipped with drainage tile, so there is never any question of perfect drainage of excess water.

The frames or flats should be deep enough so that about 4 in. of rooting medium can be used. The depth should be enough so that an average-length cutting–3 to 5 inch can be inserted up to half its total length, with the end of the cutting still an inch or more above the bottom of the frame. The rooting medium should be watered thoroughly before the cuttings are inserted, which should be as soon as possible after they are prepared. It is very important that the cuttings be protected from drying at all stages during their preparation and insertion.

After a section of the rooting bench or a flat is filled with cuttings, it should be well watered to settle the rooting medium around the cuttings.

Care of Cuttings During Rooting

Hardwood stem cuttings or root cuttings started out-of-doors in the nursery require only the usual care given to other crop plants, such as adequate soil moisture, freedom from weed competition, and insect and disease control. The best results with most species are likely to be obtained if the nursery is established in full sun where shading and root competition from large trees or shrubbery do not occur.

Leafy softwood or semi-hardwood stem cuttings and leaf-bud or leaf cuttings being rooted under high humidity require close attention throughout the rooting period. The temperature should be controlled carefully. The cuttings must not be allowed to show wilting for any length of time. Glass-covered frames, exposed even for a few hours to strong sunlight, will build up excessively high and injurious temperatures, owing to the heat accumulating under the glass. Such equipment should always be protected by cloth screens, whitewash on the glass, or some other method of reducing the light intensity.

If bottom heat is provided, thermometers should be inserted in the rooting medium to the level of the base of the cutting and checked at frequent intervals, especially at first. A temperature of 65° to 75°F (18° to 24°C) is desirable. Excessively high temperatures in the rooting medium even for a short time are likely to result in death of the cuttings.

It is very important to maintain humidity conditions as high as possible in rooting leafy cuttings to reduce water loss from the leaves to a minimum. Syringing the leaves with a spray nozzle at frequent intervals is a common practice especially during hot weather. Although more time consuming, several light sprinklings with water each day are better than heavy soakings at less frequent intervals. A nozzle should be used that breaks the water into a fine spray. A drop of the humidity to a low level with a consequent pronounced wilting of the cuttings, if prolonged for any length of time, may so injure the cuttings that rooting will not occur, even though high-humidity conditions are subsequently resumed. However most commercial nursery operations use intermittent-mist propagating beds to overcome such problems.

Although high humidity in the propagating frame is important while roots are forming, it is also necessary that adequate drainage be provided so that excess water can escape and not cause the rooting medium to become soggy and waterlogged. When peat or sphagnum moss is used as a component of the rooting medium, it is especially important to see that it does not become excessively wet. It is also necessary to maintain sanitary conditions in the propagating frame. Leaves that drop should be removed promptly, as well as any obviously dead cuttings. Organisms find ideal conditions in a humid, closed propagating frame with low light intensity and, if not controlled, can destroy thousands of cuttings overnight.

Disease problems under mist propagation conditions have not been serious. Probably the frequent washing of the leaves by the aerated water removes spores before they are able to germinate. The greater light intensity and air movement also decrease the incidence of diseases. If insects, such as red spiders, aphids, or mealy bugs, appear on the leaves of the cuttings, immediate control measures are necessary.

Handling Cuttings After Rooting

Hardwood Cuttings

Rooted hardwood cuttings in the nursery row are usually dug during the dormant season after the leaves have dropped. With fast growing species, the cuttings may be sufficiently large to dig after one season's growth. Slower growing species may require 2 or even 3 years to become large enough to transplant.

The digging should take place on cool, cloudy days, when there is no wind. If possible, digging should not be done when the soil is wet, especially if it has a high clay content. Most of the soil should drop readily from the roots after the plants are removed. After the plants are dug, they should be quickly heeled-in in a convenient location or else replanted immediately in their permanent location, "Heeling-in" is to place dug, barerooted deciduous nursery plants close together in trenches with the roots well covered. This is a temporary provision for holding the young plants until they can be set out in their permanent location.

Commercial nurseries often store quantities of deciduous plants for several months in cool, dark rooms with the roots protected by damp wood shavings, shingle two, or some similar material. Nursery stock to be kept for extended periods should be held under refrigerated conditions at 32° to 35°F (0° to 2°C).

If only a few plants are to be removed from the nursery row, they can be dug with shovels, but in large-scale nursery operations some type of mechanical digger is generally used. This digger "undercuts" the plants. A sharp U-shaped, blade travels 1 to 2 ft below the soil surface under the nursery row, cutting through the roots. Sometimes a horizontal "lifting" blade is attached to, and travels behind, the cutting blade. This slightly lifts the plants out of the soil, making them easy to pull by hand.

Unless very small, plants of evergreen species usually cannot be successfully handled bare-root, as is done with dormant and leafless deciduous plants. The presence of leaves on evergreen plants requires the continuous contact of the roots with soil. Therefore, large, salable plants of broad-or narrow-leaved evergreens, and occasionally deciduous plants, are either grown in containers or dug and sold "balled and burlapped." By the latter method, the plants are removed from the soil by carefully digging a trench around each individual. The soil mass around the roots is sometimes tapered at top and bottom, resulting in a half of soil in which the roots are embedded. It is important that the soil be at the proper moisture level–not too wet and not too dry–otherwise it will fall apart. The ball is tipped gently onto a large square of burlap, which is then pulled tightly and sewed in place with heavy twine. If properly done, this adequately holds the soil to the roots, and the plant can then be moved safely of considerable distances and replanted successfully.

To properly prepare field-grown nursery stock for transplanting, either by hand "ball and burlapping" or by machine digging prior root pruning is necessary. This is to produce a compact, fibrous root system and should be started when the young plants (*liners*) are first set in the field. Long or curled and twisted roots should be cut back to a few inches from the crown of the liner. Root pruning the second year should also be done, either by a tractor-powered U-blade cutter for large-scale operations, or with a sharp shovel where only a few plants are involved. This procedure helps confine the roots to the soil mass which will be taken with the plant during the digging and balling operation.

The production of balled and burlapped nursery stock is gradually being replaced, especially in areas with mild winters, by the production of plants in containers primarily because of the lower labour costs involved and the greater opportunities for mechanization.

Softwood, Herbaceous, Semi-Hardwood, Leaf-Bud, Leaf Cuttings

Cuttings such as these, rooted with leaves attached and under conditions of high humidity, require considerable care in being removed from the rooting medium. After rooting has started, the humidity should be lowered and ventilation of the bed provide. They should be dug as soon as a substantial root system with secondary roots has formed. Many propagators have experienced rapid rooting of cuttings only to have them die when they are dug and potted. Sometimes this trouble may be overcome by leaving the cuttings in the rooting bed longer, until after the first-formed primary roots have branched to develop a dense, fibrous secondary root system to which the rooting medium clings in a ball.

Most of the rooting media used contain little or no mineral nutrients available to the plants, hence the cuttings are dependent upon the nutrients already stored in the stems and leaves at the time the cuttings were made. This is generally sufficient to maintain the cuttings during the period of root formation. With some species, however, it may be helpful to water the cuttings with a nutrient solution about 10 days before they are to be removed, especially if digging has been delayed until a secondary root system has developed.

If for any reason the cuttings are to be kept for a prolonged period in the rooting medium, they should be watered several times with a nutrient solution.

In digging, the cuttings should be lifted gently from the rooting medium with a trowel or some similar device, taking care not to break off the roots. It is desirable for the cuttings to be lifted out with a mass of the rooting medium still adhering to the roots. This can be done if some material such as peat moss or vermiculite is included in the rooting medium. The cuttings are ready for potting when most of the roots are 1 to 2 in. long.

The rooted cuttings should be watered thoroughly shortly after potting. It is very important that potted cuttings be moved gradually from the protected conditions under which they were rooted (high humidity, low light intensity) to out-of-door conditions (low humidity, high light-intensity, and wind). It is often best to leave the cuttings for several days after they have been potted under the same conditions in which they were rooted. If they were rooted in mist, they must be given especially close attention and moved gradually to the dryer atmosphere. Before pacing the rooted cuttings in full sun, they should be hardened-off for 12 or 2 weeks in a cold frame or lathhouse or under some partial protection from the sun.

Cold Storage of Rooted and Unrooted Leafy Cuttings

Sometimes it may be convenient to take cuttings at certain times, such as when nursery plants are being sheared and shaped, for later rooting. This was done experimentally with the Kurume type of azaleas overwintered in a cold greenhouse. Softwood cuttings were taken in the spring and held in polyethylene bags at temperatures form 31° to 40°F (–0.5° to 4.5°C) for up to 10 weeks, with subsequent rooting equal to that from unstored cuttings. Unrooted carnation cuttings have long been stored commercially for subsequent rooting. In test on the effects of storage on subsequent performance of the plants, it was found that cuttings rooted after stored at 33°F (0.5°C) gave better results than those stored after rooting. In some situations cuttings may be rooted in late summer or early fall for planting outdoors the following spring.

Several studies have shown that it is possible to hold rooted cuttings of some species in polyethylene bags for prolonged periods under cold storage temperatures of 35° to 40°F (1.5° to 4.5°C). In one test, rooted cuttings of certain plants were safely stored for about 5 months at temperatures of 34° to 39°F (1° to 4°C) when placed in polyethylene bags. Packing material around the roots was of no advantage. Other studies on cold storage of rooted cuttings of 31 different species for 6 months at two temperatures (32° and 40°F) showed better survival with many of the species at 32° than at 40°, although no difference was noted with other species. With some species, dusting prior to storage with 5 per cent Captan increased survival, but in others this made no difference. Apparently some species survive cold-storage treatments much better than others.

3

Propagation by Layering

Layering is the development of roots on a stem while it is still attached to the parent plant. The rooted stem is then detached to become a new plant growing on its own roots. A layered stem is known as a *layer.* This may be a natural means of reproduction, as in black raspberries and trailing blackberries, or it may be induced by the "artificial" methods described in this chapter.

Factors Affecting the Propagation of Plants by Layering

Root Formation during layering is stimulated by various stem treatments which cause an interruption in the downward translocation of organic materials–carbohydrates, auxin, and other growth factors–from the leaves and growing shoot tips. These materials accumulate near the point of treatment, and rooting occurs in this general area even though the stem is still attached to the parent plant.

Water and minerals are supplied to the layered shoot, because the stem is not severed and the xylem, remains intact. Thus layering does not depend upon the length of time during which a severed shoot (cutting) can be maintained before rooting occurs. This is an important reason for layering being more successful with many plants than propagation by cuttings.

Etiolation is another means by which the internal conditions of the developing shoot can be modified during layering to stimulate rooting. It is accomplished in mound or trench layering by covering the newly developing shoot as it grows so that the basal part of the layered shoot is not exposed to the light. This apparently accounts in large measure for the success with which shy-rooting plants are rooted by these two methods.

Applying root-promoting substances, such as indolebutyric acid, during layering is sometimes beneficial, as it is with cuttings, although the method of application may be somewhat different. Applying the material to girdling cuts as a powder, in lanolin or as a solution in 50 percent alcohol can be utilized effectively.

Root formation on layers depends upon continuous moisture, good aeration, and moderate temperatures in the rooting zone. These conditions are best provided by a rooting medium such as sawdust. Prolonged dry spells and compact, heavy soils hinder root development, particularly during the initial stages of rooting. The addition of granulated peat moss to the soil mounded around apple and quince stock plants has promoted rooting. Excessively high temperatures in the upper layers of soil during the spring and summer may reduce the moisture content and cause compaction, not only inhibiting but rooting injuring the shoots as well.

Characteristics and Uses of Layering

The principal advantage of layering is the success with which plants can be rooted by this method. Many clones which will not root easily by cuttings can be propagated by layering, enabling the plant to be established on its own roots. Most methods of layering are relatively simple to perform and can be practiced out-of-doors in the nursery or garden. When small numbers of plants are involved, layering can give a high degree of success with somewhat less skill, effort, and equipment than is necessary with cuttings. With those few kinds of plants in which layering occurs naturally, it is a simple and economical method of propagation.

In some cases, a larger plant can be produced in a shorter time than it would if started as a cutting. However, since transplanting becomes increasingly difficult as the size of the layer increases, special precautions are necessary to establish successfully the larger plant on its own roots.

On the other hand, layering is an expensive method of propagation and does not lend itself to the large-scale techniques of mechanization used in modern nurseries. Part of the increased cost of propagation is due to the additional hand labour required. A layered plant requires a certain amount of individual attention, depending upon the particular method in use, even though the operations involved are in themselves simple. Also, the number of salable plants from a given number of stock plants is smaller than with cuttings, buds, or scions. The methods tend to be cumbersome, and the stock plants take up a considerable area which is difficult to cultivate and maintain free of weeds.

Layering is usually limited by American nurserymen to those plants which propagate naturally in this manner–e.g., black raspberry, trailing blackberries, and those plants too difficult or impossible to propagate by other methods, yet of sufficient value to justify the cost. For instance, the filbert (*Corylus* sp), Muscadine grape (*Vitis rotundifolia*), and litchi (*Litchi chinensis*) are propagated commercially in this manner. Layering is used in propagating certain clonal rootstocks, such as the Malling apple stocks which are not easily rooted as cuttings or which require special equipment, such as mist installations.

Some ornamental trees and shrubs are propagated by layering particularly in Europe. Established blocks of plants to be used for layering (known as *stool beds*) have been in production for many years in some cases, and nurserymen are skilled in their management.

Layering is perhaps best utilized by the amateur horticulturist who wishes to propagate a relatively small number of plants, or by specialists involved with reproducing certain kinds of plants. In these cases, expense per plant and the individual attention necessary would not be factors in the choice of method.

PROCEDURES IN LAYERING

Tip Layering

In tip layering, rooting takes place near the tip of the current season's shoot which is bent to the ground. The shoot tip begins to grow downward into the soil but recurves to produce a sharp bend in the stem, from which roots develop. This natural method of reproduction is characteristic of trailing blackberries, dewberries, and black and purple raspberries.

Stems of these plants are biennial in that the canes are vegetative during the first year, fruitful the second, and pruned out after fruiting. "Summer topping" new canes by pinching off 3 to 4 in of the tip after growth of 18 to 30 in. encourages lateral shoot production. This will increase the number of potential tip layers, and also next year's fruit crop. By late summer, the canes begin to arch over, and their tips assume a characteristic appearance in that the terminal ends become elongated and the leaves small and curled to give a "rat-tail" appearance. The best time for layering is when

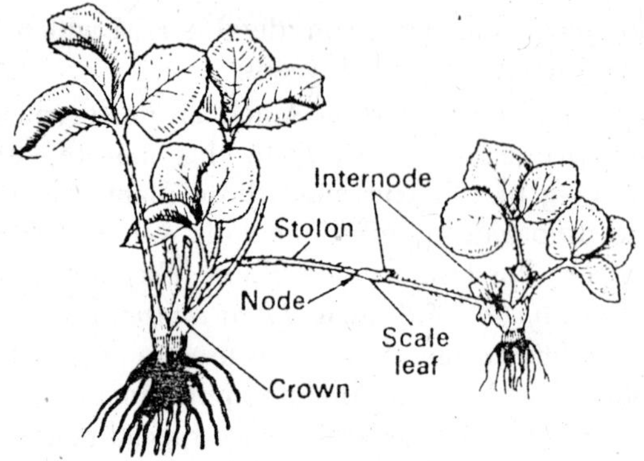

Fig. 3.1. A stolon.

only part of the lateral tips have attained this appearance. If the operation is done too soon, the shoots may continue to grow instead of forming a terminal bud. If it is done too late, the root system will be small.

The tips are preferably layered by hand, a spade or trowel being used to make a hold 3 to 4 in. deep. The end of the shoot is inserted into the hole and covered with soil. For large scale operations, a shallow furrow may be plowed along the row into which the tips are placed. Rooting takes place rapidly, and the plants are ready for digging by the end of the same season. The rooted tip consists of a terminal bud, a large mass of roots, and 6 to 8 in. of the old cane to serve as a "handle" and to mark the location of the new plant. Since the tip layers are tender, easily injured and subject to drying out, digging should be done preferably just before replanting.

Rooted tip layers are planted in the late fall or early spring. New canes develop rapidly during the first season.

Simple Layering

Simple layering is performed by bending a branch to the ground and covering it partially with soil or rooting medium, but leaving the terminal end exposed. The end of the branch is sharply bent to an upright position about 6 to 12 in. back from the tip. The sharp bending of the shoot may be all that is necessary to induce rooting, although additional benefit may be gained by twisting to loosen the bark. Cutting or notching the underside of the stem is often practiced. The bent part of the shoot is next inserted into the soil so that it can be covered to a depth of 3 to 6 in. A wooden peg, bent wire, or stone may be used to hold the layer in place, and a vertical wooden stake should be inserted beside the layer to support the exposed shoot and hold it upright. If the branch is relatively inflexible and hard to bend, the tension may be lessened by notching the upper side of the shoot on the highest part of the bend back from the layer itself.

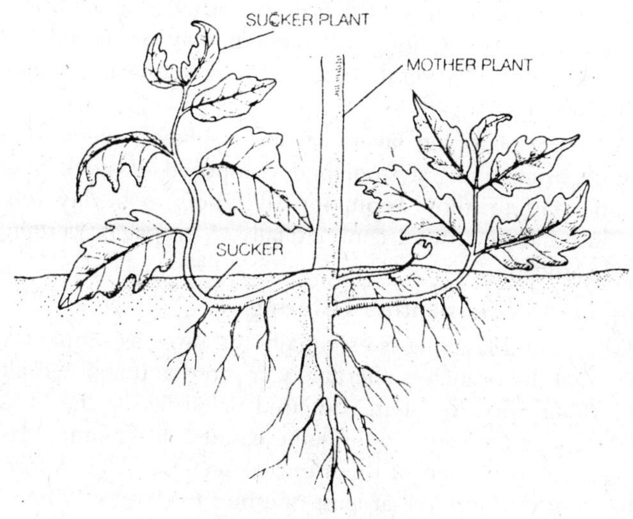

Fig. 3.2. A sucker of Chrysenthemum.

The usual time for layering is in the early spring, and dormant, 1-year-old shoots are used. Low, flexible branches of the plant which can be bent easily to the ground are chosen. In some cases, the suckers produced near the crown of the plant may serve as a source of shoots for layering. Layering could also be delayed until later in the growing season after the current season's shoots have attained sufficient length and become hardened. This timing would be used, perhaps, with some broad leaved evergreens such as *Rhododendron* and *Magnolia.* As a general rule, shoots older than one year are not satisfactory for layering.

Shoots layered in the spring will usually be adequately rooted by the end of the first growing season and can be removed either in the fall or in the next spring before growth starts. Mature shoots layered in summer should be left through the winter and either removed the next spring before growth begins or left until the end of the second growing season. When the rooted layer is removed from the parent plant, it is treated essentially in the same way as a rooted cutting of the same plant. Evergreen plants should be potted and kept humid and cool for a time. It should be possible to plant a well-rooted layer of a dormant deciduous plant directly into a nursery row or permanent location if the top is reduced to a size corresponding to the root system. Rooted layers removed in the fall can be planted in a cold frame or shaded greenhouse in a peat-sand mixture. Although defoliation may occur, root activity often continues for several weeks, and by spring the root system may be well developed.

A supply of rooted layers can be produced over a period of years by establishing a stool bed composed of stock plants far enough apart to allow room for all shoots to be layered. This procedure has been used commercially to propagate certain hard-to-root shrubs.

Compound or Serpentine Layering

Compound layering is essentially the same as simple layering, except that the branch is alternately covered and exposed along its length. Generally, the stem is injured or girdled at the lower part of the stem and covered in the same manner as for simple layering. Roots develop at each of these buried sections. The exposed part of the stem should have at least one bud to develop a new shoot. After rooting takes place or at the end of the growing season, the branch is cut in sections made up of the new shoot and the portion containing roots. Several new plants are thus possible from a single branch. This method is used for propagating plants which have long, flexible shoots, such as Muscadine grape. Ornamental vines, such as *Wisteria* and *Clematis*, can also be rooted this way, although such efforts are generally limited to use by amateurs rather than by commercial propagators.

Fig. 3.3. Compound layering of a Philodendron vine.

Air Layering

In air layering, roots form on the aerial part of a plant after the stem is girdled or slit at an angle and enclosed in a moist rooting medium at the point of injury. This method has been used for more than a thousand years, being known in various parts of the world as *Chinese layerage, pot layerage, circumposition, marcottage,* and *gootee.*

The principal limiting factor in air layering has been the difficulty in keeping the rooting medium properly moistened. The ancient Chinese gootee or marcottage method consists of plastering a ball of clay or other soil mixture about the ring or girdle, which is then covered with moss or fiber to hold it together. Above the gootee is a receptacle of water (perhaps a bamboo joint) from which extends a string leading to and wrapped about the gootee to keep it continuously moist. Various modifications of this technique have been made, but all require that the operation be carried out in greenhouses or in regions of high humidity.

Enclosures used to surround the rooting medium have included metal or wooden boxes, split flower pots, paper cones, and rubber sheeting. Wrapping with moss alone may be adequate if the humidity is sufficiently high and daily syringing is practiced. These devices have been replaced to a large extent by sheets of polyethylene film, which have the properties of high permeability to gases (carbon dioxide and oxygen), low transmission of water vapour, and sufficient durability to withstand weathering for long periods. By this means it is possible to keep the enclosed rooting medium damp from the plant's natural moisture for long periods of time without continuous attention. Not all plastic films have the same degree of the necessary characteristics; only *polyethylene* film should be used.

Air layering is used to propagate a number of tropical and subtropical trees and shrubs. In Florida, the litchi and the Persian lime (*Citrus aurantifolia*) are propagated commercially by this method. With polyethylene film for wrapping the layers, it is possible to extend the method of air layering out-of-door plants from the tropics to the temperature zones. However, its greatest application is for the amateur horticulturist or the propagator of selected plants for cultural or for scientific purposes rather than for the commercial nurseryman. Air layers are made in the spring on wood of the previous season's growth or, in some cases, in the late summer with partially hardened shoots. Wood older than one year can be used in some cases, but rooting is less satisfactory and the larger plants produced are somewhat more difficult to handle after rooting. With tropical greenhouse plants, layering should be done after several leaves have developed during a period of growth.

The first step in air layering is to girdle or cut the bark of the stem at a point 60 to 12 in. or more from the tip end. A strip of bark ½ to 1 in. wide, depending upon the kind of plant, is completely removed from around the stem. Scraping the exposed surface to insure complete removal of the phloem and cambium is desirable to retard healing. Another procedure is to make a slanting cut about 2 in. long up and to the center of the stem, keeping the two surfaces apart by sphagnum or a piece of wood. Application of a root promoting material, such as indolebutyric acid to the exposed stem, especially to the upper edges or between the two exposed surfaces of a cut, may be beneficial. About two handfuls of only *slightly* moistened sphagnum moss are placed around the stem to enclose the cut surfaces. If the moisture content of the sphagnum moss is too high, decay of the stem tissues will occur.

A piece of polyethylene film 8 to 10 in. square is wrapped carefully about the branch so that the sphagnum moss is completely covered. The ends of the sheet should be folded (as in wrapping meat) with the fold placed on the lower side. The two ends must be twisted to make sure that no water can seep inside. Adhesive tape, such as electrician's waterproof tape, serves well to wrap the ends; the winding should be started well above the end of the plastic film to enclose the ends, particularly the upper one, securely. Budding rubbers and florist's ties are other materials which can be used for this purpose.

The time for removal of the layer from the parent plant is best determined by observing root formation through the transparent film. In some plants, rooting occurs in 2 to 3 months or less. Layers made in spring or early summer are best left until the shoots become dormant in the fall, and are removed at that time. Holly, lilac, *Azalea*, and *Magnolia* should be left for two seasons. In general, it is desirable to remove the layer at a time when it is not actively growing. The period between the removal of the rooted layer from the plant and establishment on its own roots is very critical, and the plant may be lost at that time even though it rooted well. The root system at this time is small in proportion to the top and is incapable of supporting the plant without special precautions. The difficulties of transplanting increase in proportion to the size of the top. Severe pruning usually is advisable to bring the top more in line with the size of the roots, but it may not be imperative if the following precautions are followed. The rooted layer should be potted into a suitable container and placed under cool, humid conditions, such as an enclosed frame, where the plants can be frequently syringed. If this operation is done in the fall, a sufficiently large root system may develop by spring to permit successful growth in the open. Placing the rooted layers under mist for several weeks, followed by gradual hardening-off, is probably the most satisfactory procedure.

Mound (Stool) Layering

Mound or stool layering involves cutting a plant back to the ground during the dormant season and mounding soil or other media around the base of the newly developing shoots in the spring to encourage roots to form on them. Covering with soil keeps the shoots etiolated and encourages root formation. Plants with stiff branches that do not bend easily and which are capable of producing an abundance of shoots from the crown year after year are particularly adapted to this method. Plants commonly propagated in this manner include the clonal apple stocks, certain Mazzard cherry stocks, currants, and gooseberries.

A stool bed can remain in use for 15 to 20 years; therefore it should be on loose, fertile, well-drained soil and established 1 year

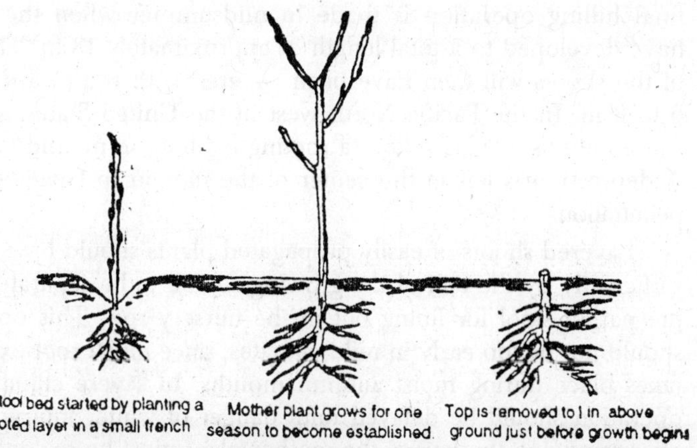

Fig. 3.4. Mound (stool) layering.

before the operations begin. The mother plants should be set 12 to 15 in. apart in the row, but the spacing between rows varies under different conditions and types of nursery equipment. Width between rows should be sufficient to allow for cultivation and hilling operations during spring and summer. In England 3½-ft rows are recommended, but in New York a minimum of 8 ft was found desirable. Planting in a shallow trench will permit the shoots to arise from the crown at a low level. Plants are cut back to 15 to 18 in. from the ground and left to grow unchecked for a season, the space between the rows being kept cultivated.

Before new growth starts the next spring, all plants are cut back to an inch above the ground level. Two to five new shoots usually develop from the crown the second year, more in later years. When these shoots have grown 3 to 5 in., loose soil or sawdust is drawn up around each shoot to one-half of its height. When the shoots have grown to a total height of 8 to 10 in., a second hilling operation takes place. Additional rooting medium is again added, to be mounded around the bases of the shoots but not to more than half their total height. In England, with a spacing of 3½ ft between the rows, a ridging plow is drawn between the rows so that two ridges of soil are produced about 15 in. apart. Each row of plants will thus have a ridge on each side. The trench made by the plow is kept cultivated, and the loose soil is shoveled around and between the shoots. The shoots arise close together and, to prevent crowding, should be spread apart by placing a shovelful of soil in the middle of a cluster of shoots. A third and final hilling operation is made in midsummer when the shoots have developed to a total length of approximately 18 in. The base of the shoots will then have been covered with soil to a depth of 6 to 8 in. In the Pacific North west of the United States, sawdust is used almost exclusively as a rooting medium in mound layering. A depression is left in the center of the mound to facilitate water penetration.

Layered shoots of easily propagated plants should have rooted sufficiently by the end of the growing season to be separated from the parent stool for lining out in the nursery row. This operation should not be too early in mild climates, since much root extension takes place during moist autumn months. In severe climates, the operation should be delayed until danger of winter injury is past; after removing the layers, the mother plant must be re-covered by several inches of soil. The rooted layers are cut as close to their base as possible to keep the height of the stool plant low. After the cutting away of the shoots, the mother stool remains exposed until new shoots again reach a height of 3 to 5 in., and the operation is underway again. Shoots which do not root or which root poorly often can be cut off at the same time as the well-rooted ones and treated as hardwood cuttings ready for planting in the nursery row.

Several points which are important to the proper utilization of a stool bed should be kept in mind. To prolong the life of the stool bed and to maintain the plants in a vigorous condition, good fertility should be maintained by annual fertilizer application. Disease, insects, and weeds must be controlled. Supplementary overhead irrigation is essential in some areas to maintain proper moisture conditions in the rooting zone. Soil must be added around the base of the shoots while they are still soft and succulent, not delayed until the shoots have hardened. With more difficult plants, such as certain plum varieties, it is best to add soil before the shoot begin to grow so as to induce etiolation at their bases. Girdling the bases of the shoots by wiring about 6 weeks after they begin to grow will stimulate rooting in many plants. To gradually rejuvenate old stool beds, new layers may be planted in spaces between old ones.

Trench Layering

Trench layering consists of growing a plant or a branch of a plant in a horizontal position in the base of a trench and filling in soil around the new shoots as they develop. Roots develop from the bases of these new shoots.

The first step in this procedure involves the establishment of the mother bed which, as in mound layering, can be used over a period of years. Rooted layers or 1-year-old nursery-budded or grafted trees are planted 18 to 30 in. apart at an angle of 30 to 45 deg. down a row. The rows should be 4 to 5 ft apart–wide enough to allow for cultivation and to draw soil up around the plants to a height of 6 in. The plants are then cut back to a uniform length–18 to 24 in.–and left to grow one season. In some cases- in layering walnuts, for instance–the plants are planted horizontally in the trench and the developing shoots layered the first year.

Before the beginning of growth in the spring, the parent layers are bent over and laid flat on the bottom of a trench dug along the row, about 2 in. deep and wide enough to receive the entire layer. Weak lateral branches are cut to ½ in. in length and strong laterals merely tipped back. Wooden pegs may be used to hold the mother layer in place. Short lengths of wire bent to form a U, or a longer single wire rolled on the end and set at an angle, will be useful to hold down the small shoots. It is important that all parts of the shoot be completely flat on the floor of the trench.

Before the buds swell, the entire layer is covered with 1 to 2 in. of fine soil or other rooting medium such as peat moss, sawdust, or wood shavings. Another inch of medium is added when the developing shoots have pushed through the first soil layer but before their leaves have expanded. Several more additions of soil are made the first 2 or 3 weeks of the growing season to ensure the etiolation of the basal 2 to 3 in. of the shoot. When the shoots have grown an additional 3 to 4 in. soil is again added to half the height of the exposed shoot. Further additions are made at intervals until the bases of the shoots are finally covered to a depth of 6 to 8 in. by midsummer. Roots form on the base of the current season's shoot.

At the end of the growing season after the plants have become dormant–or the next spring–the soil is removed from around the layered shoots, and rooted layers are cut off from the original layered stock as close to the base as possible. Unrooted shoots can be left to be pegged down the nexy year to fill in gaps or to rejuvenate an old bed. New shoots may arise from adventitious buds along the old original layer under ground. However, some of the vigorous shoots growing from the mother plant can be pegged down each year to produce a new supply of rooted layers.

Trench layering is primarily a nursery method for propagating particular fruit tree rootstocks difficult to propagate by other methods. It could also be practiced on established shrubs or trees by bending long, flexible shoots or vines to the ground under the same conditions as is done in simple layering. The shoot is covered along its entire length, but the tip is left exposed. New shoots which develop from the buds along the stem grow upward through the soil, with roots forming at their base. This latter procedure is sometimes known as *continuous layering.*

Plant Modifications Suitable for Natural Layering

Some plants exhibit modifications of their vegetative structure or method of growth which lead to their natural vegetative increase.

Those listed below could be considered natural forms of layering and often can be utilized for propagation.

Runners

A *runner* is a specialized stem which develops from the axil of a leaf at the crown of a plant, grows horizontally along the ground, and forms a new plant at one of the nodes. The strawberry is a typical plant propagated in this way. Other plants propagated by runners include bugle (*Ajuga*) and the strawberry geranium (*Saxifraga sarmentose*). Plants of these species grow as a typical rosette or crown.

In most strawberry varieties, runner formation is related to the length of day, runners being produced under day lengths of 12 to 14 hours or more. New plants are produced at alternate nodes. These take root but remain attached to the mother plant for some time. New runners may in turn be produced by the daughter plants. The connecting stems die in the late fall and winter, and each daughter plant becomes separate from the others.

In propagating by runners, the rooted daughter plants are dug when they have become well rooted, and then transplanted to the desired locations.

Stolons

Stolon is a term used to describe various types of horizontally growing stems that produce adventitious roots when in contact with the soil. Specifically these are prostrate or sprawling stems, as found in *Cornus stolonifera*, for instance. The underground stem of the potato that terminates in the tuber is a stolon. The term is sometimes used to describe the tip layers of black raspberry and blackberry or the runners of strawberry. In such plants, the stolon could be treated as a naturally occurring rooted layer, cut from the parent plant and transplanted.

Offsets

An *offset* is a characteristic type of lateral shoot or branch which develops from the base of the main stem in certain plants. This term is applied generally to a shortened, thickened stem of rosette-like appearance. Many bulbs reproduce by producing typical offset bulblets from their base. The term offset (or *offshoot*, as is sometimes used) also applies to lateral branches arising on stems of monocotyledons. The date palm produces lateral shoots from the base of the plant by which it is propagated. The pineapple is also propagated by offsets, although in commercial culture these are termed ratoons, suckers, or slips, depending upon the location on the plant where they are produced. Lateral shoots arising from rhizomes, as in the banana or orchid, are also offsets or offshoots.

Offsets are removed by cutting them close to the main stem with a sharp knife. If it is well rooted, the offset can be potted as is done with any rooted cutting. If insufficient roots are present, the shoot is placed in a favourable rooting medium and treated as a leafy stem cutting.

In cases in which offset development is meager, cutting off the main rosette may stimulate the development of offsets from the old stem just as removing the terminal bud stimulates lateral shoots in any other type of plant. For instance, in *Echeveria* the main stem may elongate so that the plant becomes a fleshy rosette borne on top of a fleshy, bare stem. The rosette, with a short piece of stem, may be removed and rooted, and new offsets will develop from buds at the base. It is desirable that this operation be carried out while the stem is somewhat soft and succulent, rather than allowing it to become hard and woody. Offshoots of the date palm do not root readily if separated from the parent plant. They are usually layered for a year prior to removing.

Suckers

A *sucker* is a shoot which arises on a plant from below ground. The most precise use of this term is to designate a shoot which arises from an adventitious bud on a root. However, in practice, shoots which arise from the vicinity of the crown are also referred to as suckers even though originating from stem tissue. Nurserymen generally designate any shoot produced from the rootstock below the bud union of a budded tree as a sucker and refer to the operation of removing them as "suckering". In contrast, a shoot arising from a latent bud of a stem several years old, as, for instance, on the trunk or main branches, should be termed a *watersprout*.

The tendency to "sucker" is a characteristic possessed by some plants and not by others. The ability of a plant to sucker and the ability of a plant to grow from root cuttings are closely related. In such cases, however, nurserymen prefer to use root cuttings rather than depend on naturally produced suckers. Since suckers arise from adventitious buds, they may show juvenile characteristics.

Suckers are dug out and cut from the parent plant. In some cases part of the old root may be retained, although most new roots arise from the base of the sucker. It is important to dig the sucker out rather than pull it, to avoid injury to its base. Suckers are treated essentially as a rooted layer or as a cutting, in case few or no roots have formed. They are usually dug during the dormant season.

Crowns

The term *crown* as generally used in horticulture designates that part of a plant at the surface of the ground from which new shoots are produced. In trees or shrubs with a single trunk, the crown is principally a points of location near the ground surface marking the general transition zone between stem and root. In herbaceous perennials, the crown is the part of the plant from which new shoots arise annually. The crown of many herbaceous perennials consists of many branches, each being the base of the current season's stem, which originated from the base of the preceding year's branch. These new shoots are stimulated to grow from the base of the old stem as it dies back after blooming. Adventitious roots develop along the base of the new shoots. These new shoots eventually flower either the same year they are produced or the following year. As a result of the annual production of new shoots and the drying back of old shoots, the crown may become extensive within a period of a relatively few years.

In certain plants for example, the strawberry or the African violet (*Saintpaulla*)–the stem is a short and thickened structure from which the leaves are produced in a rosette-like arrangement. The entire body of the plant is often referred to as the crown. Lateral shoots or offsets are produced from the base of the crown. An old plant may be composed of a number of "crowns" or "crown divisions" which have been produced in this manner.

Multibranched woody shrubs may develop extensive crowns. Although an individual woody stem may persist for a number of years, new, vigorous shoots are continuously produced from the crown, and eventually crowd out the older shoots. If left undisturbed, such shrubs develop into extensive thickets. Under normal handling, the older shoots are regularly removed by pruning to give way to the younger, more vigorous shoots.

Division of the crown is an important method of propagation for herbaceous perennials, and to some extent for woody shrubs, because of its simplicity and reliability. Such characteristics make this method particularly useful to the amateur or professional gardener who is generally interested in only a modest increase of a particular plant. Many herbaceous perennials must be divided every 2 to 3 years to prevent the plants from becoming overcrowded.

Crowns of outdoor herbaceous perennials are usually divided in the spring just before growth begins or in late summer or autumn at the end of the growing season. As a general rule, those plants which bloom in the spring and summer and produce new growth after blooming should be divided in the fall. Those which bloom in summer and fall and make little or no new growth until spring should be divided in early spring. Potted plants are divided when they become too large for the particular container in which they are growing.

In crown division, plants are dug and cut into sections with a knife. In herbaceous perennials such as the Shasta daisy or aster, where an abundance of new rooted offshoots are produced from the crown, each may be broken from the old crown and planted separately, the older part of the plant clump being discarded. If a large clump is desired, then a section of the old crown bearing a number of new shoots from its base may be used. In the case of plants in which the crown consists of a number of rosettes or offsets, as occurs in the African violet, division should be made between each of the rosettes, but roots should be present on each section.

Shrubs may be divided in the same manner with a shovel or hatchet. Such an operation should be carried out at a time when the plant is dormant. The top should be cut back and the roots trimmed at the time of divisions, and each section planted as a new shrub.

4

PROCESSES OF BUDDING

In contrast to grafting, in which the scion consists of a short detached piece of stem tissue with several buds, budding utilizes only one bud and a small section of bark, with or without wood. Budding is often termed "bud grafting", since the physiological processes involved are the same as in grafting.

The commonly used budding methods depend upon the bark "slipping". The term indicates the condition when the bark can easily separated from the wood. It denotes the period of year when the plant is in active growth, when the cambium cells are actively dividing, and newly formed tissues are easily torn as the bark is lifted from the wood. Beginning with new growth in the spring, this period should last until the plant ceases growth in the fall. However, adverse growing conditions, such as lack of water, defoliation, or low temperatures, may lead to a tightening of the bark and can seriously interfere with the budding operation. Getting the stock plants in the proper condition for budding is an important consideration. Of the commonly used methods described here, only one–the chip bud can be done when the bark is not slipping.

The budding operation, particularly T-budding, can be performed more rapidly than the simplest method of grafting, some rose budders inserting as many as 2000 or 3000 or more T-buds a day if the tying is done by helpers. If performed under the proper conditions, the percentage of successful unions in T budding is very high–90 to 100 percent. Budding is widely used in producing nursery stock of rose and fruit tree cultivars, where hundreds of thousands of individual plants are propagated each year. Therefore,

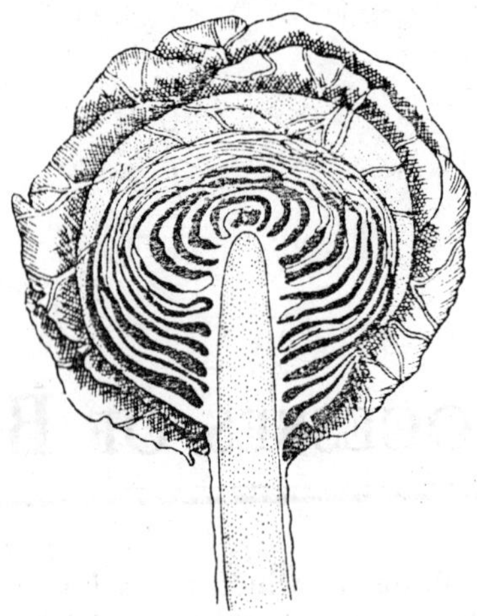

Fig. 4.1. A terminal bud of cabbage.

for propagation operations involving large numbers of plants, where speed and low mortality are essential, budding upon selected rootstocks is likely to be chosen. The use of budding is confined generally to young plants or the smaller branches of large plants

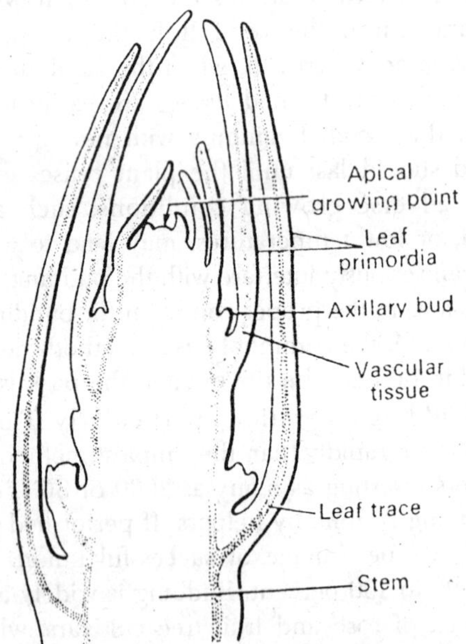

Fig. 4.2. Terminal and axillary buds in L.S.

where the buds can be inserted into shoots which are from 1/4 to 1 in. in diameter. Topworking young trees by top-budding is quite successful. Here the buds are inserted in small, vigorously growing branches in the upper portion of the tree.

Budding may result in a stronger union, particularly during the first few years, than is obtained by some of the grafting methods, and thus the shoots are not as likely to blow out in strong winds. Budding makes more economical use of propagating wood than grafting, each bud potentially being capable of producing a new plant of the desired variety. This may be quite important if propagating wood is scarce. In addition, the techniques involved in budding are simple and can be easily performed by the amateur.

Rootstocks for Budding

In propagating nursery stock of the various fruit and ornamental species by budding, a rootstock plant is used. It should have the desired characteristics of vigor, growth habit, and disease resistance, as well as being easily propagated. This rootstock plant may be a rooted cutting, a rooted layer, or more commonly a seedling. Usually, one year's growth in the nursery row before budding is to be done is sufficient to produce a rootstock plant large enough to be budded, but seedling of slow growing species may require two seasons.

To produce nursery trees free of harmful pathogens (viruses, *fungi*, or bacteria) it is essential that the rootstock plant, as well as the budwood, be free of such organisms.

Time of Budding–Fall, Spring, or June

The important budding methods are used at seasons of the year when the stock plant is in active growth and the cambial cells actively dividing so that the bark separates readily from the wood. It is also necessary that well-developed buds of the desired variety be available at the same

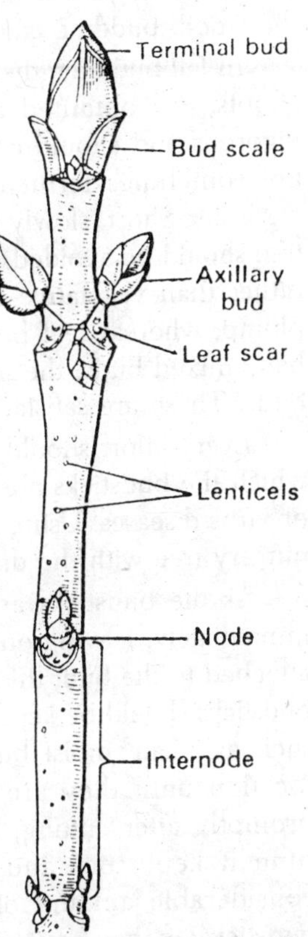

Fig. 4.3. Terminal and axillary buds.

time. These conditions exist for most plant species at three different times during the year. In the Northern Hemisphere, these periods are late July to early September (*fall budding*), March and April (*spring budding*), and late May and early June (*June budding*).

Fall Budding

This is the most important time of budding in the propagation of fruit tree nursery stock, although actually the budding is done mostly in late summer rather than fall. The rootstock plants are usually large enough by late summer to accommodate the bud, and the plants are still actively growing, with the bark slipping easily. Once growth has stopped and the bark adheres tightly to the wood, budding can no longer be done.

In fall budding, the budsticks, consisting of the current season's shoots, are obtained at the time of budding. They should be vigorous and should contain vegetative or leaf buds. Such shoots are sometimes termed *watersprouts*, especially if they are very vigorous. Short, slowly growing shoots on the outer portion of the tree should be avoided, because they may have chiefly flower buds rather than vegetative buds. Flower buds are usually around and plump, whereas leaf buds are smaller and pointed. Some species have mixed buds, the node containing both vegetative and flower buds. These are satisfactory for use in budding.

Every effort should be made to make sure that the trees from which the budsticks are obtained are free of any bacterial, fungus, or virus diseases. Using infected budsticks can infect every budded nursery tree with the disease.

As the budsticks are selected the leaves should be removed immediately, leaving only a short piece of the leaf stalk or petiole attached to the bud; this will aid in handling the bud later on. The budsticks should be kept from drying by wrapping in some material such as clean, moist burlap and keeping them in a cool, shady location until they are needed. The budsticks should be used promptly after cutting, although they can be stored for a short time if kept cool and moist. It is best, if possible, when a considerable amount of budding is being done, to collect the budsticks as they are being used, a day's supply at a time.

The best buds to use on the stick are usually those in the middle and basal portions. Buds on the succulent terminal portion of the shoot should be discarded. In certain species, such as the

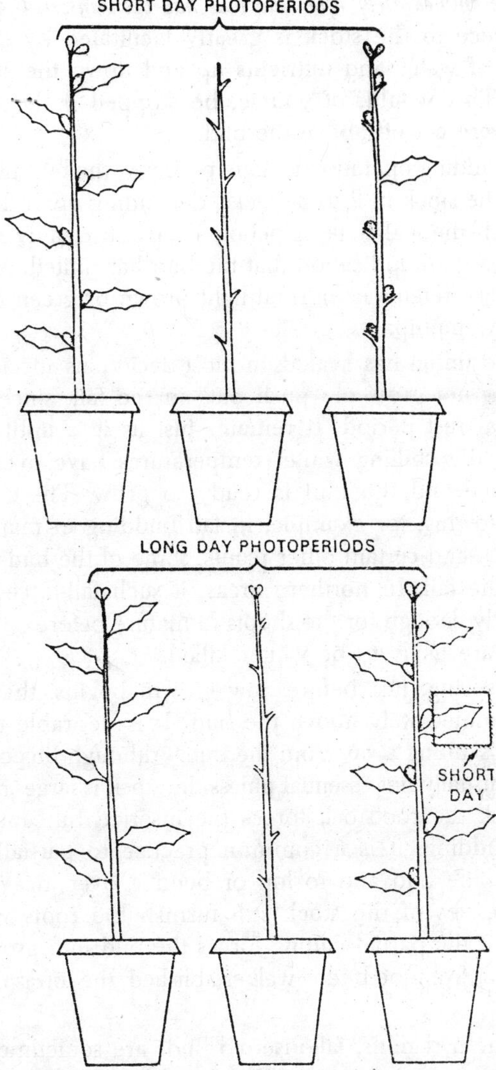

Fig. 4.4. Experiments on Xanthium plants indicate that photoperiodic stimulus is perceived by leaves.

sweet cherry, buds on the basal portion of the shoots are flower buds, which of course should not be used.

In fall budding, after the buds have been inserted, there is nothing more to be done until the following spring. *Although eventually the rootstock is to be cut off above the bud, in no case should this be done immediately after the bud has been inserted.* Healing of the bud piece to the stock is greatly facilitated by the normal movement of water and nutrients up and down the stem of the rootstock. This would, of course, be stopped if the top of the rootstock were cut off above the bud.

If the budding operation is done properly, the bud piece should unite with the stock in 2 to 3 weeks, depending upon the growing conditions. If the leaf stalk or petiole drops off cleanly next to the bud, this is a good indication that the bud has united, especially if the bark piece retains its normal light brown or green colour and the bud stays plump.

The bud union has healed, in most deciduous species the bud usually does not grow or "push out" in the fall, since it is in a physiological rest period. It remains just as it is until spring, at which time the chilling winter temperatures have overcome the rest influence and the bud is ready to grow. There are some exceptions to this; for example, in fall budding of maples, roses, honey locust, and certain other plants, some of the buds may start growth in the fall. In northern areas, if such fall-forced buds do not start early enough for the shoots to mature before cold weather starts, they are likely to be winter killed.

In the spring, just before new growth begins, the rootstock is cut off, immediately above the bud. It is desirable to make a sloping cut, slanting away from the bud. Although this cut may be waxed, it is usually not essential unless the stock is large in diameter. Cutting back the rootstock forces the inserted bud into growth. In citrus budding, it is a common practice to partially cut the stock above the bud and to lop or bend it over away from the bud. The leaves of the stock still furnish the roots with some nutrients, but the partial cutting forces the bud into growth. After the new shoot from the bud is well established, the top is completely removed.

In northern regions, fall-inserted buds are sometimes covered with soil during the winter until danger of frost has passed and are then uncovered and topped-back in late spring.

Where strong winds occur and in species in which the new shoots grow vigorously, support for the newly developing shoot may be necessary. One practice sometimes followed is to cut off the rootstock several inches above the bud, using this projecting

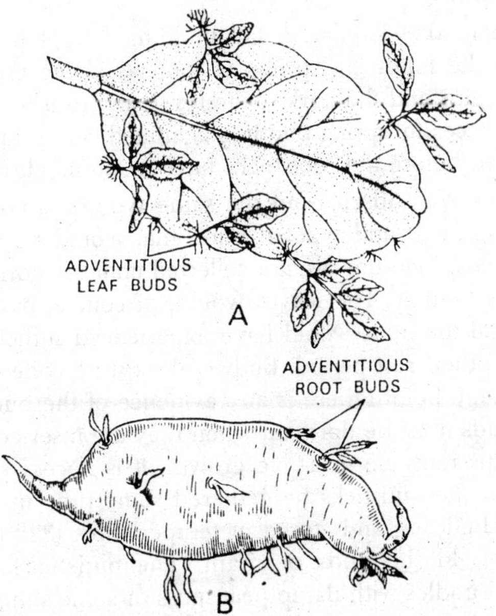

Fig. 4.5. A–Bryophyllum–Adventitious leaf buds, B–Sweet potato–Adventitious root buds.

stub as a support on which to tie the tender young shoot arising from the bud. This stub is removed after the shoot has become well established. In another procedure, stakes may be driven into the ground next to the stock to which the developing shoot is tied at intervals during its growth.

Cutting back to force the main bud to grow also forces many latent buds on the rootstock into growth. These must be rubbed off as soon as they appear, or they will soon choke out the desired inserted bud. It may be necessary to go over the budded plants several times before these "sprouts" stop appearing. Nurserymen refer to this procedure as "suckering".

The shoot arising from the inserted bud becomes the top portion of the plant. After one season's growth in the nursery, with favourable conditions of soil, water and nutrients, temperature, and insect and disease control. Such a tree would have a 1-year-old top and a 2-or perhaps 3-year-old root, but it is still considered a "yearling" tree. If the top makes insufficient growth the first year, it can be allowed to grow a second year, and is then known as a 2-year-old tree. However, abnormally slow-growing trees should be discarded.

Spring Budding

This method is similar to fall budding except that the work takes place the following spring as soon as active growth of the rootstock begins and the bark separates easily from the wood. The period for successful spring budding is limited, and budding should be completed before the rootstocks have made much new growth.

Budsticks are chosen from the same type of shoots-in regard to vigor of growth and type of buds–that would be used in fall budding, except that they are not collected until the dormant season the following winter. The leaves would, of course, have fallen by this time, and the buds would have experienced sufficient chilling to overcome their rest period. Budwood must be collected while it is still dormant–before there is any evidence of the buds swelling. Since the buds must be dormant when they are inserted, and since the rootstocks must be in active growth, it is necessary in spring budding that the budsticks be gathered some time in advance of the time of budding and stored at temperatures (32° to 40°F; 0° to 4°C) to hold the buds dormant. The budsticks should we wrapped in bundles with damp peat moss or some similar material to prevent drying out.

In spring budding, the actual budding operation should be done just as soon as the bark on the rootstock slips easily. Then, about 2 weeks after budding, when the bud unions have healed, the top of the stock must be cut off above the bud to force the inserted bud into active growth. At the same time, latent buds on the rootstock begin to grow and should be removed. Sometimes it is helpful to permit such shoots from the rootstock to develop to some extent to prevent sunburn and help nourish the plant. They must be held in check, however, and eventually removed.

Although the new shoot from the inserted bud gets a later start in spring budding than in fall budding, spring buds will usually develop rapidly enough, if growing conditions are favourable, to make a satisfactory top by fall. Fall budding, however, is to be preferred for several reasons : the higher temperatures at that time promote more certain healing of the union, the budding season is longer, there is no necessity to store the budsticks, the inserted buds start growth earlier in the spring, and the pressure of other work is usually not so great for nurserymen in the late summer as in the spring. Spring budding is used sometimes on rootstocks which were fall-budded but on which the buds failed to take.

June Budding

June budding is used to obtain a "1-year-old" budded tree in a single growing season. Its principal characteristic is that budding is done in the early part of the growing season and the inserted bud forced into growth immediately during the same season. As a method of nursery propagation, June budding is confined to regions which have a relatively long growing season–in the U.S, this includes California and the southern states. In the propagation of fruit trees, June budding is used mostly in producing such stone fruits as peaches, nectarines, apricots, almonds, and plums. Peach seedlings are generally used as the rootstock, but almond seedlings and plum hardwood cuttings can also be used. Budding is done by the T-bud method. If seeds are planted in the fall or startified seeds as early as possible in the spring, the seedlings usually attain sufficient size (12 in.high and at least 1/8 in. in diameter) to be budded by mid-May or early June (in the Northern Hemisphere). Preferably, June budding should not be done much after mid-June, or a nursery tree of satisfactory size will not be obtained by fall. June-budded trees are not as large by the end of the growing season as those propagated by fall or spring budding but they are of sufficient size (about 3/8 in. to 5/8 in. caliper and 3 to 5ft tall) to be used commercially.

The budwood used in June budding consists of current season's growth, that is, of new shoots which have developed since growth started in the spring. By late May or early June, these shoots will usually have grown sufficiently to have a well developed bud in the axil of each leaf. At this time of year these buds will not have entered the rest period, so when they are used in budding they continue their growth on through the summer, producing the top portion of the budded seedling.

For June-budded trees handling subsequent to the actual operation of budding is some what more exacting than for fall-or spring-budded trees. The rootstocks are smaller and have less stored food than those used in fall or spring budding. The object behind the following procedures–is to keep the rootstock (and later the budded top) actively and continuously growing so as to allow no check in growth, while at the same time changing the seedling shoot to a budded top. The bud should be inserted high enough (5 to 6 in.) on the stem so that a number of leaves–at least three or four–can be retained below the bud. The method of T-budding with the "wood out" described later should be used. Healing of the inserted bud should be very rapid at this time of year, since temperatures are relatively high, and rapidly growing, succulent plant parts are used. By 4 days after budding, healing should have started, and the top of the root stock can be cut back somewhat–2 to 5 in. above the bud–leaving at least one leaf above the bud and several below it. This operation will force the inserted bud into growth and will check terminal growth from basal buds of the rootstock, which will produce additional leaf area. This continuous leaf area is necessary so that there always will be enough leaves to keep manufacturing food for the small plant. Ten days to two weeks after budding, the rootstock can be cut back to the bud, which should be starting to grow. If the budding rubber has not broken, if should be cut at this time. Other shoots arising from the rootstock should be headed back to retard their growth. After the inserted bud grows and develops a substantial leaf area, it can supply the plant with the necessary nutrients. By the time the shoot from the inserted bud has brown 8 to 12 in. high, it should have enough leaves so that all other shoots and leaves can be removed. Later inspections should be made to remove any shoots arising from the rootstock below the budded shoot.

Another method which works well is to partially cut the stock just above the bud and break it over. Nutrients are still able to pass from the top of the stock to the roots, but this partial blocking forces the inserted bud into growth.

June budding is often of considerable value to nurserymen who find that their regular supply of fall or spring-budded trees is in sufficient to meet their expected demand. By this method they can, as late as the end of June, still propagate trees to be ready by fall, provided that they have a supply of rootstock plants large enough to bud.

METHODS OF BUDDING

T-Budding (Shield Budding)

This method is known by both names, the "T-bud" designation arising from the T-like appearance of the cut in the stock, whereas the "shield bud" name is derived from the shield-like appearance of the bud piece when it is ready for insertion in the stock.

T-budding is by far the most common method of budding and is widely used by nurserymen in propagating nursery stock of most

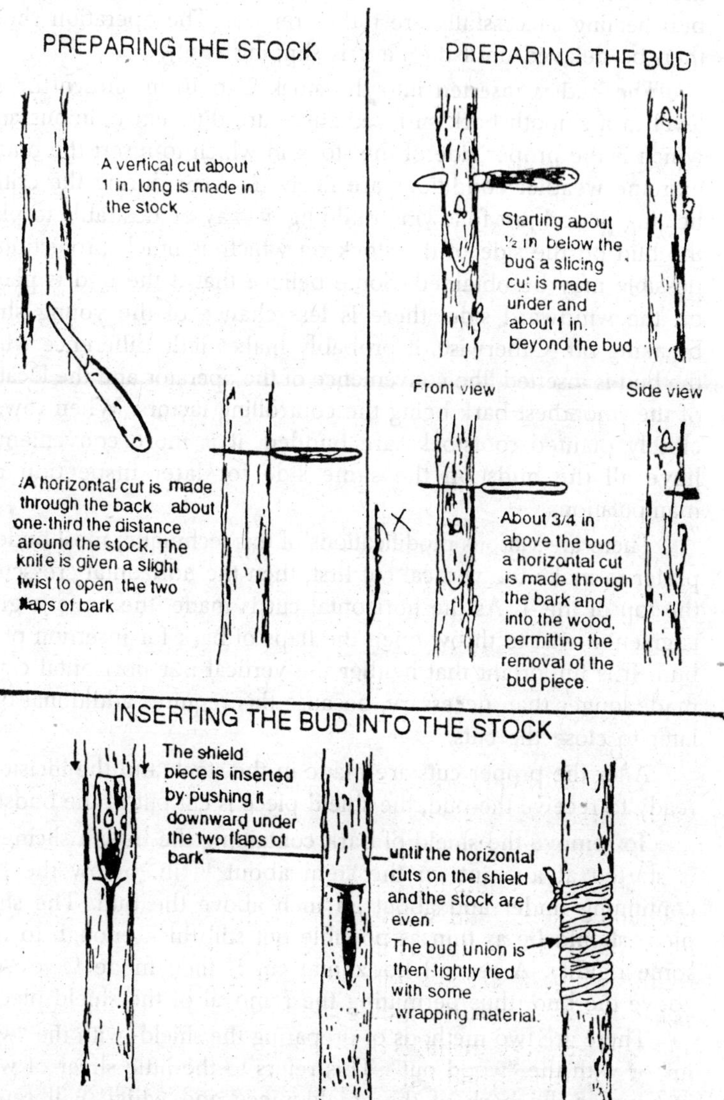

Fig. 4.6. Steps in making the T-bud (shield bud).

fruit tree species, roses, and many ornamental shrubs. Its use is generally limited to stocks which are about ¼ to 1 in. in diameter, with fairly thin bark, and which are actively growing so that the bark will separate readily from the wood. If the bark is so tight on the wood that it has to be forcibly pried loose, the chances of the bud healing successfully are rather remote. The operation should then be delayed until the bark is slipping easily.

The bud is inserted into the stock 2 to 10 in. above the soil level in a smooth bark surface. There are different opinions as to which is the proper side of the stock in which to insert the bud. If extreme weather conditions are likely to occur during the critical healing period just following budding, it may be desirable to place the bud on the side of the stock on which as much protection as possible may be obtained. Some believe that if the bud is placed on the windward side, there is less chance of the young shoot breaking off. Otherwise, it probably makes little difference where the bud is inserted, the convenience of the operator and the location of the smoothest bark being the controlling factors. When rows of closely planted rootstocks are budded, it is more convenient to have all the buds on the same side for later inspection and manipulations.

There are various modifications of this technique; most budders prefer to make the vertical cut first, then the horizontal crosscut at the top of the T. As the horizontal cut is made, the knife is given is given a twist to throw open the flaps of bark for insertion of the bud. It is important that neither the vertical nor horizontal cut be made longer than necessary, because this requires additional tying later to close the cuts.

After the proper cuts are made in the stock and the incision is ready to receive the bud, the shield piece is cut out of the budstick.

To remove the shield of bark containing the bud, a slicing cut is started at a point on the stem about ½ in. below the bud, continuing under and about an inch above the bud. The shield piece should be as thin as possible but still thick enough to have some rigidity. A second horizontal cut is then made ½ to ¾ in. above the bud, thus permitting the removal of the shield piece.

There are two methods of preparing the shield–with the "wood in" or with the "wood out". This refers to the little sliver of wood just under the bark of the shield piece and which will remain attached to it if the second horizontal cut is deep and goes through the bark and wood, joining the first slicing cut. Some professional budders believe it is best to remove this sliver of wood, but others retain it. In budding certain species, however, such as maples and walnuts, much better success is usually obtained with "de-wooded" buds. If it is desired to prepare the shield with the wood out, the second horizontal cut should be just deep enough to go through the bark and not through the wood. Then if the bark is slipping easily, the bark shield can be snapped loose from the wood (which still remains attached to the budstick) by pressing it against the budstick and sliding it sideways. A small core of wood comprising the vascular tissues supplying the bud is present, and this should remain in the bud, rather than adhering to the wood and leaving a hole in the bud. If the shield is pulled outward rather than being slid sideways from the wood, this core usually pulls out of the bud, eliminating the chances of success. In June budding of fruit trees, the shield piece is usually prepared with the wood out. In most other instances, however, the wood is left in. In spring budding, using dormant budwood, this silver of wood is tightly attached to the bark and cannot be removed.

The next step is the insertion of the shield piece containing the bud into the incision in the stock plant. The shield is pushed under the two raised flaps of bark until its upper horizontal cut matches the same cut on the stock. The shield should fit snugly in place, will covered by the two flaps of bark, but with the bud itself exposed.

No waxing is necessary, but the bud union must be wrapped, using materials to hold the two components firmly together until healing is completed. Rubber budding strips, especially made for wrapping, are widely used for this purpose. Their elasticity provides sufficient pressure to hold the bud securely in place. The rubber, being exposed to the sun and air, usually deteriorates, breaks, and drops off after several weeks, at which time the bud should be healed in place. If the budding rubber is covered with soil, the rate of deterioration will be much slower. This material has the advantage of eliminating cutting the wrapping ties, which can be a costly operation if thousands of plants have been budded. The rubber will expand as the rootstock grows, and thus there is little danger of constriction.

In tying the bud, the ends of the budding rubbers are held in place by inserting them under the adjacent turn. The bud itself should not be covered. The amount of tension given the budding rubber is quite important. It should not be too loose, or there will be too little pressure holding the bud in place. On the other hand, if the rubber is stretched extremely tight, it may be so thin that it will deteriorate rapidly and break too soon–before the bud union has taken place. Often the tying is done from the top down to avoid forcing the bud out through the horizontal cut.

Raffia (fiber-like leaf segments of certain *Raphia* species) has been widely used for wrapping buds. This material is soaked in water over-night before it is used so that it will be flexible. Raffia must be cut later–about 10 days after budding–to prevent constriction at the bud union at the plant grows. If such nonelastic ties are not promptly cut, the resultant constriction can have a very adverse affect on subsequent growth of the bud.

Plastic ties of polyvinyl chloride (PVC) film, 3/8 in. wide, are quite useful for budding. Such material is moisture-proof and elastic, and since it is transparent, it permits inspection of the buds after covering.

Inverted T-Budding

In localities where a great deal of rain occurs, water running down the stem of the rootstock will often enter the T-cut, soak under the bark, and cause decay of the shield piece. Under such conditions an inverted T-bud may give better results, since it is more likely to shed excess water. In citrus budding, the inverted T method is widely used, even though the conventional method also gives excellent results. In species which bleed badly during budding, such as chestnuts, the inverted T-bud allows better drainage and better healing.

The techniques of the inverted T-bud method are the same as those already described, except that the incision in the stock has the transverse cut at the bottom rather than at the top of the vertical cut, and in removing the shield piece from the budstick the knife starts above the bud and cuts downward below it. The shield is removed by making the transverse cut ½ to ¾ in. below the bud. The shield piece containing the bud is inserted into the lower part of the incision and pushed upward until the transverse cut of the shield meets that made in the stock.

It is important in using this inverted T-bud method that a normally oriented shield bud piece should not be inserted into an inverted incision in the stock. The bud would then have a reversed polarity. Although such upside down buds do live and grow, at least in some species, their use may not result in normal shoot development.

Patch Budding

The distinguishing feature of patch budding and related methods is that a rectangular patch of bark is removed completely from the stock and replaced with a patch of bark of the same size containing a bud of the variety to be propagated.

Patch budding is somewhat slower and more difficult to perform than T-budding, but it is widely and successfully used on thick-barked species, such as walnuts and pecans, in which T-budding gives poor results, presumably owing to the poor fit around the margins of the bud. Patch budding, or one of its modifications, is also extensively used in propagating various tropical species, such as the rubber tree (*Hevea brasiliensis*).

Patch budding requires that the bark of both the stock and budstick be slipping easily. It is usually done in late summer or early fall, but can also be done in the spring. In propagating nursery stock, the diameter of the root stock and the budstick should preferably be about the same, from ½ to 1 in. Although the budstick should not be much larger than about 1 in. in diameter, the patch can be inserted successfully into stocks as large as 4 in. in diameter, although adequate healing of such large stubs may be a problem.

Special knives have been devised to remove the bark pieces from the stock and the budstick. Some type of double bladed knife that will make two transverse parallel cuts 1 to 1-3/8 in. apart is necessary. These cuts, about an inch in length, are made through the bark to the wood in a smooth area of the rootstock several inches above the ground. Then the two transverse cuts are connected at each side by vertical cuts made with a single-bladed knife.

The patch of bark containing the bud is cut from the budstick in the same manner in which the bark patch is removed from the stock. Using the same two-bladed knife, the budder makes two transverse cuts through the bark, one above and one below the bud. Then two vertical cuts are made on each side of the bud so that the bark piece will be about an inch wide. The bark piece containing the bud is now ready to be removed. It is important that it be slid off sideways rather than being lifted or pulled off. There is a small core of wood, the bud trace, which must remain inside the bud if a successful "take" is to be obtained. By sliding the bark patch to one side, this core is broken off, and it stays in the bud. If the bud patch is lifted off, this core of wood is likely

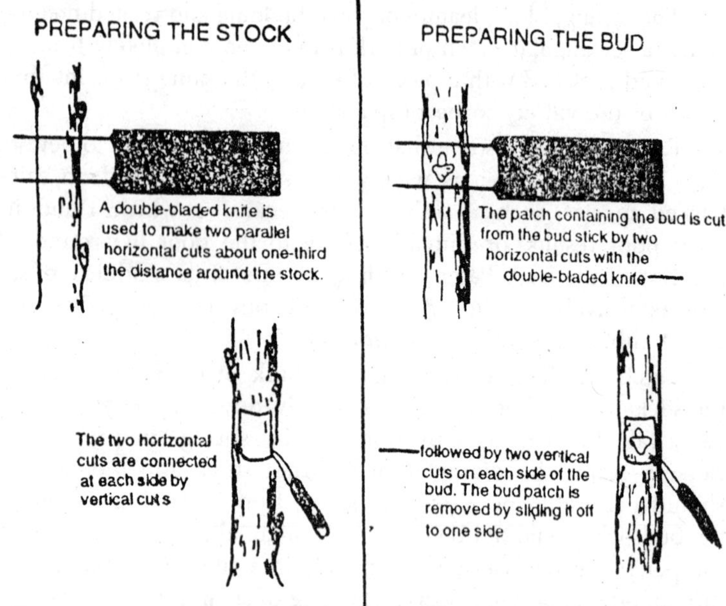

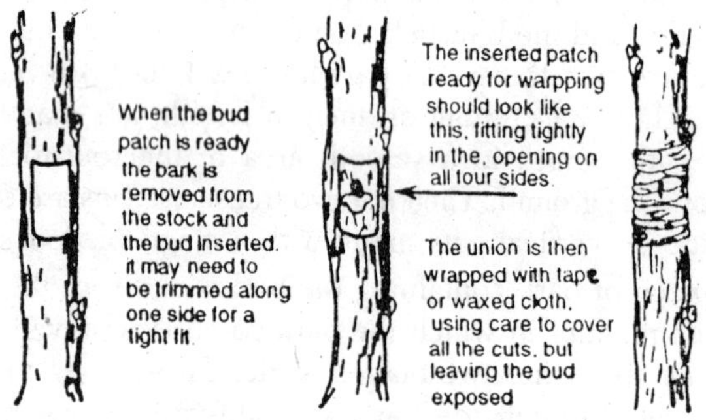

Fig. 4.7. Steps in making the patch bud.

to remain attached to the wood of the budstick, leaving a hold in the bud.

After the bud patch is removed from the budstick, it must be inserted immediately on the stock, which should already be prepared, needing only to have the bark piece removed. The patch from the budstick should fit snugly at the top and bottom into the opening in the stock, since both transverse cuts were made with the same knife. It is more important that the bark piece fit tightly at top and bottom than that it fit along the sides. The inserted patch is now ready to be wrapped. Often the bark of the stock will be thicker than the bark of the inserted bud patch so that, upon wrapping, it is impossible for the wrapping material to hold the bud patch tightly against the stock. In this case it is necessary that the bark of the stock be pared down around the bud patch so that it will be of the same thickness, or preferably slightly thinner, than the bark of the bud patch. Then the wrapping material will hold the bud patch tightly in place.

One type of tool, has a pair of vertical cutting blades in addition to the horizontal blades. With this, all four cuts of the rectangle are made at once. In cases in which difficulty is experienced in obtaining successful unions with the patch bud method, it may help to make the four cuts of the rectangle in the stock 1 to 3 weeks ahead of the time when the actual budding is to be done. This bark patch is not removed from the stock, however, until the patch containing the bud is ready to be inserted. When the cuts are made ahead of time, the wounding causes the callusing process to start, so that when the new bark patch is inserted, it heals very rapidly.

In wrapping the patch bud, a material should be used which not only will hold the bark tightly in place but will cover all the cut surfaces to prevent the entrance of air under the patch, with subsequent drying and death of the tissues. The bud itself mud not be covered during wrapping. The most satisfactory material is nurserymen's adhesive tape. Waxed cloth strips, are also satisfactory and have been used extensively. Another method is to tie the bud patch with heavy cotton string and cover the cut edges with grafting wax. It is important in patch budding, especially with walnuts, that the wrapping not be allowed to cause a constriction at the bud union. When the stock is rapidly growing, it is necessary to cut the tape about 10days after budding. A single vertical knife cut on the side opposite the bud is sufficient. The cut tape should not be pulled off.

Patch budding is best performed in late summer when both the seedling stock and the source of budwood are growing rapidly and their bark slipping easily. The budsticks for patch budding done at this time should have the leaf blades cut 2 to 3 weeks before they are taken from the tree. The petiole or leaf stalk is left attached to the base of the bud, but by the time the budstick is taken this petiole has dropped off or is easily removed.

Patch budding can also be done in the spring after new growth has appeared on the stocks and it has been determined that the bark is slipping. There is a problem, however, in obtaining satisfactory buds to use at this time of year, since it is necessary that the bark of the budstick separate readily from the wood. At the same time, the buds should not be starting to swell. There are two methods by which satisfactory buds can be obtained for patch budding in the spring. In one, the budsticks are selected during the dormant winter period and stored at low temperatures (about 36°F; 2°C) and wrapped in moist sphagnum or peat moss to prevent their drying out. Then, about 3 weeks before the time the spring budding is to be done, they are brought out into a warm room and set with their bases in a container of water, or they may be left packed in damp peat or sphagnum moss. The increased temperature will cause the cambium layer to become active, and soon the bark will slip sufficiently for the buds to be used. Although a few of the more terminal buds on each stick may start swelling in this time and cannot be used, there should be a number of buds in a satisfactory condition.

The second method of obtaining buds for spring patch budding is to take them directly from the tree which is the source of the budwood, but at the time the budding is to be done. If the trees are inspected carefully, it will be seen that not all of the buds start pushing at once. The terminal ones are usually more advanced than the basal ones. There is a period when the bark is slipping easily throughout the shoot containing the desired buds, but when only a few of the buds have developed so far that they cannot be used. The remaining buds, which are still dormant but upon bark that can be removed readily, may be taken and used immediately for budding. It is easier to obtain suitable buds from young trees which made vigorous shoot growth the previous year than it is from old trees. When the budding is done will be governed then by the stage of development of the buds. This will vary considerably with the species and variety being used. Stage of development of the rootstock is not as critical. Its bark must be slipping well, and the budding should be done before the stock plant has made much new growth.

Flute Budding

Flute budding is similar to patch budding, except that the patch of bark removed from the rootstock almost completely encircles it, leaving a narrow connection (about one-eighth of the circumference) between the upper and lower parts of the stock. In taking the bud patch from the budstick, a two-bladed knife is used, with the two transverse cuts completely encircling the budstick. A single vertical cut connects the two horizontal cuts, permitting the bud patch to be removed. In fitting this bud patch to the stock, it may be necessary to shorten its circumference by a vertical cut to remove the surplus amount of bark. If the bud patch fails to unite, the narrow connecting strip of bark on the stock keeps the top of the stock alive.

Ring or Annular Budding

By this method, a complete ring of bark is removed from the stock and a complete ring from the budstick. In order for the two to match, the size of the stock and that of the budstick should be about the same. Since the stock is completely girdled, if the bud patch fails to heal in, the stock above the ring may eventually die. This method is not as widely used as the ordinary patch bud, since it has no particular advantages and is rather cumbersome to perform.

I-Budding

In this type of budding, the bud patch is cut just as for patch budding, that is, in the form of a rectangle or square. Then, with the same parallel-bladed knife, two transverse cuts are made through the bark of the stock. These are joined at their centers by a single vertical cut to produce the figure I. The two flaps of bark can then be raised for insertion of the bud patch beneath them. It may make a better fit to slant the side edges of the bud patch. In tying the I-bud, care should be taken to see that the bud patch does not buckle upward and fail to touch the stock.

This method should be considered for use when the bark of the stock is much thicker than that of the budstick. In such cases, if the patch bud were used, considerable paring down of the bark of the stock around the patch would be necessary. This operation is not necessary in the I-bud method.

Chip Budding

This is a budding method that can be done at times when the bark is not slipping e.g., early in the spring before growth starts, or during the summer when active growth stops prematurely owing to lack of water or some other cause. It is generally used with

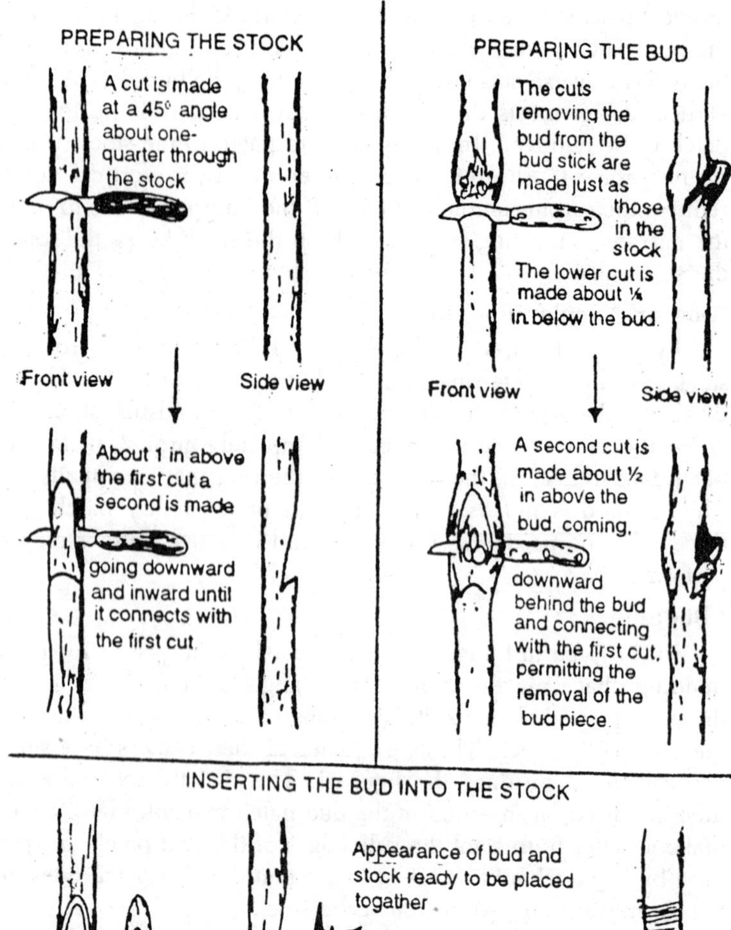

Fig. 4.8. The chip bud is really a form of grafting, a variation of the side-veneer graft, with the scion reduced to a small piece of wood containing only a single bud.

fairly small material, ½ to 1 in. in diameter. Although chip budding is quite successful, it is not as fast or simple as T-budding and is not likely to be used if conditions are favourable for the use of T-budding. Chip budding in the fall has given consistently excellent results, however, in budding grape varieties or phylloxera or nematode- resistant rootstocks. It is not commonly used on deciduous tree fruit species.

A chip of bark is removed from a smooth place between nodes near the base of the stock and replaced by another chip of the same size and shape from the budstick which contains a bud of the desired variety. The chips in both stock and budstick are cut out in the same manner. The first cut is made just below the bud and down into the wood at an angle of about 45 deg. The second cut is started about ½ in. above the bud and goes inward and downward behind the bud until it intersects the fist cut. The order of making these two cuts may be reversed. The chip is removed from the stock and replaced by the one from the budstick. If they have both been cut to the same size and shape–as they should be–a good fit is obtained. It is important that the cambium layer of the bud piece be placed so as to coincide with that of the stock, preferably on both sides of the stem, but at least on one side.

There are no protective flaps of bark to prevent the bud piece from drying out, as there are in T-budding. It is very important, then, that the chip bud be wrapped to seal the cut edges as well as to hold the bud piece tightly into the stock. Nurseryman's adhesive tape works very well for this, although string can be used if all the cut edges are covered with grafting wax. If the bud is inserted into the stock close to the ground level, as it is in grape propagation, it is sufficient to wrap the bud with budding rubber and cover the whole bud union immediately with several inches of finely pulverized moist soil, which can be removed after the bud has united. Budding rubbers under the soil level are very slow to disintegrate, so they may need to be cut or removed before constriction occurs.

In chip budding, as in the other methods, the stock is not cut back above the bud until the union is completed. If the chip bud is inserted in the fall, the stock is cut back just as growth starts the next spring. If the budding is done in the spring, the stock is cut back about 10 days after the bud has been inserted.

Top-Budding

In young trees when there is an ample supply of vigorous shoots at a height of 4 to 6 ft, top-budding provides a fast and certain method of topworking. It can be used in older trees, too, if they are cut back rather severely the year before to provide a quantity of vigorous watersprout shoots fairly close to the ground.

Depending upon the size of the tree, 10 to 15 buds are placed in vigorously growing branches 1/4 to 3/4 in. in diameter in the upper portion of the tree–about shoulder height. Although a number of buds could be placed in a single branch, usually only one will be saved to develop into secondary branches, which will then form the permanent new top of the tree. The T-bud method is used on thin-barked species, and the patch bud on those with thick bark.

Top-budding is usually done in midsummer, as soon as well matured budwood can be obtained and while the stock tree is still in active growth with the bark slipping easily. Orchard trees generally stop growth earlier in the season than young nursery trees ; therefore the budding must be done earlier. When top-budding is done at this time of year, the buds will remain inactive until the following spring. At that time, just as vegetative growth starts, the stock branches are cut back just above the buds. This forces the buds into active growth, and they should develop into good-sized branches by the end of the summer. At the time the shoots are cut back to the buds, all other unbudded branches should be removed at the trunk. It is important that the trees be inspected carefully through the summer and any shoots removed that are arising from any but the inserted buds.

Top-budding can also be done in the spring just as the tree to be top-worked is starting active growth and the bark is slipping easily.

Double-Working by Budding

In propagating nursery trees, some of the budding methods can be used in developing double-worked trees. The intermediate stock can be budded on the rootstock; then the following year the desired variety is budded on the interstock. Although quite effective, this is a rather lengthy process, taking 3 years. It is possible, to develop a double-worked tree in one operation in 1 year by the double shield bud method. A T-bud is used, but just under and below a budless shield piece of the desired interstock is inserted.

In studies comparing the two methods of "double-shield" budding–that of Garner and that of Nicolin–it appears that the Nicolin method, with a full-length intermediate piece, is simpler and more easily done, but the results by either method are satisfactory, at least for some combinations.

Micro-Budding

This type of budding is used successfully in propagating citrus trees and probably could be utilized also for other tree and shrub species. It has been of commercial importance in the citrus districts of southeastern Australia. Micro-budding is similar to ordinary T-budding, except that the bud piece is reduced to a very small size. The leaf petiole is cut off just above the bud, and then the bud is removed from the budstick by a flat cut just underneath the bud, with a razor-sharp knife. Only the bud itself and a small piece of wood under it are used. In the stock an inverted T-cut is made, and the micro-bud is slipped into this, right side up. The entire T-cut, including the bud, is covered with polyvinyl chloride (PVC) plastic budding tape (3/4 in. in width and 0.002 in. thick). The tape is allowed to remain for 10 to 14 days for spring budding and 3 weeks for all budding, after which it is removed by cutting with a knife. By this time the buds should have healed in place; subsequent handling is the same as for conventional T-budding.

5

GRAFTING AND BUDDING

Grafting is the art of joining parts of plants together in such a manner that they will unite and continue their growth as one plant. The part of the graft combination which is to become the upper portion or top of the new plant is termed the *scion* (cion), and the part which is to become the lower portion or root is termed the *rootstock*, or *understock*, or sometimes the *stock*. All methods of joining plants are properly termed *grafting*, but when the scion part is a small piece of bark (and sometimes wood) containing a single bud, the operation is termed *budding*. Most of the situations discussed in this chapter affecting grafting would therefore apply equally well to budding, although the techniques involved are somewhat different.

Natural Grafting

One occasionally sees branches of trees that have become grafted together naturally following a long period of being pressed together without disturbance. The English Ivy (*Hedera helix*) forms such grafts, and detailed studies have been made of translocation in natural grafts of this species. Not so obvious but of much greater occurrence, particularly in forest stands, is the natural grafting of roots. Such grafts are most common between roots of the same tree or between roots of trees of the same species. Grafts between roots of trees of different species are rare. In the forests living stumps sometimes occur, kept alive because their roots have become grafted to those of nearly intact, living trees. The anatomy of natural grafting of aerial roots has been studies; the initial contact is established by the formation and fusion of epidermal hairs. Such natural grafting provides a means of virus transmission from an infected tree to its neighbours. This can be important in closely set orchard and nursery plantings of fruit trees, where numerous root grafts could occur and result in the slow spread of the virus throughout the planting . Natural root grafting is a potential source of error in virus indexing procedures where virus-free and virus-infected trees are grown in close proximity. The fact that natural grafting does take place throws some doubt on the assumption usually held that each tree is an independent, discrete organism.

Formation of the Graft Union

A number of detailed studies have been made of the healing of graft unions, mostly with woody plants.

Briefly, the usual sequence of events in the healing of a normal graft union is as follows:

(a) Freshly cut scion tissue capable of meristematic activity is brought into secure, intimate contact with similar freshly cut stock tissue in such a manner that the cambial regions of both are in close proximity. Temperature and humidity conditions must be such as to promote activity in the newly exposed, and surrounding, cells.

(b) The outer exposed layers of cells in the cambial region of both scion and stock produce parenchyma cells which soon intermingle and interlock; this is called *callus tissue.*

(c) Certain cells of this newly formed callus which are in line with the cambium layer of the intact scion and stock differentiate into new cambium cells.

(d) These new cambium cells produce new vascular tissue, xylem toward the inside and phloem toward the outside, thus establishing vascular connection between the scion and stock, a requisite of a successful graft union.

The healing of a graft union can be considered as the healing of a wound. Such injury to tissue as would occur if the end of a branch were split longitudinally would heal quickly if the split pieces were bound tightly together. New parenchyma cells would be produced by abundant proliferation of cells of the cambium region of both pieces, forming callus tissue. Some of the interlocking parenchyma cells differentiate into cambium cells, which subsequently produce xylem and phloem.

If between the two split pieces, one interposed a third, detached, piece which had been cut so that a large number of its cells in the cambial region could be placed in intimate contact with cells of the cambial region of the two split pieces, proliferation of cells from all cambial areas would soon result in complete healing, with the foreign, detached piece joined completely between the two original split pieces. A graft union is essentially a healed wound, with an additional, foreign, piece of tissue incorporated into the healed wound.

This added piece of tissue, the scion, will not resume its growth successfully, however, unless vascular connection has been established so that it may obtain water and mineral nutrients. In addition, the scion must have a terminal meristematic region–a bud–to resume shoot growth.

In the healing of a graft union, the parts of the graft that are originally prepared and placed in close contact do not themselves move about or grow together. The union is accomplished entirely by cells which develop *after* the actual grafting operation has been made.

In addition, it should be stressed that in a graft union *there is no intermingling of cell contents.* Cells produced by the stock and by the scion each maintain their own distinct identity.

Considering in more detail the steps involved in the healing of a graft union, we may say that the first one listed below is a preliminary step, but nevertheless, it is essential and one over which the propagator has control.

(a) Establishment of intimate contact of a considerable amount of the cambial region of both stock and scion under favourable environmental conditions.

Temperature conditions that will cause high cell activity are necessary. Usually, temperatures from 55°to 90°F (12.8° to 32°C), depending upon the species, are conductive to rapid cell growth. The grafting operation should thus take place at a time of year when such favourable temperatures can be expected and when the plant tissues, especially the cambium, are in a naturally active state. These conditions generally occur during the early spring months.

The new callus tissue arising from the cambial region is composed of thin-walled, turgid cells which can easily become desiccated and die. It is important for the production of these parenchyma cells that the humidity in the vicinity of the cambial region of the graft union be kept at a high level. This explains the necessity of thoroughly waxing the graft union or using some other method to maintain a high degree of tissue hydration.

It is important, too, that the region of the graft union be kept as free as possible from pathogenic organisms. The thin walled parenchyma cells, at relatively high humidity and temperature, will provide favourable conditions for growth of certain fungi and bacteria which are exceedingly detrimental to the successful healing of the graft union. Prompt waxing of the graft helps prevent disease infection.

It is essential that the two original graft components be held together firmly by some means, such as wrapping, tying, or nailing, or better yet, by wedging (as in the cleft or sawkerf grafts) so that the parts will not move about and dislodge the interlocking parenchyma cells after proliferation has started.

The statement is often made that for successful grafting the cambium layers of stock and scion must be "matched." Although this is desirable, it is unlikely that complete matching of the two cambium layers is, or ever can be, attained. In fact, it is only necessary that the cambial regions be close fact, it is only necessary that the cambial regions be close enough together so that the parenchyma cells from both stock and scion produced in this region can become interlocked. It is in the region of the cambium that the essential callus production is the highest. Two badly matched cambial layers may delay union or, if extremely mismatched, prevent union. In studies of grafting monocotyledonous plants, it was found that a cambium layer is not required for a successful graft union but that any meristematic tissue could be utilized for this purpose and is capable of forming a union between stock and scion.

(b) Production and interlocking of parenchyma cells (callus tissue) by both stock and scion.

During the grafting operation the cells cut and damaged by the grafting knife turn brown and die. Underneath these dead cells new parenchyma cells arise from both stock and scion, coming from the parenchyma of the phloem rays and the immature parts of the xylem. The actual cambial layer itself seems to take little or no part in this first development of the callus. In grafting scions on established stocks, the stock produces most of the callus. These parenchyma cells, comprising the spongy callus tissue, fill the space between the two original components of the graft (the scion and the stock), becoming intimately interlocked and providing some mechanical support, as well as allowing for some passage of water and nutrients from the stock into the scion. For a time, between the callus arising from the stock and that arising from the scion, there is a more or less continuous brown line consisting of the above- mentioned dead and crushed cells remaining from the grafting cuts. This line of cells is gradually resorbed, however, and disappears. At the final stage of healing the cells of the outer layer of callus become suberized. Exposed dead cells (tracheids, vessels) are sealed off with a deposit of gum.

(c) Production of a new cambium across the callus "bridge"

At the edges of the newly formed callus mass, parenchyma cells which are touching the cambial cells of the stock and scion will differentiate into new cambium cells. This cambial formation in the callus mass proceeds farther and farther inward away from the original stock and scion cambium, and on through the callus bridge, until a continuous cambial connection between stock and scion has finally formed.

(d) Formation of new xylem and phloem from the new vascular cambium in the callus bridge

The newly formed cambial sheath in the callus bridge begins typical cambial activity, laying down new xylem and phloem, along with the original vascular cambium of the stock and scion, continuing this throughout the life of the plant.

In the formation of new vascular tissues following cambial continuity, it appears that the type of cells formed by the cambium may be influenced by the cells of the stock adjacent to the cambium. For example, xylem ray cells are formed where the cambium is in contact with xylem rays of the stock, and xylem elements where they were in contact with xylem elements.

The new xylem tissue originates from the scion rather than from the stock. This is shown by "ring grafting" (where a ring of bark from a young tree is removed and replaced by a ring of bark from another tree). Researchers found that using bark rings of the 'Scugog' apple variety, which has purple xylem, subsequent xylem growth of the tree following grafting was entirely purple in colour just under the 'Scugog' ring of bark, whereas in the remainder of the tree the xylem remained white.

This production of new xylem and phloem then establishes vascular connection between the scion and the stock. Under high transpiration rates this must occur before much new shoot growth takes place from buds on the scion; otherwise, the enlarging leaf surfaces on the scion shoots will have little or no water supply to offset that lost by transpiration, and the scion will quickly become desiccated and die. Under some conditions, however, where vascular connections fail to occur, enough translocation can take place between the parenchyma cells of the callus to permit survival. In grafts of the monocot, *Vanilla* orchid, scions survived and grew for two years with only parenchyma unions.

A somewhat different developmental sequence occurs in tobacco and cotton. Here, xylem tracheary elements or phloem sieve tubes, or both, form directly by differentiation of callus into these vascular elements. A cambium layer subsequently differentiates between the two vascular elements. Apparently parenchyma cells, which make up the callus, can differentiate into tracheid-like elements with relative ease. Buds, as would occur on the scions of grafts, seem to be effective in inducing processes of dedifferentiation and subsequent redifferentiation in the tissue on which they are grafted. This bud influence has been shown by inserting a bud into a piece of *Cichorium intybus* root consisting only of old vascular parenchyma, and observing that under the influence of the bud, the old parenchyma cells redifferentiate and dedifferentiate into groups of conducting elements.

The Healing Process in T-Budding

In T-budding, the bud piece usually consists of the periderm, cortex, phloem, and often some xylem tissue to which is attached a lateral bud. In budding, this bark piece is laid against the exposed xylem of the stock.

Detailed studies of the healing process in T-budding have been made for the rose, citrus and apple.

When the bud piece is removed from the budstick, and when the flaps of bark on either side of the "T" incision, on the stock are raised, the cambium and newly formed cambial derivatives in these tissues are usually destroyed, owing to their very tender, succulent nature.

In the apple when the bark is lifted in the budding operation the separation occurs in the young, undifferentiated xylem. The entire cambial zone remains attached to the inside of the bark flaps. After the bud shield is inserted, callus is produced, which surrounds the bud shield and holds it in place. The callus originates almost entirely from the rootstock tissue, mainly from the exposed surface of the xylem cylinder. Very little callus is produced from the sides of the bud shield. Callus starts development 2 days after budding and continues rapidly for 2 to 3 weeks until all internal air pockets are filled. Following this, a continuous cambium is established between the bud and the rootstock. The callus then begins to lignify, an isolated tracheary elements appear. Lignification of the callus is completed about 12 weeks after budding.

In the rose, about 3 days after budding the terminal cells of the broken xylem rays and adjacent cambial derivatives on the exposed surface of the stock begin to enlarge and divide, leading to the production of callus strands. In the same manner, callus strands develop from terminal cells of broken phloem rays and adjacent young secondary phloem cells on the cut surface of the inner side of the bud piece. Within 14 days the space between the stock and the bud piece is completely filled with callus, which has developed mainly from the proliferating immature secondary xylem of the stock and the immature secondary phloem of the bud piece. During the second week, short areas of cambium cells appear in this newly developed callus tissue. By the tenth day, a completed band of cambium tissue extends over the face of the stock and is joined to the uninjured cambium on either side of the bud piece.

After cambial continuity is completed, continuity of vascular tissues soon becomes established between the bud and stock. In T-budding, then, the primary union is between the surface of the phloem on the inner face of the shield and the meristematic xylem surface of the stock. A secondary type of union may occur, however, at the edges of the shield piece, as it would in chip budding.

The various stages in the healing process of the union in T-budding of citrus have been determined as follows:

Stage of Development	*Approximate Time after Budding*
1. First cell division	24 hours
2. First callus bridge	5 days
3. Differentiation :	
(a) in the callus of the bark flaps	10 days
(b) in the callus of the shield	15 days
4. First occurrence of xylem tracheids:	
(a) in the callus of the bark flaps 15 days	
(b) in the callus of the shield 20 days	

5. Lignification of the callus completed :
   (a) in the bark flaps — 25 to 30 days
   (b) under the shield — 30 to 45 days
6. First occurrence of meristematic layers in the callus between shield and bark flaps — 15 days

Factors Influencing the Healing of the Graft or Bud Union

As anyone experienced in grafting or budding knows, the results obtained in grafting are often inconsistent, an excellent percentage of "takes" occurring in some operations, whereas in others the results are very discouraging. There are a number of factors which influence the healing of graft unions.

Incompatibility

One of the symptoms of incompatibility in grafts between distantly related plants is a complete lack, or a very low percentage, of successful unions. Grafts between some plants known to be incompatible, however, will initially make a satisfactory union, even though the combination eventually fails.

Kind of Plants

Some plants are much more difficult to graft than others even when no incompatibility is involved. Difficult ones, for example, are the hickories, oaks, and beeches. Nevertheless, such plants, once successfully grafted, grow very well with a perfect graft union. In top grafting apples and pears, even the simplest techniques usually give a good percentage of successful unions, but in topgrafting certain of the stone fruits, such as peaches and apricots, much more care and attention to details are necessary. Strangely enough, topgrafting peaches to some other compatible species, such as plums or almonds, is more successful than reworking them back to peaches. Many times one method of grafting will give better results than another, or budding may be more successful than grafting, or vice versa. For example, in topworking the native black walnut (*Juglans hindsii*) to the Persian walnut (*Juglans regia*) in California, the bark graft method is more successful than the cleft graft.

Some easily grafted plants, such as the apple, form a "wound gum" plugging exposed xylem elements after the grafting operation, thus preventing excessive dessication and the death of tissues. Other plants, such as the walnut, in which graft unions heal with difficulty, form such "wound gum" very slowly, and in these desiccation and death of tissues in the area of the graft union may be extensive.

Some species, such as the Muscadine grape (*Vitis rotundifolia*), mango (*Mangifera indica*), and *Camellia reticulata*, are so difficult to propagate by the usual grafting or budding methods that "approach grafting," in which both partners of the graft are maintained for a time on their own roots, is often used. This variation among plant species and varieties in their grafting ability is probably related to their production of callus, which is essential for a successful graft union. *Camellia reticulata*, for example, is difficult to graft and is a very poor callus producer.

Temperature and Moisture Conditions During and Following Crafting

There are certain environmental requirements which must be met for callus tissue to develop.

Temperature has a pronounced effect on the production of callus tissue. In apple grafts little, if any, callus is formed below 32°F (0°C) or above about 104°F (40°C). Even around 40°F (4°C), callus development is slow and meager, and at 90°F (32°C) and higher, callus production is retarded, with cell injury becoming more apparent as the temperature increases, until death of the cells occurs

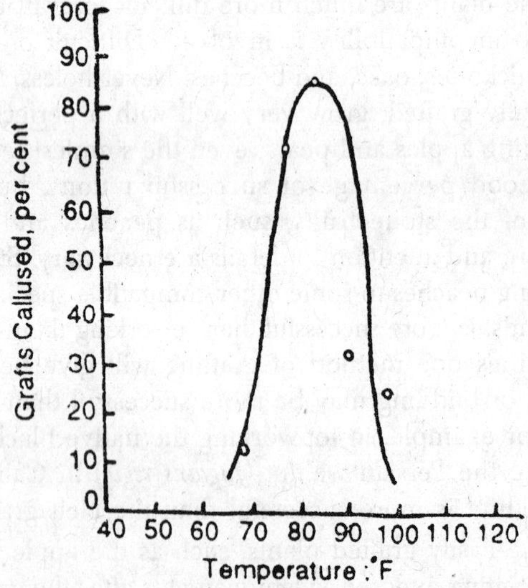

Fig. 5.1. Influence of temperature on the callusing of walnut (Juglans) grafts.

at 104°F. Between 40°and 90°F, however the rate of callus formation increases directly with the temperature. In such operations as bench grafting, callusing may be allowed to proceed slowly for several months by storing the grafts at relatively low temperatures (45° to 50°F; 7° to 10°C), or if rapid callusing is desired, they may be kept at higher temperatures for a shorter time. Since excessive callusing of root graft unions may result in the formation of undersirable "callus knots" (sometimes mistaken by plant inspectors for crown gall tumors), it can be controlled by storing well-callused grafts at reduced temperatures to prevent further callus development.

Temperature during the healing period following grafting was found to have a pronounced effect upon the degree of successful unions obtained in whip grafting of black walnuts. Studies show clearly that a temperature in the range of 77° to 86°F (25° to 30°C) during the callusing period gives much better results than either higher or lower temperatures. This was confirmed by studies in which black walnut root grafts placed in a cool nursery cellar at 41° to 56° (5°to 13°C) for callusing failed completely to unite. Another group of grafts planted in the nursery row immediately after grafting (air temperature 41° to 72°F; (5° to 22°C) resulted in only about 3 percent growing. In a third group which was allowed to callus in a warm green house (air temperature 65° to 80°, 8° to 27°C), an average of 88 percent grew.

Following bench grafting of grapes, a temperature of 70° to 75°F (21° to 24°C) is about optimum; 85°F (29°C) or higher results in profuse formation of a soft type of callus tissue which is easily injured during the planting operations. Below 70°F, callus formation is slow, and below 60°F (5°C), it almost ceases.

Grafting operations performed late in the spring when excessively high temperatures may occur, often result in failure. Tests of walnut top grafting in California during very hot weather in May showed that whitewashing the area of the completed graft union definitely promoted healing of the union. Whitewash would reflect a considerable portion of the radiant energy of the sun rather than allow it to be absorbed, thus resulting in lower bark temperatures. In addition in these tests scions placed on the north and east sides of the stub survived much better than these on the south and west. Undoubtedly this was due to the lower temperatures resulting from their shaded position. Since the parenchyma cells comprising the important callus tissue are thin-walled and tender, with no provision for resisting desiccation, it is obvious that if they are exposed to drying air for very long, they will be killed. This was found to be the case in studies of the effect of humidity on healing of apple grafts. Air moisture levels below the saturation point inhibited callus formation, the rate of desiccation of the cells increasing as the humidity dropped. In fact, the presence of a film of water against the callusing surface was much more conductive to abundant callus formation than just maintaining the air at 100 percent relative humidity. Unless a completed graft union is kept by some means at a very high humidity level, the chances of successful healing are rather remote. With most plants, thorough waxing of the graft union, which retains the natural moisture of the tissues, is all that is necessary.

Often root grafts are not waxed but stored in a moist packing material during the callusing period. Damp peat moss is an excellent material for callusing, because it provides proper moisture and aeration.

It has been shown that oxygen is necessary at the graft union for the production of callus tissue. This would be expected, since such rapid cell division and growth is accompanied by relatively high respiration, which requires oxygen. For some plants, a lower percentage of oxygen than is found naturally in air is sufficient, but for others, healing of the graft union is better if the union is left unwaxed and placed in an enclosure of water-saturated air. This may indicate that the latter plants have a high oxygen requirement for callus formation. Waxing restricts air movement to such an extent that oxygen may become limiting, and callus tissue fails to form. This situation apparently exists in grafting the grape, in which usually the union is not covered with wax or other air excluding material.

Growth Activity of the Stock Plant

Some propagation methods, such as T-budding and bark grafting, dependent upon the bark "slipping," which means the vascular cambium cells are actively dividing, producing young thin-walled cells on each side of the cambium. These newly formed cells readily separate from one another, so the bark "slips." Initiation of such cambial activity in the spring is thought to result from swelling of the buds in warm weather, since shortly afterwards cambial activity can be detected beneath each developing bud with a wave of cambial activity progressing down the stems and trunk. This stimulus is believed to be due to increased auxin levels originating in the expanding buds.

In budding seedlings in the nursery row in late summer it is important that they have an ample supply of soil moisture just before, during, and following the budding operation. If they should lack water during this period, active growth is checked, cell division in the cambium stops, and the chances of getting the buds to unite are lessened.

There is evidence to show that callus proliferation essential for a successful graft union-occurs most readily at the time of year just before and during "bud break" in the spring, diminishing through the summer and into the winter. Increasing callus proliferation takes place again in late winter, but this is not dependent upon breaking of bud dormancy.

Plants exhibiting strong root pressure (such as the walnut) will show, at certain periods of high growth activity in the spring, excessive sap flow or "bleeding" when cuts are made preparatory to grafting. Grafts made with moisture exudation around the union will not heal, and should be made at some other stage of growth; such "bleeding" can be overcome by making slanting knife cuts around the tree through the bark and into the xylem to permit the exudation to take place below the graft union. When potted rootstock plants to be grafted (in such species as *Fagus*, *Betula*, or *Acer*) show excessive root pressure, they should be put in a cool place with reduced watering until the "bleeding" stops.

On the other hand, potted rootstock plants, such as junipers or rhododendrons, when first brought into a warm greenhouse in winter for grafting, are dormant, and grafting done then would be unsuccessful. Grafting should be delayed until the rootstock plants have been held for several weeks at 60° to 65°F (15° to 18°C) and new roots start to form; then the rootstock plant is physiologically active enough for the union to heal.

When the rootstock plant is physiologically overactive (excessive root pressure and "bleeding"), or underactive (no root growth being made), some form of side graft in which the rootstock top is not at first removed, should be used. In situations in which the rootstock is neither overactive or underactive, one of the many forms of topgrafting, in which the top of the stock is completely removed at the time the graft is made, is likely to be successful.

Propagation Techniques

Even in using the standard grafting or budding methods, there are numerous possible variables which may affect the success of the operation. There are many opinions, often conflicting, about the proper techniques to use.

Sometimes the grafting technique is so poor that only a very small portion of the cambial regions of the stock and scion are brought together. Although healing occurs in this region and growth of the scion may start, when a large leaf area develops and high temperatures and high transpiration rates occur, sufficient movement of water through the limited conducting area cannot take place, and the scion subsequently dies. Other errors in grafting technique, such as poor or delayed waxing, uneven cuts, or use of desiccated scions can, of course, result in grafting failure.

Poor grafting techniques, although they may delay adequate healing for some time–weeks, or even years–do not in themselves cause any permanent incompatibility. Once the union is adequately healed, growth can proceed normally.

Virus Contamination, Insect Pests, and Diseases

Usring virus-infected propagating materials in nurseries can reduce bud "take" as well as the vigor of the resulting plant. In stone fruit propagation the use of budwood free of ring spot virus has consistently given improved percentages of "takes" over infected wood.

Topgrafting olives in California is seriously hindered in some years by attacks of the American plum borer (*Euzophera semifuneralis*), which feeds on the soft callus tissue around the graft union, result in the death of the scion. (This is easily controlled, however, by painting the completed graft, before waxing , with a 3 oz per gal, solution of 50 percent DDT in water). In England, nurserymen are often plagued with the red bud borer (*Thomasiniana oculiperda*), which eats the healing callus beneath the bud-shield in newly inserted T-buds.

Sometimes bacteria or fungi gain entrance at the wounds made in preparing the graft or bud unions. For example, it was found that a rash of failures in grafts of *Cornus florida rubra* on common *C. florida* stock was due to the presence of the fungus *Chalaropsis thielavioides.* Chemical control of such infections materially aids in promoting healing of the unions.

In South and Central American, rubber (*Hevea*) trees are propagated by a modification of the patch bud. A major cause of budding failures in these countries has been infection of the out surfaces by a fungus, *Diplodia theobromae.* Control of this infection has been obtained by fungicidal treatments.

In topgrafting mangos in Florida, control of the fungus diseases anthracnose and scab is essential for success. This is done by spraying the rootstock trees and the source of scionwood regularly with copper fungicides before grafting is attempted.

Possible Relation of Growth Substances to Healing of the Graft Union

Studies of the practical application of growth substances, particularly auxin, to tree wounds or to graft unions and their effect on promoting subsequent healing have not given consistent results; consequently such materials are not generally used for this purpose. It is known from tissue culture studies, however, that there is a definite relationship between callus production (which is

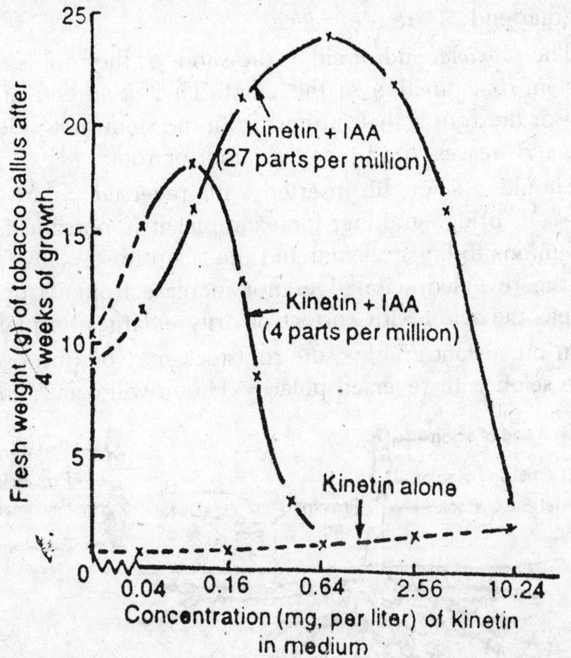

Fig. 5.2. Relationship between relative concentration of auxin (indoleacetic acid) and kinetin upon callus production.

essential for graft healing) and the levels of certain endogenous growth substances, particularly kinetin, and auxin. The relationship between auxin (indoleacetic acid) and kinetin levels in the production of callus in tobacco segments maintained under aseptic conditions. Perhaps further studies may show some practical benefit from the use of combinations of auxin and kinetin in stimulating callus formation and subsequent healing of graft unions, thereby facilitating grafting of combinations considered to be difficult.

Polarity in Grafting

Attention to proper polarity is very important if the graft union is to be permanently successful. In all commercial grafting operations correct polarity is strictly observed.

As a general rule, in grafting two pieces of stem tissue together, the morphologically proximal end of the scion should be inserted into the morphologically distal end of the stock. But in grafting a piece of stem tissue on a piece of root, as it is done in root grafting, the proximal end of the scion should be inserted into the proximal end of the root piece.

The *proximal* end of either the shoot or the root is that nearest the stem-root junction of the plant. The *distal* end of either the shoot or the root is that furthest from the stem-root junction of the plant and nearest the tip of the shoot or root.

Should a scion be inserted with reversed polarity– "upside down"–in bridge-grafting, for example, it is possible for the two graft unions to be successful and the scion to stay alive for a time. But, the reversed scion does not increase from its original size, whereas the scion with correct polarity enlarges normally.

In nurse-root grafting, the rootstock may be purposely grafted to the scion with reversed polarity. Union will occur, and the root

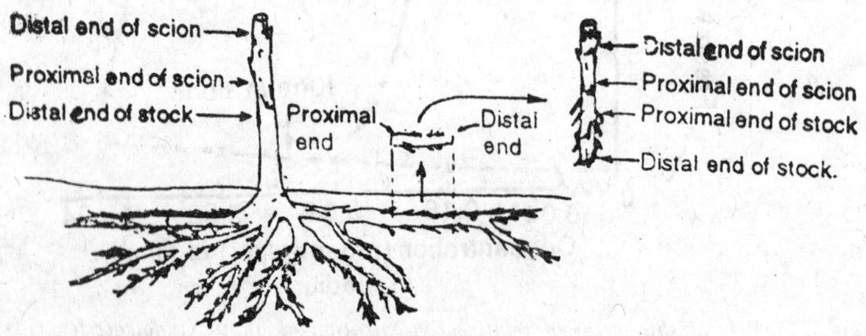

Fig. 5.3. Polarity in grafting.

will supply water and mineral nutrients to the scion. Since the scion is unable to supply necessary organic materials to the rootstock, the stock eventually dies. In this case, the graft union is purposely set well below the ground level, and the scion itself produces adventitious roots which ultimately become the entire root system of the plant.

In T-budding or patch budding, the rule for observance of correct polarity is not as exacting. Buds can be inserted with reversed polarity and still make permanently successful unions. The buds start growing downward, then the shoots recurve and start upward. In the inverted bud piece, the cambium seems capable of continued functioning and growth. However, in the vessels as well as in the fibres formed from cambial activity, there is a twisting configuration which apparently keeps translocation and water conduction oriented as in the original position.

If a complete ring of bark is removed from the trunk of a young tree, inverted, and regrafted into place, the growth of the tree will be markedly retarded. Organic materials that stimulate growth of the roots move downward through the phloem. When they reach the inverted bark piece, their movement is obstructed. Presumably, lack of these materials retards root growth subsequently dwarfing the entire tree. Such dwarfing is temporary, however; experiments have shown that within 2 years trees with such inverted bark pieces resume normal growth. This is due to a bridge of normally oriented tissues forming at the vertical seams of the ring grafts.

Limits of Grafting

Since one of the requirements for a successful graft union is the close matching of the callus- producing tissues near the cambium layers, grafting is generally confined to the dicotyledons in the angiosperms, and to the gymnosperms,. the cone-bearing plants. Both have a vascular cambium layer existing as a continuous tissue between the xylem and the phloem. In the monocotyledonous plants of the Angiospermae, which do not have a vascular cambium, grafting is more difficult and the percentage of "takes" much lower than in the dicots. There are cases of successful graft unions between the stem parts of monocots. By making use of the meristematic properties found in the intercalary tissues (located at the base of internodes), successful grafts have been obtained with various grass species as well as the large tropical monocotyledonous vanilla orchid of commerce.

Before a grafting operation is started, it should be determined that the plants to be combined are capable of uniting. There is no definite rule that can be followed exactly which will give this information. *Generally, the more closely the plants to be grafted are related botanically, the better the chances are of the graft union being successful.* This will not hold true consistently, however, since botanical classifications are based on reproductive characteristics, whereas grafting is concerned primarily with the vegetative properties of plants.

Grafting Within a Clone

A scion can be grafted back on the plant from which it came, and a scion from a plant of a given clone can be grafted to any other plant of the same clone. For example, a scion taken from an "Elberta' peach tree could be grafted successfully to any other 'Elberta' peach tree in the world.

Grafting between Clones within a Species

Almost always, different clones within a species can be grafted together without difficulty and will produce normal plants.

Grafting between Species within a Genus

For plants in different species but in the same genus, the situation is confused. In some cases grafting can be done successfully, in others it cannot. Grafting between most species in the genus *Citrus*, for example, is successfully and widely used commercially. Varieties of the almond *Prunus amydalus*), the apricot (*Prunus armeniaca*), the European plum (*Prunus domestica*), and the Japanese plum (*Prunus salicina*) all different species–are grafted commercially on the peach (*Prunus persica*),a still different species, as a rootstock. On the other hand, the almond and the apricot, both in the same genus, cannot be integrated successfully. The complexity of the situation is further illustrated by the fact that the 'Beauty' variety of Japanese plum (*Prunus salicina*) makes a good union when grafted on the almond, but another variety of *P. salicina*, 'Santa Rosa,' cannot be successfully grafted on the almond.

Compatibility between species may depend upon the particular clone or seedling used, either for stock or scion. For example, choice of the almond variety (clone) is highly important when grafting to 'Marianna 2624' plum as a rootstock, some varieties being, successful, others not. Likewise, clonal selections from different seedlings of the St.Julien plum (*Prunus insititia*), when used as rootstocks, have resulted in greatly different behaviour of the trees after top working to the same peach clone. For example, clonal selections A and C produced good peach trees with no signs of incompatibility, while trees of the same peach variety worked on selection B died within 5 years; peach trees on rootstock selection D died after 1 year.

There are some cases in which a given interspecies graft is successful, but the reciprocal combination is not. For instance, the 'Marianna plum' (*Prunus cerasifera* × *P. munsoniana* ?) on the peach (*Prunus persica*) makes an excellent graft combination, but grafts of the peach on 'Marianna plum either soon die or fail to develop normally. Although many varieties of Japanese plums (*Prunus salicina*) can be successfully grafted on the European plum (*P. domestica*), grafts of most varieties of the European plum on the Japanese are unsuccessful.

Grafting between Genera within a Family

When the plants to be grafted together are in different genera but in the same family, the chances of the union being successful become more remote. Cases can be found in which such grafts are highly successful and used commercially, but in most instances such combinations are failures.

Trifoliate orange (*Poncirus trifoliata*) is used commercially as a dwarfing stock for various species in the genus *Citrus*. The quince (*Cydonia oblonga*) has long been used as a dwarfing root stock for certain varieties of the pear (*Pyrus communis*). The reverse combination, quince on pear, though, is unsuccessful. Intergeneric grafts in the nightshade family, Solanaceae, are quite common. Tomato (*Lycopersicon esculentum*) can be grafted successfully on Jimson weed (*Datura stramonium*), tobacco (*Nicotiana tobacum*), potato (*Solanum tuberosum*), and black nightshade (*Solanum nigrum*).

Grafting between Families

Grafting such plants together is usually considered impossible, but there are reported instances in which it has been accomplished. These are with short-lived, herbaceous plants, though, for which the item involved is relatively short. Grafts, with vascular connections between the scion and stock, were successfully made, using white weet clover, *Melilotus alba* (Leguminosae), as the scion and sunflower. *Helianthus annuus* (Compositae), as the stock. Cleft grafting was used, with the scion inserted into the pith parenchyma of the stock. The scions continued growth with normal vigor for over 5 months. As far as is known, however, there are no instances in which woody perennial plants belonging to different families have been successfully and permanently grafted together.

For any successful grafting operation there are five important requirements :

(a) *The stock and scion must be compatible.* They must be capable of uniting. Usually, but not always, plants closely related, such as two apple varieties, can be grafted together. Distantly related plants, such as an oak tree and an apple tree, cannot be grafted together.

(b) *The cambial region of the scion must be in intimate contact with that of the stock.* The cut surfaces should be held together tightly by wrapping, nailing, or some other such method. Rapid healing of the graft union is necessary so that the scion may be supplied with water and nutrients from the stock by the time the buds start to open.

(c) *The grafting operation must be done at a time when the stock and scion are in the proper physiologcial stage.* Usually this means that the scion buds are dormant. For deciduous plants, dormant scion wood is often collected during the winter and kept inactive by storing at low temperature. The rootstock plant may be dormant or in active growth, depending upon the grafting method used.

(d) *Immediately after the grafting operation is completed, all cut surfaces must be carefully protected from desiccation.* This is done either by covering with grafting wax or by placing the grafts in moist material or in a covered grafting frame.

(e) *Proper care must be given the grafts for a period of time after grafting.* Shoots coming from the stock below the graft will often choke out the desired growth from the scion. Or, in some cases, shoots from the scion will grow so vigorously that they break off unless staked on tied or cut back.

Methods of Grafting

Whip, or Tongue, Grafting

This method is particularly useful for grafting relatively small material, 1/4 to 1/2 in. in diameter. It is highly successful if properly done because there is considerable cambial contact. It heals quickly and makes a strong union. Preferably, the scion and stock should

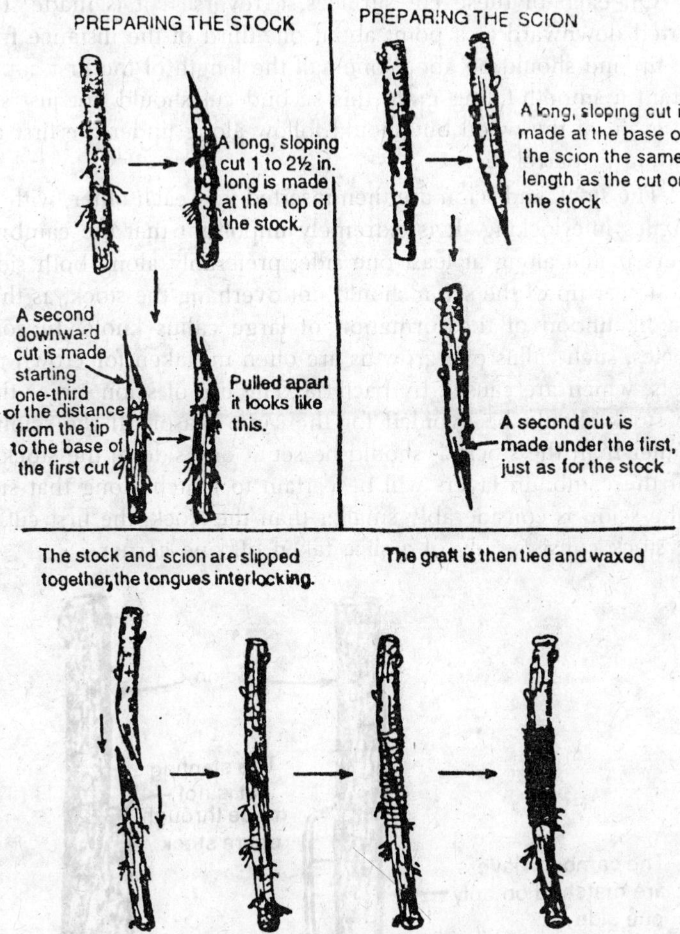

Fig. 5.4. The whip, or tongue, graft.

be of equal diameter. The scion should contain two or three buds with the graft made in the smooth internode area below the lower bud.

The cuts made at the top of the stock should be exactly the same as those made at the bottom of the scion. First, a long, smooth, sloping cut is made, 1 to 2½ in. long. The longer cuts are made when working with large material. This first cut should be made, preferably with one single stroke of the knife, so as to leave a very smooth surface. To do this, the knife must be sharp. Wavy, uneven cuts will not result in a satisfactory union.

On each of these cut surfaces, a reverse cut is made. It is started downward at a point about one-third of the distance from the tip and should be about one-half the length of the first cut. To obtain a smooth-fitting graft, this second cut should not just split the grain of the wood but should follow along under the first cut, tending to parrel it.

The stock and scion are then inserted into each other, with the tongues interlocking. It is extremely important that the cambium layers match along at least one side, preferably along both sides. The lower tip of the scion should not overhang the stock, as there is a likelihood of the formation of large callus knots. In some species, such callus overgrowths are often mistaken for crown gall knots, which are caused by bacteria. The use of scion larger than the stock should be avoided for the same reason. If the scion is smaller than the stock, it should be set at one side of the stock so that the cambium layers will be certain to match along that side. If the scion is considerably smaller than the stock, the first cut on the stock consists only of a slice taken off one corner.

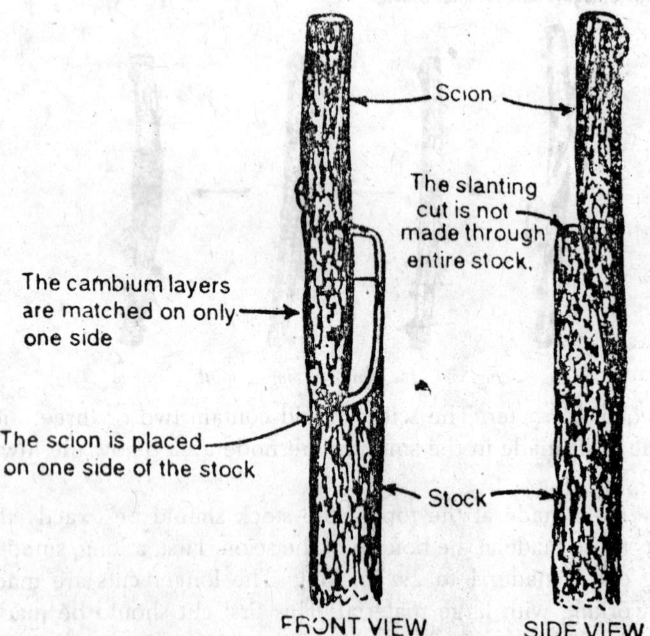

Fig. 5.5. Method of making a whip, or tongue, graft when the scion is considerably smaller than the stock.

After the scion and stock are fitted together, they should be held securely in some manner until the pieces have united. There are a number of possible ways of doing this.

(a) If the unions are very well made with a tight, snug fit, it is possible that no additional wrapping or tying is needed, but it is safer to provide some type of wrapping. If not wrapped, the grafts must be protected from drying by burying in moist sand, peat moss, or sawdust until the union has healed. Or they may be planted directely in the nursery with the union below soil level. If the whip graft is used in topworking, the exposed union must be protected in some manner.

(b) With a secure fit it may be sufficient to omit tying and merely cover the union with hot grafting wax, which will secure the pieces to some extent and give good protection against drying. This is not recommended for inexperienced grafters.

(c) A common method is to wrap the union with budding rubbers or possibly raffia or waxed string. After wrapping, the whole union can be covered with grafting wax. Waxing may be omitted if the grafts are to be protected from drying by burying in moist sand or peat moss, or if the grafts are planted immediately with the union below the soil surface. Grafts wrapped with budding rubbers and covered with soil should be inspected later, since the rubber decomposes very slowly below ground and may cause a constriction at the graft union.

(d) A practice widely used is to wrap the grafts with some type of adhesive tape. A special nurseryman's tape is available. The tape is drawn tightly around the graft union with the edges slightly overlapping. This holds the parts together very well and prevents drying, thus eliminating the need for waxing. If just one thickness of tape is used, it will decompose sufficiently fast (if the union is below ground) that no constriction of growth will develop. If used above ground, the tape should be cut after the graft has healed. The use of tight wrapping material such as this is especially recommended when difficulty is encountered with the formation of excessive callus.

(e) Plastic tapes are available for wrapping grafts. They are used just as adhesive tape, although they are not adhesive. The final turn of the tape is secured by slipping it under the previous turn. This tape has some elasticity. Also, it deteriorates more slowly below ground than above.

The whip graft can also be used in topworking young trees. The stocks in this case would be small, pencil-sized branches, well distributed around the tree. The grafts are usually tied with nurseryman's adhesive tape or with string, the latter covered with a coat of grafting wax. After the parts of the graft have united, the tying material must be cut; otherwise, the branch may be constricted as growth commences.

Splice Grafting

This method is the same as the whip, or tongue, graft except that the second or "tongue" cut is not made in either the stock or scion. A simple slanting cut of the same length and angle is made in both the stock and the scion. These are placed together and wrapped or tied as described for the whip graft. The splice graft is simple and easy to make. It is particularly useful in grafting plants that have a very pithy stem or which have wood that is not flexible enough to permit a tight fit when a tongue is made as in the whip graft.

Side Grafting

There are numerous variations of the side graft; three of the most useful are described. As the name suggests, the scion is inserted into the side of the stock, which is generally larger in diameter than the scion.

Stub Graft

This method is useful in grafting branches of trees that are too large for the whip graft yet not large enough for other methods such as the cleft or bark graft. For this type of side graft, the best stocks are branches about 1 in. in diameter. An oblique cut is made into the stock branch with a chisel or heavy knife at an angle of 20 to 30 degrees. The cut should be about 1 in. deep, and at such an angle and depth that when the branch is pulled back the cut will open slightly but will close when the pull is released.

The scion should contain two or three buds and be about 3 in. long and relatively thin. At the basal end of the scion, a wedge about 1 in. long is made. The cuts on both sides of the scion should be very smooth, each made by one single cut with a sharp knife. It is best to insert the scion into the stock at an angle as so as to obtain maximum contact of the cambium layers. The grafter inserts the scion into the cut while the upper part of the stock is

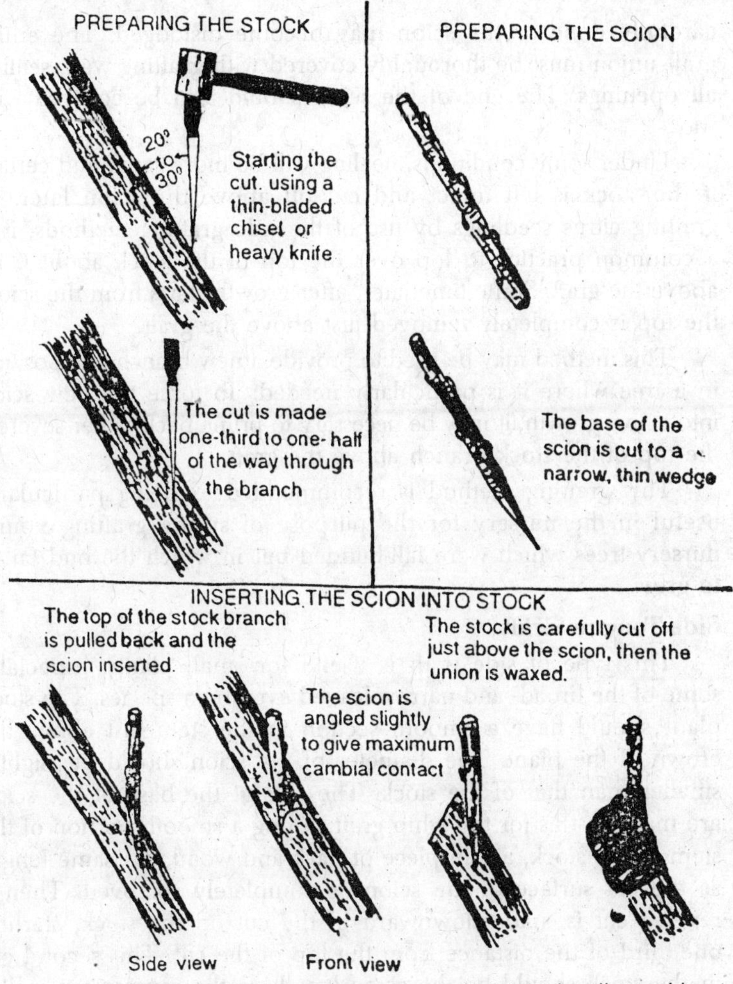

Fig. 5.6. Steps in preparing the side, or stub graft. A thinbladed chisel as illustrated here is ideal for making the cut.

pulled backward, using care to obtain the best cambium contact. Then the stock is released. The pressure of the stock should grip the scion tightly, making tying unnecessary, but if it is desired, the scion can be further secured by driving two small flat-headed wire nails (20 gauge, 5/8 in. long) into the stock through the scion. Wrapping the stock and scion at the point of union with nurseryman's tape may also be helpful. After the graft is completed, the stock may be cut off just above the union. This must be very carefully done or the scion may become dislodged. The entire graft union must be thoroughly covered with grafting wax, sealing all openings. The end of the scion should also be covered with wax.

Under some conditions, healing will be more rapid and certain if the stock is left intact and cut off above the scion later. In grafting citrus seedlings by use of the side grafting methods, it is a common practice to lop over the top of the stock about 6 in, above the graft. Some time later, after growth starts from the scion, the top is completely removed just above the graft.

This method may be used to provide a new branch at a position in a tree where it is particularly needed. To force the new scion into active growth, it may be necessary to prune back rather severely the top of the stock branch above the graft.

This grafting method is recommended as being particularly useful in the nursery for the purpose of spring grafting young nursery trees which were fall-budded but in which the bud failed to grow.

Side-Tongue Graft

This type of side graft is useful for small plants, especially some of the broad- and narrow-leaved evergreen species. The stock plant should have a smooth section in the stem just above the crown of the plant. The diameter of the scion should be slightly smaller than that of the stock. The cuts at the base of the scion are made just as for the whip graft. Along a smooth portion of the stem of the stock, a thin piece of bark and wood, the same length as the cut surface of the scion, is completely removed. Then a reverse cut is made downward in the cut on the stock, starting one-third of the distance from the top of the cut. This second cut in the stock should be the same length as the reverse cut in the scion. The scion is then inserted into the cut in the stock, the two tongues interlocking, and the cambium layer(s) matching.

The top of the stock is left intact for several weeks until the graft union has healed. Then it may be cut back above the scion gradually or all at once. This forces the buds on the scion into active growth.

Side-Veneer Graft (Spliced Side Graft)

This variation of side grafting is widely used, especially for grafting small potted plants, such as seedling evergreen. A shallow

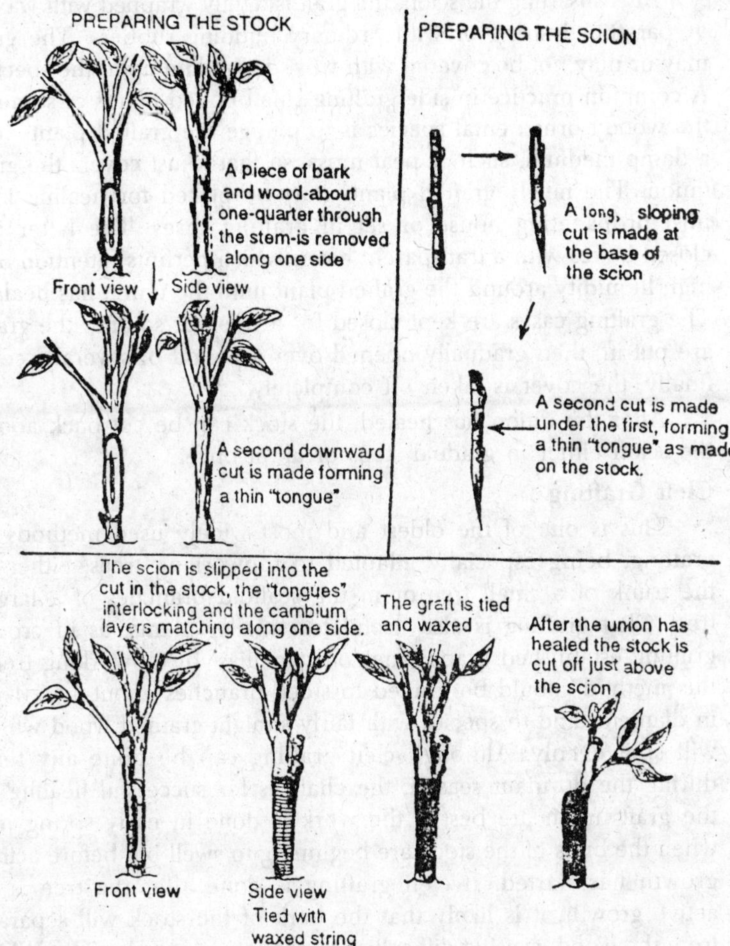

Fig. 5.7. The side-tongue graft.

downward and inward cut from 1 to 1½ in. long is made in a smooth area just above the crown of the stock plant. At the base of this cut, a second short inward and downward cut is made, intersecting the first cut, so as to remove the piece of wood and bark. The scion is prepared with a long cut along one side and a very short one at the base of the scion on the opposite side. These scion cuts should be the same length and width as those made in the stock so that the cambium layers can be matched as closely as possible.

After inserting the scion, the graft is tightly wrapped with waxed or paraffined string or with ordinary budding rubbers. The graft may or may not be covered with wax, depending upon the species. A common practice in side grafting small potted plants of some of the woody ornamental species is to plunge the grafted plants into a damp medium, such as peat moss, so that it just covers the graft union. The newly grafted plants may be placed for healing in a mist propagating house or set in grafting cases. The latter are closed boxes with a transparent cover which permits retention of a high humidity around the grafted plant until the union has healed. The grafting cases are kept closed for a week or so after the grafts are put in, then gradually opened over a period of several weeks; finally, the cover is taken off completely.

After the union has healed, the stock can be cut back above the scion either in gradual steps or all at once.

Cleft Grafting

This is one of the oldest and most widely used methods of grafting, being especially adapted to topworking trees, either in the trunk of a small tree or in the scaffold branches of a larger tree. Cleft grafting is also useful for smaller plants, as in crown grafting established grape vines or camellias. In topworking trees, this method should be limited to stock branches about 1 to 4 in. in diameter and to species with fairly straight-grained wood which will split evenly. Although cleft grafting can be done any time during the dormant season, the chances for successful healing of the graft union are best if the work is done in early spring just when the buds of the stock are beginning to swell but before active growth has started. If cleft grafting is done after the tree is in active growth, it is likely that the bark of the stock will separate from the wood, causing difficulties in obtaining a good union. When this occurs, the loosened bark must be firmly nailed back in place. The scions should be made from dormant, 1 - year-old wood. Unless the grafting is done early in the season (when the dormant scions can be collected and used immediately), the scion wood should be collected in advance and held under refrigeration until time for use.

In sawing off the branch for this and other topworking methods, the cut should be made at right angles to the main axis of the branch. In making the cleft graft, a heavy knife, such as a butcher knife, or one of several special cleft grafting tools, is used to make a vertical split for a distance of 2 to 3 in. down the center of the

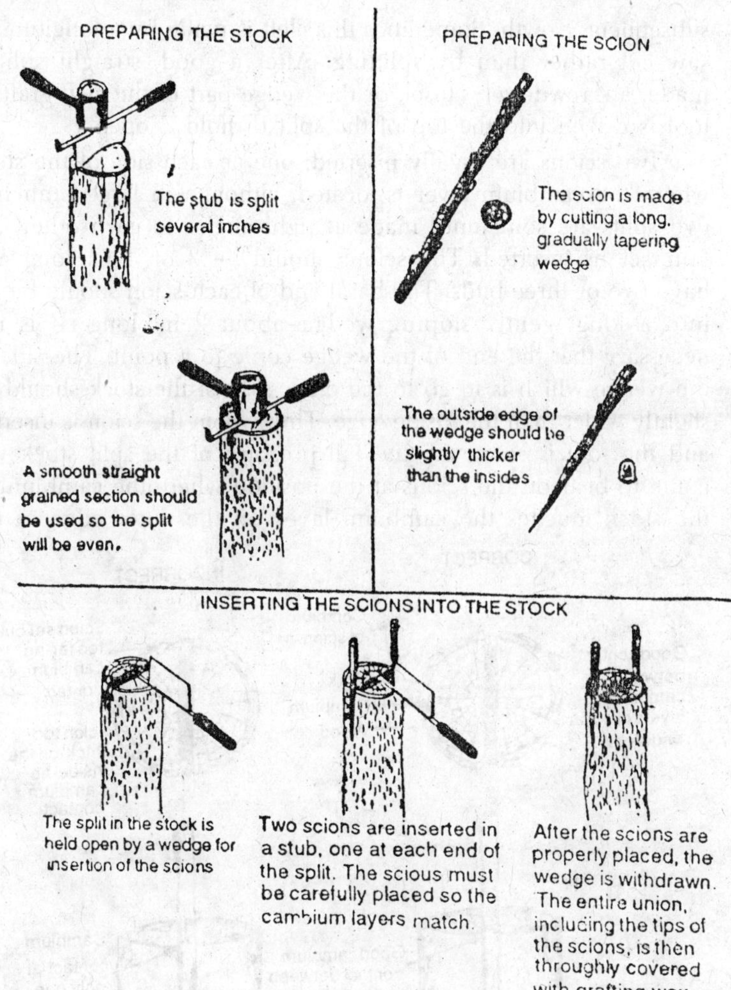

Fig. 5.8. Steps in making the cleft graft.

stub to be grafted. This is done by pounding the knife in with a hammer or mallet. It is very important to have the branch sawed off in such a position that the end of the stub which is left is smooth, straight-grained, and free of knots, for at least 6 in. Otherwise, when the split is made it may not be straight, or the wood may split one way and the bark another. The split should be in tangential rather than radial direction is relation to the center of the tree. This permits better placement of the scions for their subsequent growth. Sometimes this cleft is made by a longitudinal saw cut rather than by splitting. After a good, straight split is made, a screwdriver, chisel, or the wedge part of the cleft-grafting tool is driven into the top of the split to hold it open.

Two scions are usually inserted, one at each side of the stock where the cambium layer is located, although in large branches, two splits are sometimes made at right angles to each other and four scions inserted. The scions should be 3 or 4 in. long and have two or three buds. The basal end of each scion should be cut into a long, gently sloping wedge–about 2 in. long. It is not necessary that the end of the wedge come to a point. The side of the wedge which is to go to the outer side of the stock should be slightly wider than the inside edge. Thus, when the scion is inserted and the tool is removed, the full pressure of the split stock will come to bear on the scions at the position where the cambium of the stock touches the cambium layer on the outer edge of the

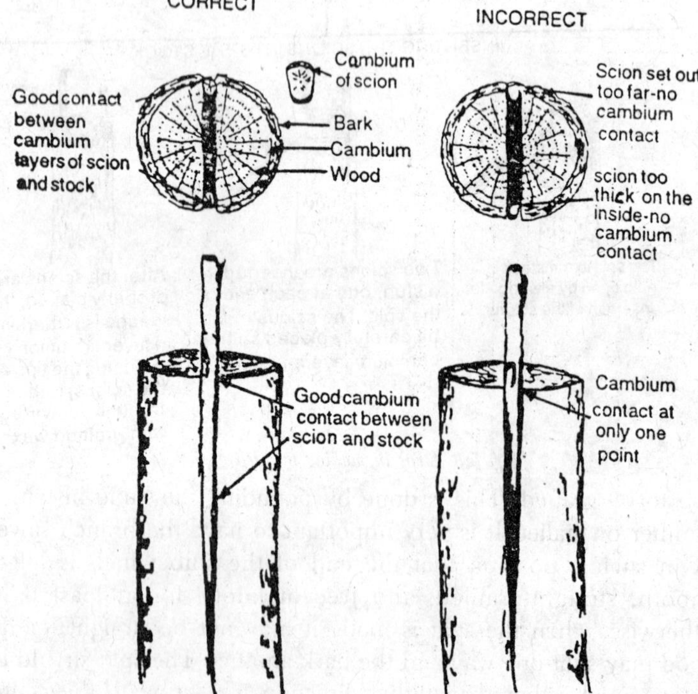

Fig. 5.9. In making the cleft graft, the proper placement of the scions is very important.

scion. Since the bark of the stock is almost always thicker than the bark of the scion, it is usually necessary for the outer surface of the scion to set slightly in from the outer surface of the stock in order to match the cambium layers.

In all types of grafting the scion must be inserted right side up. That is, the points of the buds on the scion should be pointing away from the stock. Failure to observe this means ultimate failure of the graft to grow.

The long, sloping wedge cuts at the base of the scion should be smooth, made by a single cut on each side with a very sharp knife. Both sides of the scion wedge should press, firmly against the stock for their entire length. A common mistake in cutting scions for this type of graft is to make the cut on the scion too short with too abrupt a slope, so that the only point of contact is just at the top. Shaving slightly the sides of the split in the stock will often permit a smoother contact.

After the scions are properly made and inserted, the tool is carefully withdrawn, using care not to disturb the scions. They should be held so tightly by the pressure of the stock that they cannot be pulled loose by hand. No further tying or nailing is needed unless very small stock branches have been used, in which case wrapping around the top of the stock tightly with string or waxed cloth can be done to hold the scions in place more securely.

Through waxing of the completed graft is essential. The top surface of the stub should be entirely covered, permitting the wax to work into the split in the stock. The sides of the grafted stub should be well covered with wax as far down the stub as the split has gone. The top of the scions should be waxed but not necessarily the bark or buds of the scion. Two or 3 days later all the grafts should be inspected and rewaxed where openings appear. Lack of thorough and complete waxing in this type of graft is almost certain to result in failure.

Saw-Kerf (Notch) Grafting

This can be used in place of the left graft. It is especially useful in topworking trees with branches 3 to 4 in. or more in diameter. It does not have the serious disadvantage of a deep split in the stock (as does the cleft graft), which may permit the entrance of decay-producing organisms. Sawkerf grafting can be done over a long period of time - 2 to 3 months–before growth of the stock starts in the spring. If the work is to be performed late in the season, at the time active growth is starting, it is necessary to collect the scion wood earlier and hold it under refrigeration to keep it dormant until the grafting is to be done. Curly-grained stock branches, which could not split evenly for the cleft graft, can be worked readily by the saw-kerf graft. The saw-kerf graft is somewhat more difficult for beginners to perform than most of the other types, but in the hands of experienced workers it can be done rapidly and is highly successful, especially with certain hard-to-graft species, such as the peach.

There are two types of saw-kerf grafts–deep and shallow. Usually, three scions are inserted into each stock branch. A cut with a thin-bladed, fine-toothed saw is made into the stub for each scion. This cut should extend 1 to 1½ in. toward the center of the stub and about 4 in down the side of the stub. Then, using a very sharp "round knife," the grafter widens this saw cut to fit the scion. The knife should be placed at the bottom of the saw cut and brought upward and inward to cut out thin slices of wood. Care should be taken not to get the cut in the stock too wide for use with the available scions.

The scions should be 4 or 5 in. long and contain two or three buds. The basal end of the scion is cut to a wedge shape with the outside edge of the wedge somewhat thicker than the inside edge so as to conform to the general shape of the widened saw cut as made in the stock. The wedge cut on the scions should be 1½ to 2 in. long and, as in all grafting, the cut areas should be perfectly smooth, with no wavy surfaces. The scion should be tried for a good fit as the groove in the stock is widened to the right size. It is usually easier to enlarge the cut in the stock to fit the scion rather than trying to fit the scion to the cut in the stock. After the cuts on the scion and stock are completed, the scion is tapped firmly into place. The cambium layers should cross to insure contact. Since the bark of the scion is usually thinner than that of the stock, at the correct position the scion will set slightly in from the outside of the stock. If this graft is properly made, the scion should be held very securely by tapping it into place. No further nailing or tying is necessary, but of course, as with all exposed grafts, the cut surfaces must be thoroughly waxed.

The round knife is a useful tool in making this graft. It may be adapted from a leather-worker's knife or a cook's mincing knife. The edge of the knife should be sharpened to that it is flat on one side and beveled on the other, being used so that the flat side is held against the wood in widening out the saw cut made in the stock. If a round knife of this type is not available, a large-bladed grafting knife may be substituted.

The shallow type of saw-kerf graft is made in essentially the same manner, except that the cut into the stock is quite shallow. The outer edge of the scion is considerably thicker than the inner edge. It should be cut to just fit into the opening in the stock. Care must be taken to see that the cambium layers match exactly. The scions, in this case, are held in place by nailing with flat-headed wire nails (5/8 in. long. 20 gauge). Finally, of course, the exposed cut surfaces and the end of the scions are thoroughly covered with wax.

Bark Grafting

This method is rapid, simple, readily performed by amateurs, and if properly done, gives a high percentage of "takes". It requires no special equipment and can be performed on branches ranging from 1 in. up to a foot or more in diameter. The latter size is not recommended as it is difficult to heal over such large stubs before decay-producing organisms get started. The bark graft, since it depends on the bark separating readily from the wood, can only be done after active growth of the stock has started in the spring. As dormant scions must be used, it is necessary to gather the scion wood for deciduous species during the dormant season and hold it under refrigeration until the grafting operation is done. For evergreen species, freshly collected scion wood can be used. Scions are not as securely attached to the stock as in some of the other methods and are more susceptible to wind breakage during the first year even though the healing has been satisfactory. Therefore the new shoots arising from the scions probably should be staked during the first year, or cut back to about half their length, especially in windy areas. After a few years growth, the bark graft union is as strong as the unions formed by other methods.

There are several modifications of the bark graft. Three important types are described.

Bark graft (Method No.1)

Several scions are inserted into each stub. For each scion, a vertical knife cut about 2 in, long is made at the top end of the stub through the bark to the wood. The bark is then lifted slightly along both sides of this cut, in preparation for the insertion of the scion. The scion should be of dormant wood, 4 or 5 in. long, containing two or three buds, and be 1/4 to 1/2 in. in thickness. One cut about 2 in. long is made along one side at the base of the scion. With large scions, this cut extends about one third of the way into the scion, leaving a "shoulder' at the top. The purpose of this shoulder is to reduce the thickness of the scion to minimize the separation of bark and wood after insertion in the stock. The scion should not be cut too thin, however, or it will be mechanically weak and break off at the point of attachment to the stock. If small scions are used, no shoulder is necessary. On the side of the scion opposite the first long cut, a second shorter cut is made, thereby bringing the basal end of the scion to a wedge shape. The scion is then inserted between the bark and the wood of the stock, centered directly under the vertical cut through the bark. The longer cut on the scion is placed against the wood, and the shoulder on the scion is brought down until it rests on top of the stub. The scion is then ready to be fastened in place. A satisfactory method is to nail the scion into the wood, using two nails per scion. Flat-headed nails 5/8 to 1 in. long. of 19 or 20 gauge wire, depending on the size of the scions, are satisfactory. The bark on both sides of the scion should also be securely nailed down, or it will tend to peel back from the wood. Another method commonly used with soft-barked trees, such as the avocado, is to insert all the scions in the stub and then hold them in place by wrapping with string, adhesive tape, or waxed cloth around the stub. This is more effective than nailing in preventing the scions from blowing out but probably does not give as tight a fit. Both nailing and wrapping are advisable for maximum strength. If a wrapping material is used, it may be necessary to cut this later to prevent constriction.

In cases in which the bark of the stock is quite thick and small scions are used, it may not be necessary to make the vertical slit in the bark. The scions, cut as described before, can be pushed into place between the bark and the wood. They can then be nailed or tied securely with string or waxed cloth.

After the stub has been grafted and the scions fastened by nailing or tying, all cut surfaces, including the end of the scions, should be thoroughly covered with grafting wax.

Bark graft (Method No.2)

This type of graft is similar to method No.1, except that the scion is not inserted centered on the vertical cut in the bark of the stock. The bark is lifted only along one side of the vertical cut;

the bark on the other side is not disturbed. The scion is inserted under the raised bark and held in place with two nails driven through the bark of the stock, through the scion, and into the wood of the stock. The scion is cut with a shoulder, just as described for method No.1. However, the short cut on the back of the scion, rather than being parallel to the first longer cut, is slanted to one side slightly to conform to the slope of the bark under which the scion is inserted. As in method No.1, the raised bark near the scion should be fastened securely in place with two or more nails or wrapped with tape to prevent it from peeling back. This modification of the bark graft has been widely used with good results. It has an advantage over method No.1, in that one of the edges of the scion is placed against undisturbed bark with its intact cambium cells. This tends to promote more rapid healing of the union. In method No.1, where the bark is lifted away from the scion on both sides, intact cambium cells are some distance from the scion and healing is usually slower.

Bark graft (Method No.3)

In this method, *two* knife cuts about 2 in, long are made through the bark of the stock down to the wood, rather than just one as in the other two methods. The distance between these two cuts should be exactly the same as the width of the scion. The piece of bark between the cuts should be lifted and the terminal two-thirds cut off. The scion is prepared with a smooth slanting cut along one side at the basal end. This cut should be about 2 in. long but made *without* the shoulder, in contrast to the other two methods. On the opposite side of the scion, a cut about ½ in. long is made, forming a wedge at the base of the scion. The scion should fit snugly into the opening in the bark with the longer cut inward and with the wedge at the base slipped under the flap of remaining bark. Rapid healing can be expected because both sides of the scion are touching undisturbed bark and cambium cells, which is not the case in the other two type of the bark graft.

The scion should be nailed into place with two nails, the lower nail going through the flap of bark covering the short cut on the back of the scion. If the bark along the sides of the scion should accidentally become disturbed, it must be nailed back into place.

Method No.3 is well adapted for use with thick-barked trees, such as walnuts, on which it is not feasible to insert the scion under the bark.

Approach Grafting

The distinguishing feature of approach grafting is that two independent, self-sustaining plants are grafted together. After a union has occurred, the top of the stock plant is removed above the graft

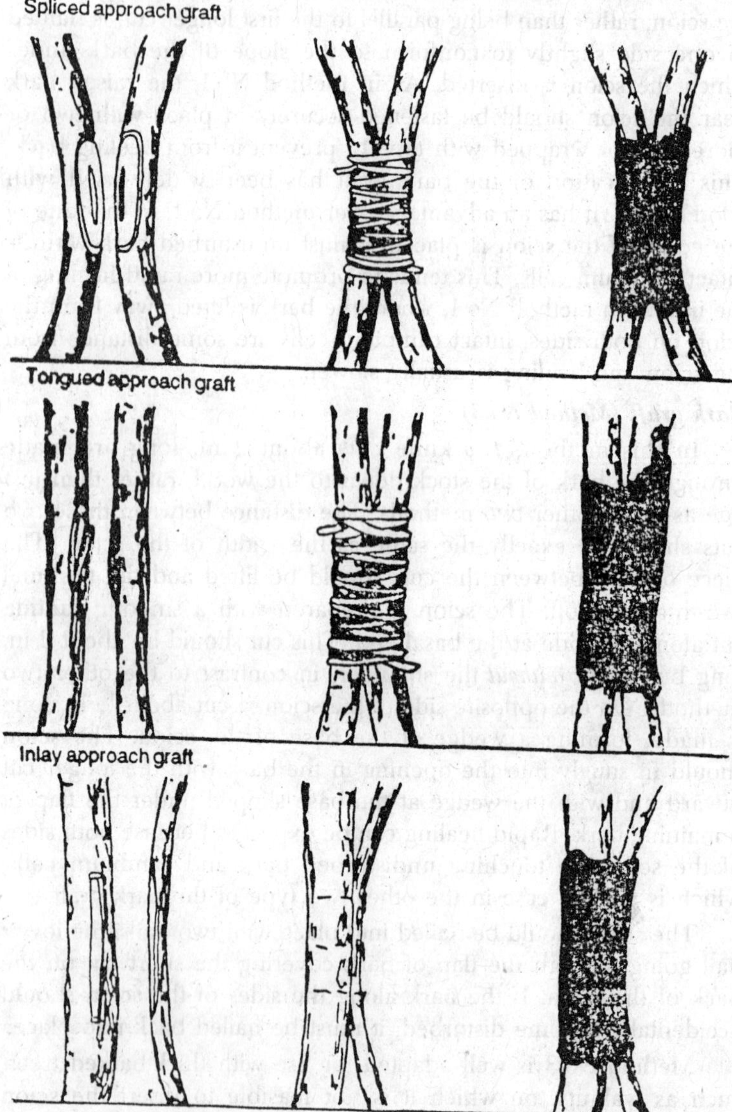

Fig. 5.10. Three methods of making an approach graft.

and the base of the scion plant is removed below the graft. Sometimes it is necessary to sever these parts gradually rather than all at once. Approach grafting provides a means of establishing a successful union between certain plants which are difficult to graft together otherwise. It is usually performed with one or both of the plants to be grafted growing in a pot or container. Use is often made of this method by placing seedling plants growing in containers under an established plant which is to furnish the scion part of the new, grafted plant.

This type of grafting can be done at any time of the year, but healing of the union is more rapid if it is performed at a season when growth is active. As in other methods of grafting, the cut surfaces should be securely fastened together, then covered with grafting wax to prevent drying of the tissues.

Spliced Approach Graft

Preferably the two stems should be approximately the same size. At the point where the union is to occur, a slice of bark and wood 1 to 2 in. long is cut from both stems. This cut should be the same size on each so that identical cambium patterns will be made. The cuts must be perfectly smooth and as nearly flat as possible so that when they are pressed together there will be close contact of the cambium layers. The two cut surfaces are then bound tightly together with string, raffia, or nurseryman's tape. The whole union should then be covered with grafting wax. After the parts are well united, which may require considerable time in some cases, the stock above the union and the scion below the union are cut, and the graft is then completed. It may be necessary to reduce the leaf area of the scion if it is more than the root system of the stock can sustain.

Tongued Approach Graft

This is the same as the spliced approach graft except that after the first cut is made in each stem to be joined, a second cut–downward on the stock and upward on the scion–is made, thus providing a thin tongue on each piece. By interlocking these tongues a very tight, closely fitting graft union can be obtained.

Inlay Approach Graft

This method may be used if the bark of the stock plant is considerably thicker than that of the scion plant. A narrow slot, 3 or 4 in, long, is made in the bark of the stock plant by making two parallel knife cut stand removing the strip of bark between. This can only be done when the stock plant is actively growing and the bark "slipping." The slot should be exactly as wide as the scion to be inserted. The stem of the scion plant, at the point of union, should be given along, shallow cut along one side, of the same length as the slot in the stock plant and deep enough to go through the bark and slightly into the wood. This cut surface of the scion branch should be laid into the slot cut in the stock plant and held there by nailing with two or more small, flat-headed wire nails. The entire union must has healed, the stock can be cut off above the graft and the scion below the graft.

Inarching

This method is similar to approach grafting in that both stock and scion plants are on their own roots at the time of grafting; it differs in that the top of the new rootstock plant usually does not extend above the point of the graft union as it does in approach grafting. Inarching is generally considered to be a form of "repair grafting," being used in cases in which the roots of an established tree have been damaged by such things as cultivation implements, rodents, or disease. It can be used to very good advantage in saving a valuable tree or improving its root system.

Seedlings (or rooted cuttings) planted beside the older, damaged tree, or suckers arising near its base, are grafted into the trunk of the tree for the purpose of providing a new root system to supplant the damaged roots. The seedlings to be inarched into the tree should be spaced about 5 or 6 in. apart around the circumference of the tree if the damage is extensive. The tree will usually stay alive for sometime after the injury occurs unless it is very severe. A satisfactory procedure for inarching is to plant seedlings of a compatible species or variety around the tree during the dormant season. Then, as active growth commences in early spring, the grafting operation can be done.

Inarching old, weakly growing trees with strong, vigorous seedling rootstocks has on some occasions proved beneficial in promoting renewed active growth of the old trees.

The seedling plants to provide the new root system are usually considerably smaller than the tree to be rapaired. The graft union is made in a manner similar to that described for method No.3 of the bark graft. The upper end of the seedling, which should be ¼ to ½ in. thick, is given a long shallow cut along the side for 4 to

6 in. This cut should be on the side next to the trunk of the tree and made deep enough to remove some of the wood, thus exposing two strips of cambium tissue. At the end of the seedling another shorter cut, about ½ in. long, is made on the side opposite the long cut, this makes a sharp, wedge-shaped end on the seedling stem. A long slot is made in the trunk of the older tree by removing a piece of bark the exact width of the seedling and just as along as the cut surface made on the seedling. A small flap of bark is left at the upper end of the slot, under which the wedge end of the seedling is inserted. Then the seedling is nailed into the slot with four or five small, flat-headed wire nails. The nail at the top of the slot should go through the flap of bark and through the end of the seedling. If any of the bark of the tree along the sides of the seedling should accidentally be pulled loosed, it is necessary to nail it back in place. After nailing, the entire area of the graft union should be thoroughly waxed.

In inarching some species, such as the walnuts, better healing of the union takes place if the top of the seedling is retained and allowed to extend above the graft union for a period of time, finally being cut off just above .the union.

Owing to the food materials translocated from the larger tree, the seedlings grow very rapidly and soon provide a considerable number of new roots for the grafted tree. If shoots arise from the inarches, they should be suppressed by cutting off their tips. The shoots should finally be removed entirely as the union becomes well established.

Bridge Grafting

This is a form of repair grafting, and is used in cases in which the root system of the tree has not been damaged but where there is injury to the bark of the trunk. Sometimes cultivation implements, rodents, disease, the winter injury will damage a considerable trunk area, often girdling the tree completely. If the damage to the bark is extensive, the tree is almost certain to die, because the roots will be deprived of their food supply from the top of the tree. Trees of some species, such as the elm, cherry, and pecan, can heal over extensively injured areas by the development of callus tissue. But trees of most species which have had the bark of the trunk severely damaged should be bridge grafted if they are to be saved.

The bridge grafting operation is best performed in early spring just as active growth of the tree is beginning and the bark is slipping easily. The scions to be used should be taken when dormant from 1-year-old growth, 1/4 to 1/2 in. in diameter, of the same or a compatible species, and held under refrigeration until the grafting work is to be done. In an emergency, one may successfully perform bridge grafting late in the spring, using scion wood whose buds have already started to grow. The developing buds or new shoots must be removed.

The first step in bridge grafting is to trim the wounded area back to healthy, undamaged tissue by removing dead or torn bark. Then every 2 or 3 in. around the injured section a scion is inserted, attached at both the upper and lower ends into live bark. It is important that the scions be inserted right side up. If they are put in reversed, they may make a union and stay alive for a year or two, but the scions will not grow and enlarge in diameter as they would if inserted correctly.

It is essentially the same as method No. 3 of the bank graft. Just above and below the injured area, a slot 2 to 3 in. long and exactly the width of the scion is cut in the bark of the trunk for each scion. The piece of bark is removed, with the exception of a flap about 1/2 in. long which is left at the end of the slot. The scions are cut to fit into these slots at each end of the wound, and long enough to bow outward slightly. This bow allows for good contact at each end and permits some swaying of the trunk in the wind without tearing the scions loose. To prepare the scions, a cut is made along both ends on the side which is to fit into the slot. The cut should be the same length as the slot and deep enough to remove some bark and wood and expose two strips of cambium tissue. Then, on the opposite side of the scion, at each end, another shorter cut, about 1/2 in. long, is made to form the end of the scion into a wedge shape. The ends of the scion should be inserted under the flaps of bark and nailed in place, using 3/4 in. 20 gauge flat-headed wire nails. One nail should go through the flap of bark at each end of the scion with enough additional nails to hold the scion securely.

After all the scions have been inserted, the cut surfaces must be thoroughly covered with grafting wax, particular care being taken to work the wax around the scions, especially at the graft unions. The Exposed wood of the injured section may also be covered with grafting wax to prevent the entrance of decay organisms and to prevent excessive drying out of the wood, which is important, being the path for upward movement of water and nutrients in the tree.

The buds on the scions will often push into growth if the grafts are successful. These shoots should be removed, because no branches would be desired in this position. The scions will rapidly enlarge in size and completely heal over the wound in a few years.

Bracing

The same type of graft union practiced in inarching and bridge grafting can often be used to establish a "natural" brace in young trees. This method is useful in supporting branches that may be in danger of breaking off or where there is a weak crotch. A small branch, about pencil size or a little larger, coming a foot or so above the weak crotch is grafted into the adjacent branch to be supported. It should be wrapped spirally and upward partly around the branch. In the region where the graft union is to be, the bark on the large branch is cut just under the small branch to its exact width. This should extend for a length of 6 in. or more, and the spiral piece of bark should be removed. Grafting such as this could only be done at a time of year when there is active growth and the bark is slipping easily. Early spring, just as the new growth is starting, is the preferable time.

After the slot in the bark is ready, the "scion"–the small branch-should be smoothly cut on the lower side through about a third of its thickness and for the length of the slot–6 in. or more. If the cuts are well made, the small branch should fit snugly into the slot. It is helpful to trim the end of the small branch to a wedge point by cutting it on the top side so that it can be inserted under the bark at the top end of the slot.

The next step is to nail the branch in place in the slot with small, flat-headed wire nails (5/8 or 3/4 in., 20 gauge) placed at 1- or 2-in, intervals. The graft union should then be thoroughly waxed to prevent drying. It is advisable to tie the two branches together temporarily with a strong cord to prevent shipping by the wind, which might put the graft union loose.

Grafting Classified According to Placement

Grafting may be classified according to the part of the plant on which the scion is placed- a root, the crown (the junction of the stem and root at the ground level), or various places in the top of the plant.

Root Grafting

In this class of grafting the rootstock seedling, rooted cutting, or layered plant is dug up, and the roots are used as the stock for the graft. The entire root system may be used (*whole-root graft*), or the roots may be cut up into small pieces and each piece used as a stock (*piece-root graft*). Both methods give satisfactory results. As the roots used are relatively small (¼ to ½ in. in diameter), the whip, or tongue, graft is generally used. Root grafting is usually performed indoors during the late winter or early spring. The scion wood collected previously is held in storage, while the rootstock plants are also dug in the late fall and stored under cool(40° to 50°F : 4° to 10°C) and moist conditions until the grafting is done.

The term *bench grafting* is sometimes given to this process, because it is often performed at benches by skilled grafters as a large-scale operation. A number of plants are propagated commercially by root grafting–apples, pears, grapes and such ornamentals as the wisteria and rhododendron.

In making root grafts, the root piece should be 3 to 6 in. long and the scions about the same length, containing two to four buds. After the grafts are made and properly tied they are bundled together in groups of 50 to 100 and stored for callusing in damp sand, peat moss, or other packing material. They may be placed in a cool cellar or under refrigeration at approximately 45°F (7°C) for about 2 months. The callusing period for apples can be shortened to around 30 days if the grafts are stored at a temperature of about 70°F (21°C) and at a high humidity. To use this higher callusing temperature the material should be collected in the fall and the grafts made before any cold weather has overcome the rest period of the scion buds. After the unions are well healed, the grafts must be stored at cool temperature–35° to 40°F (2° to 4°C)–to overcome the "rest period" of the buds and to hold them dormant until panting. The grafts are lined out in early spring to the nursery row directly from the low temperature storage conditions. For general callusing purposes, temperatures below rather than above 70°F (21°C) are the most satisfactory. By the proper regulation of temperature, callusing processes may be accelerated by increased temperature or retarded by decreased temperature so that, within reasonable limits, a desired degree of callus formation may be had within a given length of time. Provision should be made for adequate aeration of the callusing grafts.

As soon as the ground can be prepared in the spring, the grafts are lined out in the nursery row 4 to 6 in. apart. They should be planted before growth, of the buds or roots begins. If this starts before the grafts can be planted, they should be moved to lower temperatures (30° to 35° F; – 1° to 2°C). The grafts are usually planted deep enough so that the graft union is just below the ground level, but if roots are to arise only from the rootstock, the graft should be planted with the union well above the soil level. It is very important to prevent scion rooting where certain definite influences, such as dwarfing or disease resistance, are expected from the rootstock.

After one summer's growth, the grafts should be large enough to transplant to their permanent location. If not, the scion may be cut back to one or two buds, or headed back somewhat to force out scaffold branches, and then allowed to grow a second year. With the older root system a strong, vigorous top is obtained the second year.

Nurse-Root Grafting

Under certain conditions, it is desired to have a stem cutting of a difficult-to-root species on its own roots. One way this can be done is by making a root graft, using the plant to be grown on its own roots as the scion and a root of a compatible species as the stock. The scion may be made longer than usual and the graft planted deeply with the major portion of the scion below ground. In some cases scion-rooting is promoted by rubbing a rooting stimulant, such as idolebutyric acid into several vertical cuts made through the bark at the base of the scion, just above the graft union. This is done just before planting, and the grafts are set deeply so that most of the scion is covered with soil. After one or two seasons of growth many of the scions will have roots. The temporary nurse rootstock is then cut off and the top reduced in proportion to the root system. The rooted scion is replanted to grow on its own roots.

Scion rooting is often better if the nurse-root is buried deeply enough so that a new shoot of the current season's developing from a bud on the scion, will be rooted, rather than trying to root the older, lignified tissue of the original scion. This is essentially a form of mound layering. As the new scion shoot grows, soil is gradually mounded up around it to a height of 5 or 6 in. although the terminal leaves are at no time covered.

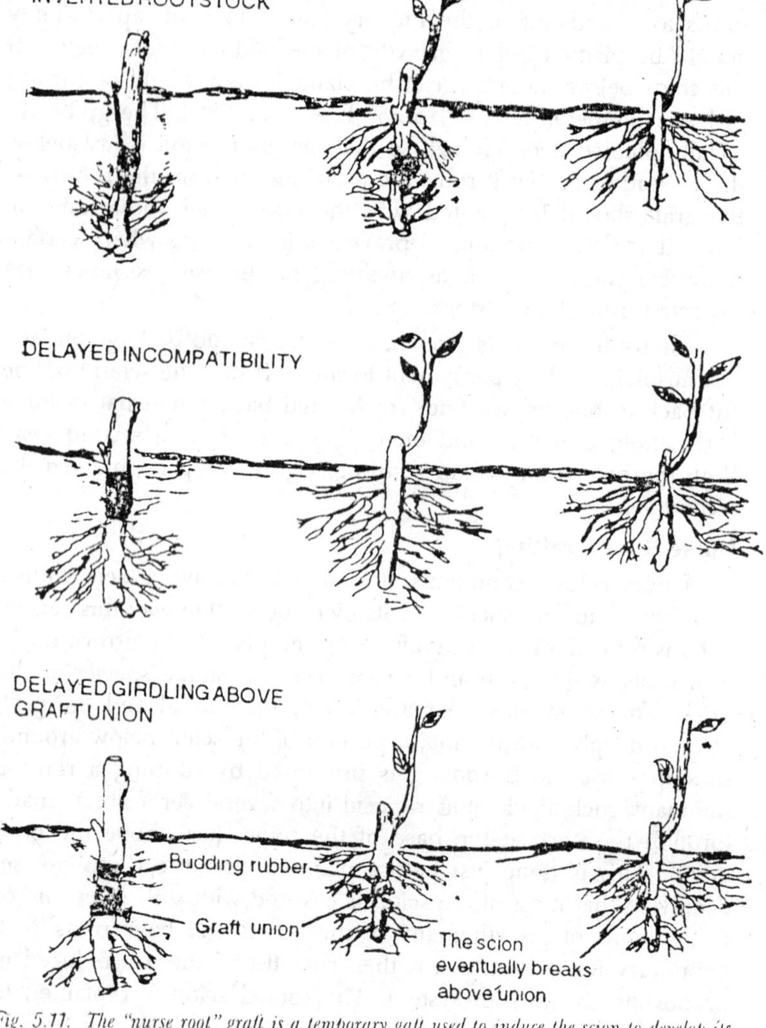

Fig. 5.11. The "nurse root" graft is a temporary gaft used to induce the scion to develop its own roots.

Several methods of handling eliminate the necessity of digging up the graft and cutting off the rootstock. The rootstock piece will eventually die if it is grafted onto the scion in an inverted position. The inverted stock piece sustains the scion until scion roots are formed, but the stock fails to receive food from the scion and eventually dies, thus leaving the scion on its own roots. In another method an incompatible rootstock is used. Hence, if the graft is planted deeply, scion roots will gradually become more important in sustaining the plant, and the incompatible rootstock will finally cease to function.

In a third method the base of the scion, just above the graft union, is bound with some type of wrapping material to eventually girdle and cut off the rootstock. Excellent results have been obtained with ordinary budding rubber strips (0.016 gauge). Budding rubbers disintegrate within a month when exposed to sun and air; when buried in the soil, they will last as long as 2 years, allowing sufficient time for the scion to become rooted. Yet, owing to the slow deterioration of the rubber below ground, the rootstock is finally girdled and cut off.

Crown Grafting

A graft union made at the rootstem transition region- the "crown" of the plant–on an established rootstock is termed a "crown graft." Several methods of grafting are commonly used in crown grafting, such as the whip, side, cleft, saw-kerf, or bark graft, depending upon the species and size of the root stock.

Crown grafting of deciduous plants is best done in late winter or early spring, shortly before new growth starts. The scions should be prepared from well-matured, dormant wood of the previous season's growth.

Since the operation is performed just below, at, or just above the soil level, it is possible to cover the graft union, or even the entire scion, with soil and thus eliminate the necessity for waxing. The union should be tied securely with string or tape to hold the grafted parts together until healing takes place.

Double-Working

A double-worked tree has three parts, usually all different genetically–the rootstock, the intermediate stock, and the scion or fruiting, top. Such a tree has two graft unions, one between the rootstock and intermediate stock, and one between the intermediate stock and the scion. The intermediate stock may be less than an inch in length or extensive enough to include the trunk and secondary scaffold branches. Double-working is used for various purposes, such as (a) overcoming graft incompatibility between a desired top variety and the rootstock, (b) providing a cold or disease-resistant trunk, (c) obtaining a dwarfing effect from the use of certain intermediate stocks, or (d) obtaining the strong trunk or crotch systems of certain varieties.

A good example of double-working is the use of an intermediate stock when the 'Bartlett' pear is grafted on dwarfing quince roots. 'Bartlett' will not, generally, form a compatible union if grafted directly on quince. Cuttings of the quince may be rooted during one summer and fall-budded to a compatible stock, such as 'Old Home' pear. This grows the following summer, after which it is again fall-budded to the 'Barlett' pear, the 'Barlett' bud being inserted in the intermediate stock. The 'Barlett' top is developed the next summer, after which the double-worked nursery tree is ready to dig and set in its permanent location.

A quicker method–which also produces a straighter trunk than double-budding. Dormant 'Bartlett' pear scions are bench-grafted onto dormant 'Hardly' pear stem pieces during late winter, using the whip or tongue grafting method. These grafts are then packed in damp material: after several weeks in a cool location, they become well-callused and have united. In early spring such grafts (scion variety on the interstock) are crown-grafted in the nursery row to the rootstock–in this case, rooted quince cuttings. After one year in the nursery row, the double-worked nursery trees are ready to be dug.

Another method of double- working all at one time is by bench grafting, using the whip or tongue grafting method, but making two graft unions rather than one. The subsequent callusing and planting procedure is the same as for a simple root graft.

A method, termed "double-shield budding" has been developed both in England and Germany for double-working. Essentially, this method consists of preparing a T-bud but inserting a thin budless shield of the intermediate variety under and below the shield piece containing the bud of the variety desired for the top. However, such short "compatibility bridges" may not always produce the desired effect. It has been shown that the degree of dwarfing is proportional to the length of the interstock.

Top-Grafting (Topworking)

One of the principal uses of grafting is to change the variety of an established plant–tree, shrub, or vine. If budding is used, then the process is termed *top-budding.*

In some cases pulling out the existing trees and planting new nursery trees may be more economical than topworking. This is especially true if the existing trees are old or diseased or if the species is relatively short-lived, as is the peach. Long-lived species, such as apples and pears, are worth topworking if the tree is in a healthy condition. Also to be considered in planning for topworking is the hazard of viruses being present in the scion wood. In the peach, for example, virus diseases such as "stubby twig." "ringspot" or "necrotic leaf spot" may be introduced by infected scions into an otherwise healthy orchard. Care should be taken to obtain scion wood for top-grafting, if possible, from known virus- tested sources. Of equal significance is the possibility that the stock trees may be infected with viruses or part of a virus complex. The value of scion wood from virus-tested sources can be nullified if the stock is virus infected.

Preparation for Top-Grafting

Any of the methods of grafting described earlier in this chapter– whip, side, cleft, saw-kerf, or bark- can be used for top-grafting, which is usually done in the spring, shortly before new growth starts. The exact time depends upon the method to be used. The cleft, side, whip and saw-kerf graft can be done before the bark is slipping. The bark graft must be done when the bark is slipping, preferably just a s the buds of the stock tree are starting growth.

It is usually advisable to obtain an ample amount of good quality scion wood prior to grafting and store it under the proper conditions.

In preparing the stock tree for topworking, one must decide, for each individual tree, which and how many scaffold branches (usually three to five) should be used. If the work is done high in the tree in the smaller, secondary scaffold branches, an earlier return to bearing will result than when fewer and larger limbs are grafted lower in the tree. The branches to be grafted should be well distributed around the tree and up and down the main trunk, avoiding branches with weak, narrow crotches. All others can be removed unless one or more nurse branches are used.

Retaining *nurse branches* is a common practice in topworking deciduous trees in regions where cold winters are experienced. In localities where the winters are mild and damage of succulent growth by winter killing is slight, nurse branches are not used. For broad-leaved evergreens, such as citrus and olives, it is customary to leave nurse branches. The presence of nurse branches, especially if they are on the south and west sides of the tree, is desirable because they protect the grafted branches from sun scald; otherwise, the branches should be whitewashed to prevent this. Retaining a large scaffold branch as a nurse limb results in less vigorous growth of the scions. They are then less likely to blow out in winds or be winter- killed than the more vigorous, succulent shoots which develop from the scions where no nurse branches are retained. Also, fewer suckers and water sprouts (which must be removed) develop when nurse branches are left . The latter should be so pruned as not to interfere with the growth of the young grafts.

A practice which is often recommended, especially for older trees, is to top work them during a 2-year period, grafting perhaps two main branches on the northeast side of the tree the first year and retaining two branches on the southwest as nurse branches. The second year these are topworked.

Topworking is most successful when done on relatively young trees where the branches to be grafted are to larger than 3 or 4 in. in diameter and are relatively close to the ground. When attempting to topwork large, old trees, it is often necessary to go high up in the trees to find branches with a diameter as small as 4 in. If the grafting is done on such branches, the new top is inconveniently high for the various orchard operations, such as thinning and harvesting. The other alternative is cutting off the branches or main trunk close to the ground and inserting the scions into wood one or more feet in diameter. Although many scions can be inserted around the tree between the bark and the wood by the bark graft method and may grow well for several years, it is quite likely, that wood rot will develop in the center of the stub before the growth of the scions can heal it over. Also, some scions are not mechanically held in place very securely and may be blown out by strong winds after they reach considerable size.

It is important that the branch to be topworked be cut off in such a location that the region just below the cut is smooth and free from knots or small branches, so that there will be a satisfactory place for inserting the scions. The branches are best cut off about 9 to 12 in. from the main trunk to keep the tree headed low. The branch should not be cut off more than a few hours before the grafting is to be done.

In Preparation for Topworking (Top-Grafting)

(a) Do the work in the spring when the trees are dormant or shortly after growth starts.

(b) Select for grafting three to five well-placed scaffold branches which are not larger than about 4 in. in diameter and which are conveniently close to the ground.

(c) Retain nurse branches for broad-leaved evergreen trees and for deciduous trees where the winters are severe.

(d) Cut off the branches properly so that the bark is not torn down the trunk.

Grafting on a cool, overcast day with no wind blowing offers the most protection from drying of the cut surfaces of the scion and stock until they can be covered with grafting wax. Grafting on hot, sunny, and windy days should be avoided. During grafting, the scion wood must not dry out by being exposed to the sun. It should be kept moist and cool in some container or be wrapped in moist burlap.

The need for prompt and thorough coverage of all cuts, including the tip end of the scions, cannot be stressed too strongly. The wax should be worked into the bark of the stock, sealing all small cuts or cracks where air could penetrate in and around the cut surfaces where healing tissue is expected to develop.

Subsequent Care of Topworked Trees

After the actual top-grafting (or top-budding) operation is finished, much important work needs to be done before the topworking is successfully completed. A good grafting job can be ruined by failure to care for the trees properly.

If the grafting has been done in late spring when growth is active, trees of some species, such as the walnut, will "bleed" to a considerable extent from the grafted stub, even though it has been covered with grafting wax. This flow of sap around the scions can be so heavy as to interfere with the normal healing processes at the graft union. If this condition appears, it can often be corrected by making several slanting cuts with a knife through the bark in the trunk of the tree several feet below the grafted stubs. The bleeding will then take place at these cuts rather than around the graft union. This extensive sap flow is not particularly harmful to the tree and will usually stop within a few days. In 3 to 5 days after grafting, the trees should be carefully inspected and the graft unions rewaxed if cracks or holes appear in the wax.

A problem needling immediate attention is the prevention of sunburn on the portions of the trunk and large branches which are exposed to the direct rays of the sun by the removal of the protecting top foliage. This is especially important if the grafting has been done late in the season when not weather can be expected, and when no protecting nurse branches have been retained. The radiant energy from the sun absorbed by the dark-coloured bark can raise the temperature of the living cells below the bark to a lethal level.

It is generally advisable to whitewash the trunk, branches and scions of the grafted trees, unless the grafting is done in late winter or early spring when the days are still cold. Various cold-water paints, some made especially for this purpose, are available. The white colour reflects a considerable portion of the sun's radiant energy, thus keeping the temperature of the living tissues within safe limits. Interior, water-base house paints (both latex and acrylic) prevent sunburn for one season. Exterior paints give longer protection but are more likely to cause injury to the tree.

To Prepare Whitewash for Grafted Trees

Formula I. Quicklime, 5 lb; salt, ½ lb; sulphur, ¼ lb. Add water to the lime to start it slaking. Then add the salt and sulphur. This mixture should be prepared in a crockery or wood container. Allow to age for several days, they dilute to a consistency just thick enough to apply with a paint brush. Dilute to a thinner consistency for spray application.

Formula II. Hydrated lime, 25 lb; zinc sulphate, 2lb; water, 50gal. Add water first to a power sprayer with agitator running, then add other components. Spray mixture on trees.

One way to protect the scions from the heat of the sun when the grafting has been done late in the season is to cover the end of the grafted stub and the scions with a large paper bag, the corners of which have been cut to allow for ventilation. The bag is tied securely on the stub with heavy string. However, in a test, of various methods of protecting scions of grafted walnut trees, including the use of paper bags, whitewashing proved the best.

Another help in preventing sunburn is to retain some of the watersprouts which soon start growth along the trunk and branches of the grafted tree. They must be kept under control or they will quickly shade out the developing scions. Rather than removing the watersprouts completely, the grafter can head them back to several inches in length, and they will shade the bark underneath. An additional benefit is that the food manufactured by this leaf area will help sustain the tree until the new scions develop sufficient foliage to do so.

It should be emphasized that, especially during the first summer after topworking the sucker and water sprout growth arising from below the grafted branches must be kept pruned back so as not to interfere with the growth of the scions. If it has been customary to fertilize the trees with nitrogen, this should be withheld for a year or two after grafting, because no stimulation of growth is usually needed. If the trees have been under irrigation, their water requirement will be much less, owing to the removal of a considerable amount of leaf area when the tops were cut back for grafting.

Two to four scions are generally inserted in each stub. If all the scions grow, they should all be retained during the first year, because they will help heal over the stub. However, just one branch, from the best placed and strongest growing scion, should be retained permanently. Growth from the remaining scions should be retarded by rather severe pruning, keeping them alive to help heal the branch, but allowing the permanent scion to become the dominant one. To keep two or more scions for permanent branches at one point will undoubtedly result in a weak crotch, which will eventually break. The first year, then, the best practice is to retain all scions to help heal the stub but, by pruning, to retard the growth of all but the best one. If two shoots arise from the scion to be retained permanently, select the best of these and remove the other. After the stub has healed over, the temporary scions can be removed completely. This may be in the second or third year.

If the permanent scion grows rather vigorously, the danger arises of its becoming top heavy and breaking off during winds. This may be handled in two ways–either by retarding the growth of the permanent branch by pruning it back or by nailing a lath or other type of stick onto the tree and tying the new branch to this stick. *When tying a cord around a branch, always make a loop so that there is no chance of the branch being girdled as it grows.*

Should only one scion grow at each stub, the problem of securing adequate healing off the stub on the side opposite the living scion may prove to be serious. Healing may be helped by sawing off the stub at an angle away from the surviving scion. The cut surface of the stub can be covered with grafting wax to retard wood rotting . It may take a number of years for the single scion to heal over the stub, and it is possible that wood rotting may start before it can do so.

When none of the scions grow in a stub, there are still some possibilities for getting it topworked. One is by allowing several well-placed water sprouts arising just below the cut surface to grow, then top-budding them during the summer. Or the watersprouts may be allowed to grow so as to keep the branch alive and healthy, then the grafting operation repeated the following year, making a fresh cut a foot or so below the original cut.

If nurse branches have been used, they should be cut back and away from the scions at intervals throughout the summer so that scion shoots are always fully exposed to the sun and have sufficient space to grow. Nurse branches should be removed entirely or grafted by the second or third year.

When the top of the tree has been finally worked over to the new variety, it will grow vigorously for a few years. Good pruning practices are needed to prevent badly placed branches from developing.

Frameworking

In this method of changing the variety of fruit trees, all the main scaffold branches are retained on the tree but most of the small laterals throughout the top of the tree are replaced by a large number of scions of the replacement variety. These are inserted by the side graft method with the cuts made very close to the origin of lateral branches (about 1 in. in diameter) arising from the secondary scaffolds. Long scions having 7 or 8 buds are often used. In frame working the tree quickly returns to fruiting, but considerable labour is needed for grafting, and much aftercare is required in keeping growth from the original tree below the scions cut away so that only the replacement variety is maintained.

Herbaceous Grafting

Grafting herbaceous types of plants is used for various purposes, such as studying virus transmission, stock-scion physiology, and grafting compatibility, as well as for the commercial greenhouse production of certain cucurbitaceous crops, particularly in Europe. Usually such grafts are made while the plants are quite small, the stock being grafted shortly after seed germination. Such material is generally very soft, succulent, and susceptible to injury. In one technique a simple splice graft is used with a diagonal cut made through the seedling stock, just above the cotyledons. A piece of thin-walled polyethylene tubing of the proper size to give a snug fit is slipped over the cut end of the stock. The basal end of the scion receives a diagonal cut similar in length and angle to that given the stock. The scion is then slipped down into the plastic tubing so that the two cut surfaces make intimate contact. The tubing holds the graft in place until healing occurs–about 12 days after grafting. Then the tubing may be slipped off over the scion if there are no leaves and if the bud has not expanded; otherwise, it may be cut off with a razor blade.

In another procedure used in grafting older herbaceous stock plants, several leaves are retained below the graft union. The cleft graft is used (but with only one scion), and the graft union is bound with raffia, budding rubbers, or adhesive latex tape. To prevent drying out, the entire plant–following grafting–is covered with a supported polyethylene bag. The grafted plant is then set in the shade until the graft has healed; then the plastic cover can be removed.

A method of establishing *Vitis vinifera* grapes on phylloxera or nematode-resistant rootstocks is by "greenwood" grafting, in which the grafting is done in the spring, placing the scion taken from new growth–on a new green shoot which has developed from the rootstock plant. A simple splice graft is made with the sloping cuts 1 to 1½ in. long. The stock and scion pieces must be the same diameter. The scion has only one bud. The cuts are matched as closely as possible, and the graft union is completely covered by wrapping with a budding rubber. The graft union should be healed and the bud growing by two weeks after grafting. Since these grafts are quite fragile, they must be tied to a stake.

Nurse-Seed Grafting

Germination of some large-seeded woody species, such as the chestnut, is hypogeal ; that is, the cotyledons remain below ground in the seed coats with the shoot tip appears above ground. In a grafting procedure using certain species with seeds of this type, when the seeds have just germinated, the petioles of the cotyledons are cut off transversely just at the seed, leaving the cotyledons inside. A knife point is inserted into the seed between the cut petioles, making an opening for the scion. The scion, which is prepared from dormant wood of the previous season's growth, is

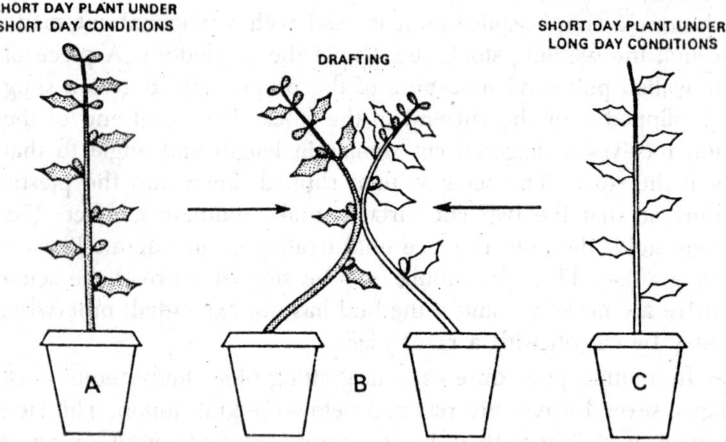

Fig. 5.12. The stimulus can translocate from one plant to another through graft union.

cut to a wedge shape at the base, as for a cleft graft. The scion is inserted into the cut between the cotyledons so that the exposed cambium surfaces of the scion are in close contact with the cut surfaces of the cotyledons. The "seed-grafts" are then lined out in the rooting medium with the union about 1 ½ in. below the surface. A graft union takes place, and roots arise from the cut cotyledon petioles. This type of grafting is done in early spring, using properly stratified seed and dormant scion wood. Chestnuts, avocados, and camellias have been grafted successfully by this method; in fact grafted camellia plants of the desired cultivar can be obtained in the same length of time it would take for the seedlings to become large enough to graft by the conventional cleft graft method.

In a modification of the nurse-seed graft, which has worked well in propagating chestnuts, pecans, walnuts, and oaks, the actual graft union is made into the seedling hypocotyl at the point of attachment of the cotyledons. The scion is cut to a thin wedge and inserted into a split made in the hypocotyl. After insertion of the scion, the graft is wrapped firmly with rubber strips or waxed string.

Cutting-Grafts

In this type of graft, a leafy scion is grafted onto a leafy, unrooted stem piece (which is to become the rootstock), and the combination is then placed in a rooting medium under intermittent mist for simultaneous healing of the graft union and rooting of the stock. This procedure was utilized many years ago in studying stock- scion physiology in citrus. More recently it has been used in commercial propagation of various types of citrus on clonal dwarfing rootstocks. It is also of value in propagating certain conifer species that are difficult to start as cuttings.

For citrus a simple splice graft is used. The slope of the cut is at 15-deg, angle, and the union is died with a rubber band. The base of the stock is dipped into a root-promoting material, such as indolebutyric acid and then the grafts are placed under mist, or in a closed case, in flats of the rooting medium over bottom heat. After healing of the union and rooting of the stock, the grafts are allowed to harden by discontinuing the mist and bottom heat for about 2 weeks. Then the grafts are ready for planting in 1-gal. cans or other containers.

In using this method for conifers the difficult-to-root scion is grafted in early spring to the unrooted (but easily rooted) stock by a side graft. The graft union is made about 2 in. above the base of the stock. The stock and scion are both 10 to 20 in. long. The base of the stock is treated with an indolebutyric acid rooting compound, and then the combination is inserted into a flat of a well drained rooting medium, such as perlite, so that the medium covers the graft union. The flats containing the grafts are kept in a cool greenhouse under intermittent mist. About 8 weeks later the cuttings should be rooted and the graft union healed. Then the top of the stock is cut off just above the graft union, and the grafted plant is transferred to an appropriate-sized container of soil.

6

Propagation by Specialized Stems and Roots

This chapter deals with propagation by specialized vegetative structures–*bulbs, corms, tubers, tuberous roots, rhizomes,* and *pseudobulbs.* These organs are primarily modified plant parts specialized for food storage. Plants possessing them are invariably herbaceous perennials in which the shoots die down at the end of a growing season, and the plant lives over in the ground as a dormant, fleshy organ which bears buds to produce new shoots the next season. Such plants are admirably suited it withstanding periods of adverse growing conditions in their yearly growth cycle. The two principal climatic cycles for which such performance is geared are the warm cold cycle of the temperate zones and the wet-dry cycle of tropical and subtropical regions. The second function of these specialized organs is the vegetative reproduction.

Bulbs

Definition and Structure

Bulbs are produced by monocotyledonous plants in which the usual plant structure is modified for storage and reproduction. A *bulb* is a specialized underground organ consisting of a short, fleshy, usually vertical stem axis (*basal plate*) bearing at its apex a growing point or a flower primordium enclosed by thick, fleshy scales.

Most of the bulb consists of *bulb scales,* which morphologically are the continuous, sheathing leaf bases. The outer bulb scales are generally fleshy and contain reserve food materials, whereas the

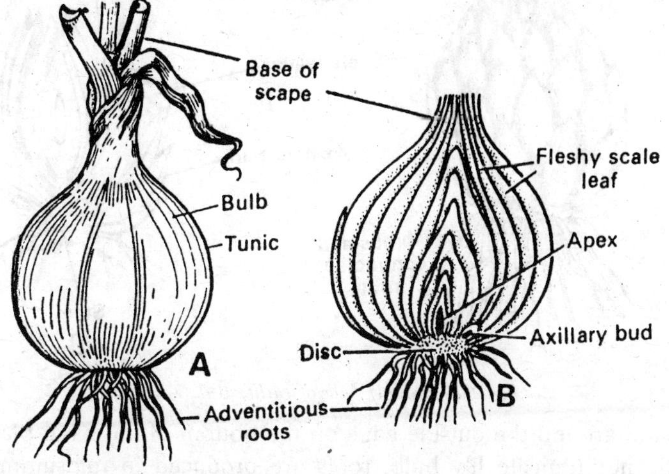

Fig. 6.1. A–Tunicated bulb of onion; B–L.S. of bulb.

bulb scales toward the center function less as storage organs and are more leaf-like. In the centre of the bulb, there will be either a vegetative growing point or an unexpanded flowering shoot. Growing points develop in the axil of these scales to produce miniature bulbs, known as *bulblets*, which when they grow to full size are known as *offsets*. In various species of lilies, bulblets may form in the leaf axils either on the underground portion or on the aerial portion of the stem. Aerial bulblets are called *bulbils*.

There are two types of bulbs;

(a) *Tunicate (laminate)* bulbs, represented by the onion and tulip. These bulb have outer bulb scales which are dry and membranous. This covering, or *tunic*, provides protection from drying and mechanical injury to the bulb. The fleshy scales are in continuous, concentric layers, or *lamina*, so that the structure is more or less solid.

(b) *Non-tumicate (scaly)* bulbs, represented by the lilies. These bulbs do not possess the enveloping dry covering. The scales are separate and give the bulb a scaly appearance. In general, the non-tunicate bulbs are easily damaged and must be handled more carefully than the tunicate bulbs; they must be kept continuously moist because they are injured by drying.

Roots are not present on a dormant, tunicate bulb but develop adventitiously at the beginning of a growth period in a narrow

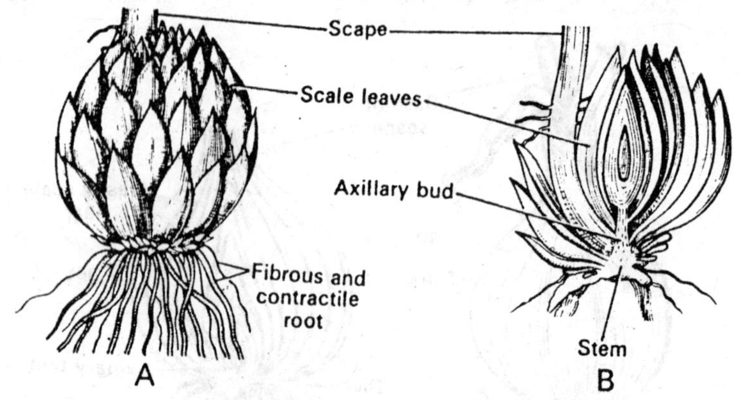

Fig. 6.2. A–Scaly bulb of garlic; B–L.S. of bulb.

band around the outside edge on the bottom of the basal plate. In the non-tunicate lily bulb, roots are produced in midsummer or later, and persist through the following year. In most lily species, roots also form on the stem above the bulb.

Contractile roots are produced in many species that appear to "pull" the bulb or corm to a given level in the ground.

Growth Behaviour

An individual bulb goes through a characteristic cycle of development, beginning with its initiation as a growing point and terminating in flowering and seed production. This general developmental cycle is composed of two stages: (a) vegetative and (b) reproductive. In the vegetative stage the bulblet grows to flowering size and attains its maximum weight. The subsequent reproductive stage includes the induction of flowering, differentiation of the floral parts, elongation of the flowering shoot, and finally flowering and (sometimes) seed production. Various bulb species have specific environmental requirements for the individual phases of this cycle which determine their seasonal behaviour, environmental adaptions, and methods of handling. They can be grouped into classes according to their time of bloom and method of handling.

Spring-Flowering Bulbs

Important commercial crops included in this group are the tulip, daffodil, hyacinth, and bulbous iris, although other kinds are grown in gardens.

Bulb formation

The vegetative stage begins with the initiation of the bulblet on the basal plate in the axil of a bulb scale. In this initial period, which usually occupies a single growing season, the bulblet is insignificant in size, since it is present within another growing bulb and can be observed only if the bulb is dissected. Its subsequent pattern of development and the time required for the bulblet to attain flowering size are somewhat different for different species. The bulbs of the tulip and the bulbous iris, for instance, disintegrate upon flowering but leave a cluster of new bulbs and bulblets which were initiated the previous season. The largest of these may have attained flowering size at this time, but smaller ones require several additional years of growth. The flowering bulb of the daffodil, on the other hand, continues to grow from the center year by year, producing new offsets which may remain attached for several years. The hyacinth bulb also continues to grow year by year, but because the number of offsets produced is limited, artificial methods of propagation are usually used.

The size and quality of bloom is directly related to the size of the bulb and the amount of stored food. Only bulbs greater than a

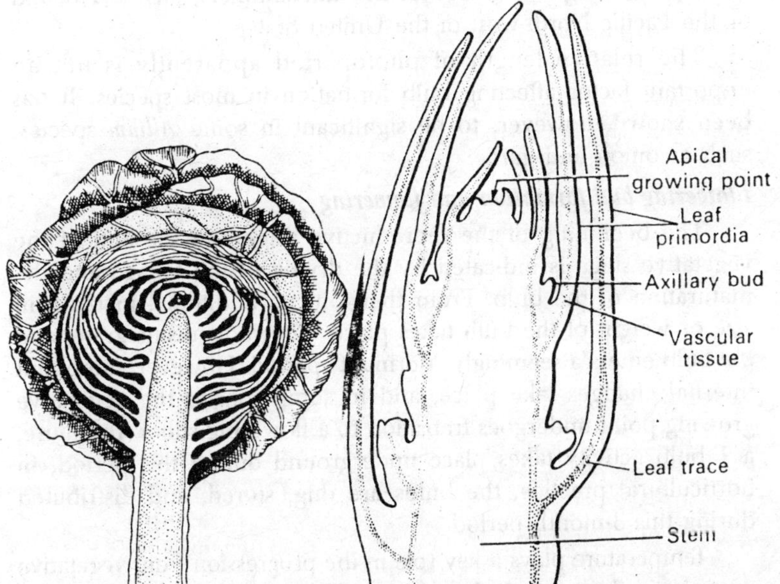

Fig. 6.3. A–Terminal bud of cabbage; B–Terminal and axillary bulb in L.S.

certain minimum size are capable of initiating flower buds. Commercial value is largely based on bulb size although the condition of the bulb and freedom from disease are also important quality factors.

The greatest increase in size and weight of the developing bulb take place in the period during and (mostly) after flowering, as long as the foliage remains in good condition. Cultural operations that encourage vegetative growth include irrigation; weed, disease, and insect control; and fertilization. Their greatest benefit, however, is to the next year's flower, because larger bulbs are produced. Conversely, adverse situations, such as poor growing conditions, removal of foliage, and premature digging of the bulb, result in smaller bulbs and inferior flower production.

Moderately low temperatures tend to prolong the vegetative period, whereas higher temperatures may cause the vegetative stage to cease and the reproductive stage to begin. Thus a shift from cool to warm conditions early in the spring, as occurs in mild climates, will shorten the vegetative period, result in smaller bulbs, and consequently produce inferior blooms the following year. Commercial bulb-producing areas for hardy spring-flowering bulbs are largely in regions of cool springs and summers, such as Holland or the Pacific North-west of the United States.

The relative length of photoperiod apparently is not an important factor affecting bulb formation in most species. It has been shown, however, to be significant in some *Allium* species, such as onion and garlic.

Flowering bud formation and flowering

The beginning of the reproductive stage and the end of the vegetative stage is indicated by the drying of the foliage and the maturation of the bulb. From then on, no additional increase in size or weight of the bulb takes place. The roots disintegrate, and the bulb enters a seemingly "dormant" period. However, important internal changes take place, and in some species the vegetative growing point undergoes transition to a flowering shoot. In nature, a 1 bulb activity takes place underground during this period; in horticultural practice, the bulbs are dug, stored, and distributed during this 3-month period.

Temperature plays a key role in the progression from vegetative growth to flower bud formation. The optimum temperature–determined as the shortest time in which the bulb would flower–

has been established for various developmental phase. In general, a deviation from these temperatures or an increase in the period of time at any one temperature results in an increase in the time required to bloom. Too high a temperature may prevent flower development entirely.

In the tulip, the differentiation of the flower parts takes place after the bulb becomes dormant in the summer. The optimum temperature for this process is 68°F (20°C). After 3 weeks of storage at this temperature, the optimum drops abruptly to 48°F (9°C), and remains there for 10 to 12 weeks, the lower temperatures being necessary to overcome the rest period and to stimulate the flower shoot to elongate. The optimum temperature then gradually increases as the flower shoot elongates and a flower is produced.

According to the studies from which the above information is quoted, the hyacinth has somewhat more exacting requirements. The optimum temperature immediately upon digging was quite high (90° to 95°F; 32° to 35°C) for a week or two. When the first flower primordium on the raceme had appeared, this shifted to 78°F (26°C). When the last flower primordium had appeared it had shifted to 63°F (17°C), and 3 weeks later it had shifted to F(55°F (13°C), where it remained for about 12 weeks. Maintaining the bulbs at a temperature above 86°F (30°C) or below 55°F (13°C) during this period inhibits flower bud development entirely. For the hyacinth, a storage temperature of 78°F (26°C) is generally recommended for the flower forming period. The daffodil ('King Alfred') has an optimum temperature of 62°F (17°C) for flower formation followed by a period of about 8 weeks at 48°F (19°C).

Holding the bulbs continuously at high temperatures (86° to 90°F; 30° to 32°C; or more) or at temperature near freezing will stop bulb development and can be used to increase the period required for flowering. With a shift to more optimum temperatures, flower bud development will continue unimpaired. Such storage conditions (below or above optimum) can be used when shipping bulbs from the Northern to the Southern Hemisphere.

In bulbous iris, a flower bud is not initiated within the bulb. It begins to appear when the shoot is a few inches out of the ground.

Lilies

A non-tunicate bulb, e.g., the lily exhibits certain important differences from a tunicate bulb. Different species of lily have different methods of bulb reproduction, some examples being given in the following section on propagation. After flowering, the old bulb may divide or form offsets which can be used for propagation. In the subsequent vegetative period during summer the new bulb for next year grows in size and accumulates reserve materials. Lily bulbs do not go dormant in the same manner as other bulbs, and the roots do not disintegrate. The bulb should be out of the ground as little as possible and, when it is for the purpose of propagation, should be handled carefully and kept from drying. The commercial value of the bulb depends not only on its size and weight but also on the condition of the fleshy roots and the freedom of the basal plate from decay.

In the Easter lily (*Lillium longiflorum*), transition of the growing point to a flowering shoot does not take place in the bulb, occurring after the shoot protrudes through the "nose" of the bulb. Storing the mature bulbs at 45° to 50°F (7° to 10°C) for 6 weeks and then growing them at temperatures of 60°F (16°C) or more (night) will produce blooms within 80 to 100 days. Storing the mature bulbs at room temperatures or at low (31°F; −1°C) temperatures will keep the bulbs dormant and delay flowering. In the latter case, the method of bulb storage is also important. If they dry out, they will deteriorate. If they absorb moisture from the packing material, they may decay.

Tender, Winter-Flowering Bulbs

There are a number of flowering bulbs from tropical areas whose growth cycle is geared to a wet-dry climatic cycle rather than a cycle related to cold-warm conditions.

The amaryllis (*Hippeastrum vittata*) is an example of this kind. Its bulb is perennial, growing continuously from the center, the outer scales disintegrating. New leaves are produced continuously from the center during the vegetative period extending from late winger to the following summer. In the axil of every fourth leaf (or scale) that develops, a growing point is initiated. Thus throughout the vegetative period a series of vegetative offsets are produced. By fall the leaves mature and the bulb goes dormant, during which time the bulb should be dry. In this period, the *fourth* growing point from the center and any external to it differentiate into flower buds, and the shoots begins to elongate slowly. After 2 to 3 months of dry storage, the bulbs are watered, the flowering shoots elongate rapidly, and flowering takes place in midwinter. Maximum foliage development and bulb growth are essential to produce a bulb large enough to form a flowering shoot.

PROPAGATION

Offsets

Offsets provide a simple and reliable method for propagating many kinds of bulbs. This method is sufficiently rapid for the commercial production of tulip, daffodil, balbous iris, and grape hyacinth bulbs, but in general it is too slow for the lily, hyacinth, and amaryllis.

If undisturbed, the offsets may remain attached to the mother bulb for several years. They can also be removed at the time the bulbs are dug and replanted into beds or nursery rows to grow into flowering sized bulbs. This may require several growing seasons, depending upon the kinds of bulb and size of the offset.

Tulip

Bulb planting takes place in the fall. Two systems of planting are used : the bed system, used mostly in Europe, and the row or field system, used mostly in the United States. Beds are usually 3 ft wide and separated by 12 to 18-in., paths. The soil is removed to a depth of 4 in. the bulbs set in rows 6 in. apart, and the soil replaced. In the other system, single or double rows are placed wide enough apart to permit the use of machines. To improve drainage, two or three adjoining rows may be planted on a ridge. Bulbs are spaced one to two diameters apart, with small bulbs scattered along the row. A mulch may be applied after planting but removed the following spring before growth.

Planting stock consists of those bulbs of the minimum size for flowering (9 to 10 cm in circumference or smaller). Since the time required to produce flowering sizes varies with the size of the bulb, the planting stock is graded so that those of one size can be planted together. For instance, an 8-cm or larger bulb normally requires a single season to become of flowering size; a 5- to 7-cm bulb, two seasons; and those 5 cm or less, 3 years.

During the flowering and subsequent bulb growing period of the next spring, good growing conditions should be provided so that the size and weight of the new bulbs will be at a maximum. Foliage should not be removed until it dries or matures. Important cultural operations include removal of competing weed growth, irrigation, fungicidal sprays to control *Botrytis blight* and fertilization.

It is desirable to remove the flower heads at blooming time, because they may serve as a source of *Batrytis* infection and can lower bulb weight.

Bulbs are dug in midsummer when the leaves have turned yellow or the outer coats of the bulb have become dark brown in colour. In the Pacific Northwest of the United States, where temperatures are cool and the leaves remain green for a longer period, digging may take place before the leaves dry. If the bulbs are dug too early or if warm weather causes early maturation, the bulbs may be small in size. The bulbs are dug by machine or by hand with a short-handled spade. After the loose soil is shaken from the bulbs, they are placed in trays in well–ventilated storage houses for drying, cleaning, sorting, and grading. General storage temperatures are 65° to 68°F (18° to 20°C). For early flowering, the bulbs should be held at 68°F for 3 weeks and then placed at 48°F (9°C) for 8 weeks. Later flowering will be produced by holding bulbs at 72°F (22°C) for 10 weeks. For shipment to the Southern Hemisphere, the bulbs can be held at 31°F (–1°C) until late December, when they are shifted to a higher temperature (78°F; 25.5°C).

Daffodil

Daffodil bulbs are perennial and produce a new growing point at the center every year. Offsets are produced which grow in size for several years until they break away from the original bulb, although they are still attached at the basal plate. An offset bulb, when it first separates from the mother bulb, is known as a "split," "spoon," or "slab", and can be separated from the mother bulb and planted. Within a year it become a "round", or "single-nose", bulb containing a single flower bud. One year later a new offset should be visible, enclosed within the scales of the original bulb, indicating the presence of two flower buds. At this stage the bulb is known as a "double-nose". By the next year the offsets split away, and the bulb is known as a "mother bulb". Grading of daffodil bulbs is principally by age, that is, as splits, round, double-nose, and mother bulbs. The grades marketed commercially are the round and the double-nose. The mother bulbs are used as planting stock to produce additional offsets, and only the surplus is marketed. Offsets, or splits, are replanted for additional growth.

Storage should be at 55° to 60°F (13° to 16°C) with a relative humidity of 75 percent. For earlier flowering, they can be stored at

48°F (9°C) for 8 weeks. To delay flowering, store at 72°F (22°C) for 10 weeks. For shipment from the Northern to the Southern Hemisphere, bulbs can be held at 86°F (30°C) until October, then stored at 31°F (–1°C) until late December, and then at 77°F (25°C).

Hot water treatment plus a fungicide, as formalin, for stem and bulb eelworm control is important. A three or four-hour treatment at 110°F (43°C) is used, but that temperature must be carefully maintained or the bulbs may be damaged.

Lilies

Lilies increase naturally, but except for a few species this increase is slow and of limited propagation value except in home gardens. Several methods of bulb increase are found among the different species. For instance, *Lilium concolor, L. hansoni, L. henryi* and *L. regale* increase by bulb splitting. Two to four lateral bulblets are initiated about the base of the mother bulb, which disintegrates during the process, leaving a tight cluster of new bulbs. *Lilium bulbiferum, L. canadense, L. paradalinum, L. parryi, L. superbum,* and *L. tigninum* multiply from lateral bulblets produced from the rhizome-like bulb. This is sometimes called "budding-off".

Bulblet Formation on Stems

The production of *underground stem bulblets* is the usual commercial method of propagating the Easter lily (*Lilium longiflorum*) and some other lily species. Flowering of the Easter lily occurs in early summer. Bulblets form and increase in size from spring through summer. Between mid August and mid-September in the Northern Hemisphere the stems are pulled from the bulbs and stacked upright in the field. Periodic sprinkling keeps the stems and bulblets from drying out. Similarly, the base of the stem can be "heeled in" the ground at an angle of 30 to 45 deg. or laid horizontally in trays at high humidity.

About mid-October the bulblets are planted in the field 4 in. deep and an inch apart in double rows which are spaced 36 in apart. Here they remain for the following season. They are dug in September as yearling bulbs and again replanted, this time 6 in deep and 4 to 6 in apart in single rows. At the end of the second year, they are dug and sold as commercial bulbs.

Digging is done in September after the stem is pulled. The bulbs are graded, packed in peat moss, and shipped. Commercial bulbs have a minimum size of 7 to 10 in, in circumference. Lily bulbs must be handled carefully so they will not be injured and must be kept from drying out. The fleshy roots should also be kept in good condition. For long-term storage that will prevent flowering, the bulbs should be packed in polyethylene-lined cases containing peat moss at 30 to 50 percent moisture and stored at 31°F (-1°C). To produce flowering in the shortest period of time, they should be stored at 45° to 50°F (7° to 10°C) for 6 weeks and then grown at temperatures of 60°F (16°C) (night) or more.

Lilies are attacked by certain viruses, fungus diseases, and nematodes which can be carried in or on the bulbs. Their control during propagation is an important consideration in bulb production. Methods of control include using disease free stocks for propagation, growing plants in disease-free locations with good sanitary procedures, and treating the bulbs with fungicides. A pre-planting dip in 2 Ib of pentachloronitrobenzene (PCNB) plus 2 Ib of ferbam in 100 gal. of water has been found useful.

To obtain commercial planting stock of lilies free of *Rhizoctonia*, stem and bulb nematode, and root-lesion nematode, cure bulbs at 05°F (95°C) and 95 percent relative humidity for 1 to 2 weeks, presoak for 2 days in cool water, soak for 2 hours at 115°F (46°C) in water plus 37 percent formaldehyde diluted 1:200; aftersoak the bulbs in a fungicide such Puratized Agricultural Spray at 1:1000. Scale the bulbs, dust the dulblets with ferbam, and place in vermiculite for bulblets to form. Plant new bulblets in treated soil.

Aerial stem bulblets, commonly known as *bulbils*, are formed in the axil of the leaves of some lily species, such as *Lilium bulbiferum*, *L. sargentiae*, *L. sulphureum*, and *L. tigrinum*. Bulbils develop in the early part of the season and fall to the ground several weeks after the plant flowers. They are harvested shortly before they fall naturally and are then handled in essentially the same manner as underground stem bulblets. Increased bulbil production can be induced by disbudding as soon as the flower buds have formed. Likewise, some lily species which do not form bulbils naturally can be induced to do so by pinching out the flower buds and a week later cutting off the upper half of the stems. Species which respond to the latter procedure include *Lilium candidum*, *L. chalcedonicum*, *L. hollandicum*, *L. maculatum* and *L. testaceum*.

Stem Cuttings

Lilies may be propagated as stem cuttings. The cutting is made shortly after flowering. Instead of roots and shoots forming on the cutting, as would occur in other plants, bulblets form at the axils of the leaves and then produce roots and small shoots while on the cutting.

Leaf-bud cuttings, made with a single leaf and a small heel of the old stem, may be used to propagate a number of lily species. A small bulblet will develop in the axil of the leaf. It is handled in the same manner as for the other methods described here.

Bulblet Formation on Scales (Scaling)

An efficient method for producing lily bulbs is the technique known as *scaling*, in which individual bulb scales are separated from the mother bulb and placed in growing conditions so that adventitious bulblets from at the base of the each scale. Three to five bulblets will usually develop from each scale. This method is particularly useful for rapidly building up stocks of a new variety or to establish disease-free stocks. Almost any lily species can be propagated by scaling.

Scaling is usually done soon after flowering in midsummer, although it might be done in late fall or even in midwinter. The bulbs are dug, the outer two layers of scales are removed and the mother bulb is replanted for continued growth. It is possible to remove the scales down to the core, but this will reduce subsequent growth of the mother bulb. The scales should be kept from drying and handled so as to avoid injury. Scales with evidence of root should be discarded and the remaining ones dusted with a fungicide. Naphthaleneacetic acid will stimulate bulblet formation. Mix 1000 parts of thiram or ferbium with 1 part of naphthalene acetic acid and apply as a dust. The PCNB-thiram mixture is also an effective fungicidal treatment.

Scales can be planted directly in the field in open beds or frames, planted no more than 2½ in. deep. Somewhat better results are obtained if they are placed in trays or flats of moist sand, peat moss, sphagnum moss, or vermiculite for 6 weeks at 65° to 70°F (18° to 21°C). The scales are inserted vertically to about half their length. Small bulblets and roots should form at the base within 3 to 6 weeks. The scales are transplanted either into the open ground or into pots or flats of soil, and then planted in the field the following spring. Subsequent treatment is the same as described for underground bulblets.

Basal Cuttage

The hyacinth is the principal plant propagated by this method, although others such as *Scilla* can be handled in this way. Specific methods include "scooping", "scoring", and "coring". Mature bulbs which have been dug after the foliage has died down and are 17 to 18 cm or more in circumference are used. In *scooping*, the entire basal plate is scooped out with a special curve-bladed scalpel, a round-bowled spoon, or a small-bladed knife. Adventitious bulblets develop from the base of the exposed bulb scales. Depth of cutting should be enough to destroy the main shoot. In *scoring*, three straight knife cuts are made across the base of the bulb, each deep enough to go through the basal plate and the growing point. Growing points in the axils of the bulb scales grow into bulblets. In *coring*, the growing point in the center of the bulb is removed entirely with an apple corer or a cork borer 3/8 to ½ in. in diameter. All of the growth potential is concentrated in the development of the bulblets which grow from the basal plate.

To combat decay that may develop during the later incubation period, any infected bulbs should be discarded; the tools disinfected frequently with an alcohol, formalin, or mild carbolic acid solution, and the cut bulbs dusted with a fungicide. Most important is to callus the bulbs at about 70°F (21°C) for a few days to a few days to a few weeks in dry sand or soil or in open trays, cut side up. After callusing, the bulbs are incubated in trays or flats, in dark or diffuse light, at 70° to 90°F (21° to 32°C) and high humidity for 2½ to 3 months.

The mother bulbs are planted about 4 in. deep in nursery beds in the fall. The next spring bulblets produce leaves profusely. Normally the mother bulb disintegrates during the first summer. Annual digging and replanting of the graded bulblets is required until they reach flowering sizes. Bulbs for greenhouse forcing should be 17 cm (6¾ in.) or more in circumference; bulbs for bedding should be 14 to 17cm (5½ to 6¾ in.) in circumference. On the average, a scooped bulb will produce 60 bulbets, but 4 to 5 years will be required to produce flowering sizes; a scored bulb will produce 24, requiring 3 to 4 years ; and a cored bulb will produce 10, requiring 2 to 3 years.

Leaf Cuttings

This method has been reported successful for blood lily (*Haemanthus*), grape hyacinth (*Muscari*), hyacinth, and *Lachenalia* although the range of species is probably wider.

Leaves are taken at a time when they are well developed and green, generally near full bloom. An entire leaf is cut from the top of the bulb and may in turn be cut into two or three pieces. Each section is placed in a rooting medium with the basal end several inches below the surface, as described for rooting cuttings. The leaves should not be allowed to dry out, and bottom heat is desirable. Within 2 to 4 weeks small bulblets form on the base of the leaf and roots develop. At this stage the bulblets are planted in soil.

Bulb Cuttings

Among those plants which have been reported to respond to this method of propagation are the *Albuca*, *Chasmanthe*, *Cooperia*, *Haemanthus*, *Hippeastrum*, *Hymenocallis*, *Lycoris*, *Narcissus*, *Nerine*, *Pancratium*, *Scilla*, *Spreakelia*, and *Urceolina*.

A mature bulb is cut into a series of 8 to 10 vertical sections, each containing a part of the basal plate. These sections are further divided by sliding a knife down between each third or fourth pair of concentric scale rings and cutting through the basal plate. Each of these fractions makes up a bulb cutting consisting of a piece or basal plate and segments of three or four scales.

The bulb cuttings are planted vertically in a rooting medium, such as peat moss and sand, with just their tips showing above the surface. The subsequent technique of handling is the same as for ordinary leaf cuttings. A moderately warm temperature, slightly higher than for mature bulbs of that kind, is required. New bulblets develop from the basal plate between the bulb scales within a matter of a few weeks, along with new roots. At this time they are transferred to flats of soil to continue development.

Corms

Definition and Structure

A *corm* is the swollen base of a stem axis enclosed by the dry, scale- like leaves. In contrast to the bulb, which is predominantly leaf scales, a corm is a solid stem structure with distinct nodes and internodes. The bulk of the corm consists of storage tissue composed of parenchyma cells. In the mature corm, the dry leaf bases persist at each of these nodes and enclose the corm. This covering, known as the *tunic*, protects it against injury and water loss. At the apex of the corm is a terminal shoot bud which will develop into the leaves and the flowering shoot. Axillary buds are produced at each of the nodes. In a large corm, several of the upper buds may develop into flowering shoots, but those nearer the base of the

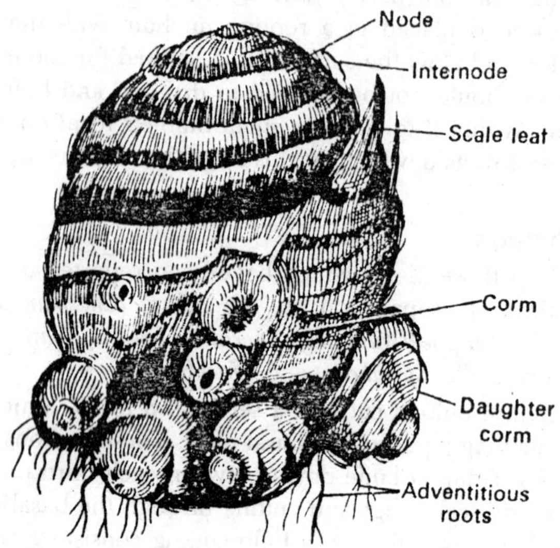

Fig. 6.4. Corm of Colocasia.

corm are generally inhibited from growing. However, should something prevent the main buds from growing, these lateral buds would be capable of producing a shoot.

Two types of roots are produced from the corm; a fibrous root system developing from the base of the old corm and enlarged, fleshy contractile roots developing from the base of the new corm.

Growth Behaviour

Gladiolus and crocus are typical cormous plants. The gladiolus is a semi-hardy to tender plant which, in areas with severe winters, must be stored over winter and replanted in the spring. At the time of planting, the corm is a vegetative structure. New roots develop from the base of the corm, and one or more of the buds begin to develop leaves. Differentiation of the inflorescence takes place within a few weeks after the bud begins to grow. At the same time the base of the shoot axis thicken, and a new corm for the succeeding year begins to form above the old corm. Root-like structures bearing miniature corms or *cormels* on their tip develop from the base of the new corm. The new corm continues to enlarge, and the old corm begins to shrivel and disintegrate as its contents are utilized in flower production. After flowering, the foliage continues to manufacture food materials, which are stored in the new corm. At the end of the summer, when the foliage dries, there are one or more new corms and perhaps a great number of little cormels. The corms are dug and stored over winter until they are planted the following spring.

PROPAGATION

New Corms

Propagation of cormous plants is principally by the natural increase of new corms. Flower production in corms, as in bulbs, depends upon food materials stored in the corm the previous season, particularly during the period following bloom. In gladiolus, cool nights and long growing periods are favourable for production of very large corms. Fertilization and other good management practices during bloom have their greatest effect on the next year's flowers. Plants are left in the ground for 2 months following blooming, or until frost kills the tops. After digging, the plants are placed in trays with a screen or slat bottom arranged to allow air to circulate between them, and cured at 95°F (35°C) at 80 to 85 percent relative humidity. Then the new corms, old corms, cormels, and top can be easily separated. The corns are graded according to size, sorted to remove the diseased ones, treated with a fungicide (Spergon wettable powder), and returned to a 95°F temperature for an additional week. This process suberizes the wounds and helps combat *Fusarium* infection. The corms are then stored at 40°F (5°C) with a relative humidity of 70 to 80 percent in well-aerated rooms to prevent excessive drying. It is also desirable to treat them immediately before planting with ½ percent Lysol solution for 2 hours or with 1/8 percent solution of New Improved Ceresan for 15 minutes.

Cormels

These are miniature corms which develop between the old and the new corms. One or 2 years growth is required for them to reach flowering size. Shallow planting of the corms, only a few inches deep, results in greater production of cormels; increasing the depth of planting reduces cormel production.

Cormels are separated from the mother corms and stored over winter for planting in the spring. Dry cormels become very hard and may be slow to start growth the following spring, but if they are stored at about 40°F (5°C) in slightly moist peat moss, they will stay plump and in good condition. Soaking dry cormels in warm water for 2 to 5 days just prior to planting will hasten the onset of growth.

Disease-free cormels can be obtained by hot-water treatment. Treatment should be done between 2 and 4 months after digging. The cormels are soaked in water at air temperature for 2 days, then placed in a 1:200 dilution of commercial 37 percent fromaldehyde for 4 hours, and then immersed in a water bath at 135°F (57°C) for 30 minutes. The temperature should be maintained within 1°F, plus or minus, of this temperature. At the end of the treatment, the cormels are cooled quickly, dried immediately, and stored at 40°F (5°C) in a clean area with good air circulation. Dusting the dry cormels with a fungicide at this time is also desirable.

The cormels are planted in the field in furrows about 2 in. deep in the manner of planting large seeds. Only grass-like foliage is produced the first season. The cormel does not increase in size but produces a new corm from the base of the stem axis, in the manner described for full-sized corms. At the end of the first growing season, the beds are dug and the corms separated by size. A few of the corms may attain flowering size, but most require an additional year of growth.

Size grades in gladiolus are determined by diameter. There are seven grades, the smallest 3/8 to ½ in. in diameter, the largest 2 in. or more.

Division of the Corm

Large corms may be cut into sections, retaining a bud with each section. Each of these should then develop a new corm. Segments should be dusted with a fungicide because of the great likelihood of decay of the exposed surfaces.

TUBERS

Definition and Structure

A *tuber* is a modified stem structure which develops below ground as a consequence of the swelling of the subapical portion of a stolon and subsequent accumulation of reserve materials.

The most notable example of a plant which produces tubers by which it is propagated is the Irish potato (*Solanum tuberosum*). The *Caladium*, grown for its striking foliage, is also propagated by tubers, as is the Jerusalem artichoke (*Helianthus tuberosus*).

A tuber has all the parts of a typical stem. The "eyes", present in regular order over the surface, represent nodes, each consisting of one or more small buds subtended by a leaf scar. The

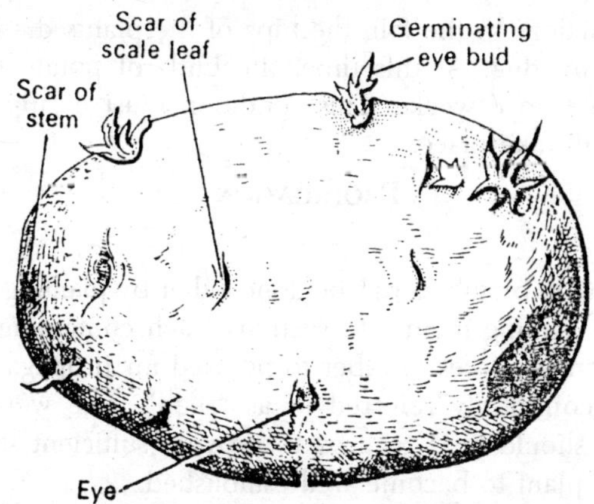

Fig. 6.5. Tuber of potato.

arrangement of the nodes is a spiral, beginning with the terminal bud on the end opposite the scar resulting from the attachment to the stolon. The terminal bud is at the apical end of the tuber, oriented farthest (distally) from the crown of the plant. Consequently it shows the same apical dominance as any stem.

Growth Behaviour

A tuber is a storage and propagative organ that is produced in one growing season, remains dormant during the winter, and then produces new shoots the following spring to start a new cycle. Most information on growth cycles in tubers is from studies of the potato. At the base of the main shoot adventitious roots are initiated, and lateral buds grow out horizontally to become the stolons. Continued growth of the stolon takes place during long photoperiods and is associated with the presence of auxin and a high gibberellin level. Tuberization begins with inhibition of terminal growth and the intermediate day lengths, reduced temperatures (particularly at night), good light, low mineral content, and a reduction in gibberellin levels in the plant.

Tuberization has been explained as being caused by the production of a tuber-inducing substance (perhaps a gibberellin inhibitor) that is produced in the leaves and the mother tuber. It seems to be necessary for the stolon tip to have attained a particular physiological age. Continued tuber enlargement is dependent on a continuing adequate supply of photosynthate. Conditions which favour rapid and luxurious plant growth above ground, such as an abundance of nitrogen, or high temperatures, are not conducive to tuber production. In the fall, the tops of the plants die down and the tubers are dug. At this time, the buds of potato tubers are dormant for 6 to 8 weeks. This condition must disappear before sprouting will take place.

Propagation

Division

Propagation by tubers can be done either by planting the tubers whole or by cutting them into sections, each containing a bud or eye. These small pieces of tuber to be used for propagation of the potato are commonly referred to as "seed". The weight of the tuber piece should be 1 to 2 oz to provide sufficient stored food for the new plant to become well established.

Division of tubers is done with a sharp knife shortly before planting. The cut pieces should be stored at warm (68°F; 20°C) temperatures and relatively high humidities (90 percent) for 2 to 3 days prior to planting. During this time, the cut surfaces heal (*suberization*) and the "seed" piece is effectively protected against drying and decay. Treatment of potato tubers prior to cutting for the control of *Rhizoctonia* and scab may be desirable. Caladium tubers are produced commercially in Florida. The tubers are cut into sections, usually two buds per piece. These are planted 3 to 4 in. deep, 4 to 6 in. apart in rows 18 to 24 in. apart. Harvest begins in November. After harvest, the tubers are dried in open sheds for 6 weeks or artificially dried for 48 hours. Further storage should be at a temperature not below 60°F (16°C).

Tubercles

Begonia evansiana and the cinnamon vine (*Dioscorea batatas*) produce small aerial tubers, known as *tubercles*, in the axils of the leaves. These tubercles may be removed in the fall, stored over winter, and planted in the spring.

Tuberous Roots

Definition and Structure

Certain herbaceous perennials produce thickened tuberous roots which contain large amounts of stored food. Although the appearance of such roots may vary considerably from one species to another, they have the internal and external features of a typical root. Thus they differ from the true (stem) tuber in that they lack nodes and internodes, buds are present only at the crown or stem (proximal) end, fibrous roots are commonly produced toward the opposite (distal) end. The polarity of the tuberous root is consequently the reverse of that of the true tuber.

Growth Behaviour

The sweet potato and dahlia produce swollen sections on lateral roots. In the latter case, these are borne in a cluster, each tuberous root attached to the crown of the plant. These roots are biennial. They are produced in one season, after which they go dormant as the herbaceous shoots die. The following spring, buds from the crown produce new shoots which utilize the food materials from the old root during their initial growth. The old root then disintergrates, and new roots are produced, which in turn maintain the plant through the following dormant period.

In the tuberous begonia, on the other hand, the primary taproot becomes a single enlarged tuberous root. Buds are produced at the proximal end (the crown). Fibrous roots are produced from the distal portion of the swollen root. This tuberous root is perennial and lives for a number of years, continuing to increase in size and produce additional buds from the crown. In the tuberous begonia reduced day length results in the enlargement of the tuberous root and the cessation of flowering.

PROPAGATION

Adventitious Shoots

The fleshy roots of a few species of plants such as sweet potato have the capacity to produce adventitious shoots if subjected to the proper conditions of temperature and moisture. The roots are laid in sand so that they do not touch one another and covered to a depth of about 2 in. The bed is kept moist. The temperature should be about 80°F (27°C) at the beginning and about 70° to 75°F (21° to 24°C) after sprouting has started. As the new shoots, or *slips*, come through the covering, more sand is added so that eventually the stems will be covered for 4 to 5 in. Adventitious roots develop from the base of these adventitious shoots. After the slips are well rooted, they are pulled from the parent plant and transplanted into the field. An increase in slip production takes place if sweet potato roots are cut in half and the pieces are subjected to 110°F (43°C) for about 26 hours. This treatment overcomes the apical dominance and also controls nematodes and fungus diseases.

Division

Most plants with fleshy roots must be propagated by dividing the crown so that each section bears a shoot bud. This is necessary in the dahlia, for example. The plant is dug with its cluster of roots intact, dried for a few days, and stored at 40° to 50°F (4° to 10°C) in dry sawdust, dry peat, or dry shavings. Storage in the open may result in shriveling. The cluster of roots is divided in the late winter or spring shortly before planting. In warm, moist conditions the buds begin to grow, and division can be carried out with better assurance that each section will have a bud.

The perennial tuberous root of the tuberous begonia can also be divided as long as each section has a bud. To combat decay, the cut surface should be dusted with a fungicide and each section dried for several days after cutting and before placing in a moist medium.

Vegetative propagation in such plants as these can often be better carried out with stem, leaf, or leaf-bud cuttings. The cuttings will develop tuberous roots at their base. This process can be stimulated if the stem cutting initially includes a small piece of the fleshy root.

Rhizomes

Structure

A *rhizome* is a specialized stem structure in which the main axis of the plant grows horizontally just below or on the surface of the ground. A number economically important plants, such as bamboo, sugar cane, banana and many grasses, as well as a number of ornamentals, such as *Iris* and lily-of- the-valley, have rhizome structures. Most are monocotyledons, although a few dicotyledons–for example, low bush blueberry (*Vaccinium angustifolium*)–have analogous underground stems classed as rhizomes. Many ferns and lower plant groups have rhizomes or rhizome-like structures.

The stem appears segmented because it is composed of nodes and internodes. A leaf-like *sheath* is attached at each node; it encloses the stem and, in an expanded form, becomes the foliage leaves. When the leaves and sheaths disintegrate and scar is left at the point of attachment identifying the node and giving a segmented appearance. Adventitious roots and lateral growing points develop in the vicinity of the node. Upright-growing, above-ground shoots and flowering stems (*culms*) are produced either terminally from the rhizome tip or from lateral branches.

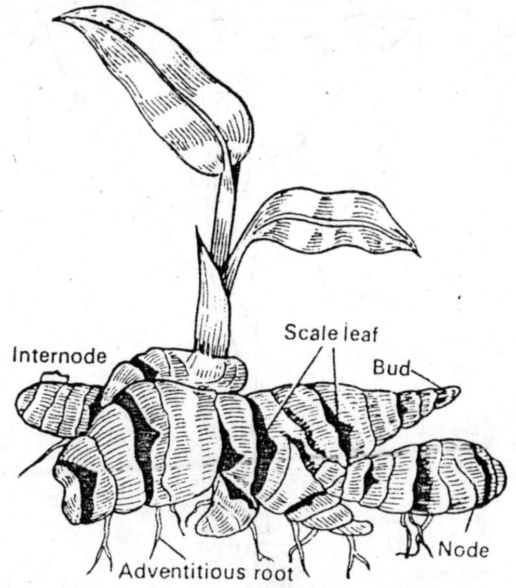

Fig. 6.6. Rhizome of ginger.

Two general types of rhizomes are found. The first (the (*pachymorph*) is illustrated by the *Iris.* The rhizone is thick, fleshy, and shortened in relation to length. It appears as a many-branched clump made up of short, individual sections. It is determinate; that is, each clump terminates in a flowering stalk, growth continuing only from lateral branches. The rhizome tends to be oriented horizontally with roots arising from the lower side.

The second type (the *leptomorph*) is illustrated by the lily-of-the-valley. The rhizome is slender with long internodes. It is indeterminate; that is, it grows continuously in length from the terminal apex and from lateral branch rhizomes. The stem is symmetrical and has lateral buds at most nodes, nearly all remaining dormant. This type does not produce a clump but spreads extensively over an area.

Intermediate forms between these two types also exist. These are called *mesomorphs.*

Growth Behaviour

Rhizomes grow by elongation of the growing points produced at the terminal end and on later branches. Length also increases by growth in the intercalary meristems in the lower part of the internodes. As the plant continues to grow and the older part dies,

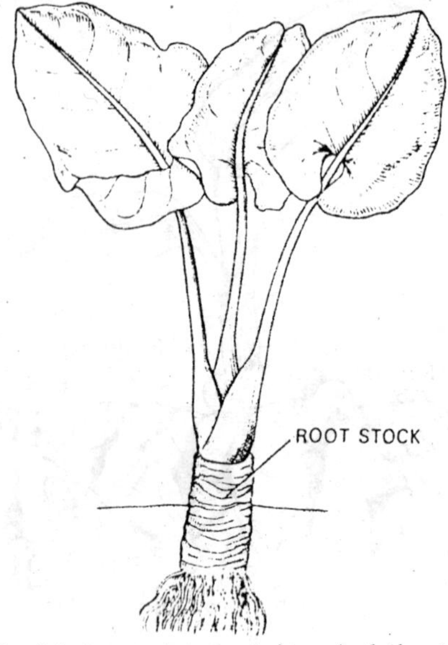

Fig. 6.7. Root stock (vertical rhizome) of Alocasia.

the several branches arising from one plant may eventually become separated to form individual plants of a single clone.

Rhizomes exhibit consecutive vegetative and reproductive stages, but the growth cycle differs somewhat in the two types described. In the pachymorph rhizome of *Iris*, a growth cycle begins with the initiation and growth of a lateral branch on a flowering section. The flowering stalk dies, but these new lateral branches produce leaves and grow vegetatively during the remainder of that season. Continued growth of the under ground stem, storage of food, and the production of a good flower bud at the conclusion of the vegetative period depend upon photosynthesis in the leaves. Consequently foliage should not be removed during this period. A flowering stalk is produced the following spring and no further terminal growth can take place. In general, plants with this structure flower in the spring and grow vegetatively during the summer and fall.

Plants with a leptomorph habit as a general rule (with exceptions) grow vegetatively during the beginning of the growth period and flower later in the same period. The length of time during which an individual rhizome remains vegetative varies with different kinds of plants. An individual branch in the lily-of-the-valley for instance, is vegetative three years before a flower bud forms. Some bamboo species remain vegetative for many years, but then they change abruptly and the entire plant produces flowers.

PROPAGATION

Division of Clumps

Clump division is the usual procedure for propagating many plants from rhizomes, but the procedure may vary some what with the two types. In pachymorph rhizomes, individual sections(or culms) are cut off at the point of attachment to the rhizome, the top is cut back, and the piece is transplanted to the new location. Leptomorph rhizomes can be handled in essentially the same way by removing a single lateral "off-shoot" from the rhizome and transplanting it. The tip of the lily-of-the-valley rhizome bearing a flower bud called a "pip" is removed along with the rooted section below and transplanted.

Division is usually carried out at the beginning of a growth period (as in early spring) or at or near the end of a growth period (i.e., in late summer or fall).

Division of Rhizomes

Dividing a rhizome is done by cutting the rhizome into sections, being sure that each piece has at least one lateral bud, or "eye", it is essentially a stem cutting. Bananas, for instance, are propagated in this way. This general method works well for the leptomorph rhizomes, in which a dormant lateral growing point is present at most nodes. The rhizomes are cut or broken into pieces, and adventitious roots and new shoots develop from the nodes. Rhizomes producing turf grasses, for instance, are cut up into sections and the individual "sprigs" transplanted. New plants can be established readily by this method.

Culm Cuttings

In large rhizome-bearing plants, such as bamboos, the aerial shoot, or culm, may be used as a cutting. These may be whole culm cuttings, in which the entire aerial shoot is laid horizontally in a trench. New branches arise at the nodes. A more common alternative procedure is to cut the stem into three- or four-node sections and place them vertically in the ground as for an ordinary stem cutting.

Pseudobulbs

Definition and Structure

A *pseudobulb* (literally "false bulb") is a specialized storage structure produced by many orchid species, consisting of an enlarged, fleshy section of the stem made up of one to several nodes. In general appearance the pseudobulb varies with different species of orchids, the differences being sufficiently characteristic to aid in identification.

Growth Behaviour

These pseudobulbs arise during the growing season on upright growths which develop laterally or terminally from the horizontal rhizome. Leaves and flowers from either at the terminal end or at the base of the pseudobulb, depending upon the species. During the growth period, they accumulate stored food materials and water and assist the plants in surviving the subsequent dormant period.

Propagation

Offshoots

In a few orchids, such as the *Dentrobium* species, the pseudobulb is long and jointed, being made up of many nodes. Offshoots develop at these nodes. From the base of these offshoots roots develop. The rooted offshoots are then cut from the parent plant and potted.

Division

Most important commercial species of orchids, including the *Cattleya*, *Laelia*, *Miltonia*, and *Odontoglossum*, may be propagated by dividing the rhizome into sections, the exact procedure used depending upon the particular kind of orchid. Division is done during the dormant season, preferably just before the beginning of a new period of growth. The rhizome is cut with a sharp knife back far enough from the terminal end to include four to five pseudobulbs in the new section, leaving the old rhizome section with a number of old pseudobulbs, or "back bulbs", from which the leaves have dehisced. The section is then potted, whereupon growth begins from the bases of the pseudobulbs and at the nodes.

The removal of the new section of the rhizome from the old part stimulates new growth, or "back breaks", to occur from the old parts of the rhizome. These new growths grow for a season and can be removed the following year.

An alternate procedure is to cut partly through the rhizome and leave it for 1 year. New back breaks will develop which can later be removed and potted.

Back Bulbs and Green Bulbs

"Back Bulbs" (i.e., those without foliage) are commonly use to propagate clones of *Cymbidium.* These are removed from the plant, the cut surface is painted with a grafting compound, and they are placed in a rooting medium for a new shoots to develop. The back bulb can be repropagated and a second shoot developed from it.

"Green bulbs"(i.e., those with leaves) can also be used in *Cymbidium* propagation. Treatment with indolebutyric acid, either by soaking or by painting with a paste, has been shown to be beneficial.

7

Asexual Reproduction in Fungi

The formation of sexual spores by Fungi fulfils two purposes:

(i) The production (normally) of vast numbers of propagating units which ensure that the organism .can be widely and quick disseminated.

(ii) To provide a mechanism of resistance to unfavourable environmental conditions.

The diversity in type and morphology of spores, and the structures that bear them, is enormous and adds to the fascination of the fundamental question regarding the process that control and are involved in their formation. Studies on the environmental parameters affecting asexual spore production are found from the earliest days of fungal physiology. More recently observers have taken advantage of electron microscopic techniques and advanced culture techniques to provide some answers to aspects of this basic question.

Types of Asexual Spore

Two main types of asexual spore are produced by fungi, sporangiospores and conidia. They are distinguished by the morphology of the structure (sporohore) that produces them and by the mechanisms by which they are formed.

Sporangiospores are produced and retained within a sporangium. This structure normally develops at the tip of an aerial hypha but may also arise at an intercalary site. During its development the sporangial wall develops from the hyphal wall and cytoplasm within

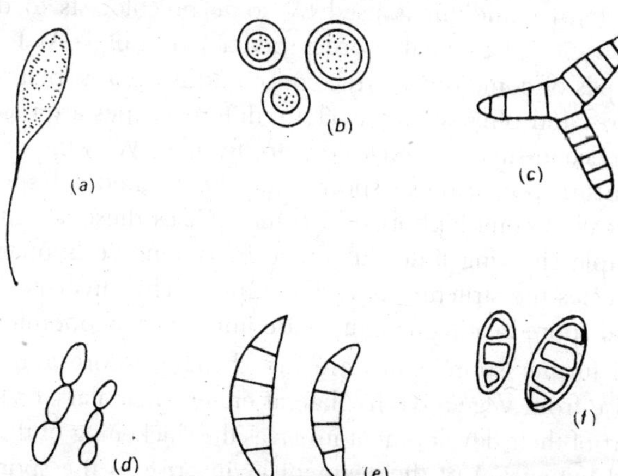

Fig. 7.1. Some examples of diversity in morphology found in asexual spores. (a) Zoospore of Allomyces. (b) Sporangiospores of Pilobolus. (c-f) Conidia of an aquatic Hyphomycete, Penicillium and Helminthosporium, respectively.

it becomes cleaved into the individual spores. The latter may or may not have a wall–walled spores are non-motile and called aplanospores, naked spores are motile zoospores. Release of the spores from the sporangium may involve a variety of mechanisms.

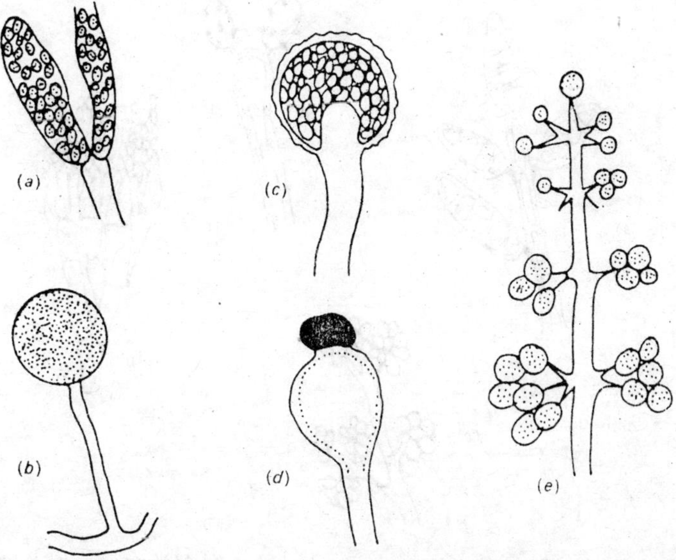

Fig. 7.2. Diversity in morphology of the sporangium; (a) Saprolegnia, (b) Pythium, (c) Rhizopus, (d) Pilobolus, (e) Mortierella.

The term conidium is used by some mycologists to describe all other types of asexual spore produced by fungi. At least four major types of conidium are recognised arthrospores, blastospores, porospores and phialoconidia. These different types are recognised by the mechanisms by which they are formed. With the exception of the arthrospores, these spores may be formed on specialised hyphae called conidiophores. In some species these structures are very simple showing little difference from somatic hyphae but in other species the structure is very complex. This diversity in spore types and spore beating structures are important taxonomic criteria.

Two further spore types are the chlamydospore and oidium. Both arise from vegetative hyphae at either intercalary or terminal sites. Part of their development involves the thickening of the hyphal wall and expansion of the segment giving rise to the spore.

The biological roles of these different spores may be considered in relation to the way in which they are produced and the number of spores involved. Sporangiospores and conidia are formed in large numbers and are used to spread the fungus and allow it to colonies new environments. Chlamydospores and oidia are generally more

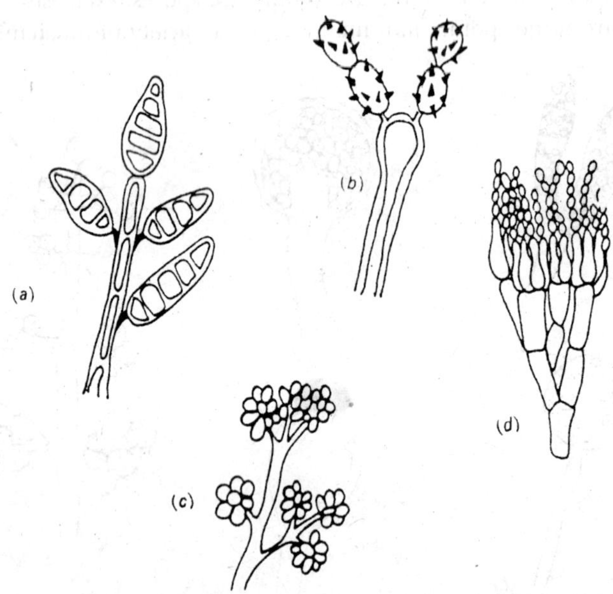

Fig. 7.3. Morphology of conidiophores; (a) Alternaria, (b) Cladosporium, (c) Botrytis, (d) Penicillium.

resistant, lacking a mechanism for liberation or dispersal, and have a role in maintenance of the species under adverse environmental conditions.

PATTERNS OF SPORE FORMATION

Our knowledge of the events that occur in the differentiation of sporangiospores and conidiospores is now quite extensive and stems from many detailed studies using the scanning and transmission electron microscopes. Undoubtedly the information regarding changes at the ultrastructural level will be of even greater interest in the future as we learn more of the associated biochemical changes.

Development of Zoosporangiospores

The differentiation of sporangiospores has been investigated in several Phycomycete fungi and in all of them the cleavage of sporangial cytoplasm into the individual spore units is brought about by the activity of vesicles. What is different in spore development in these species is the origin of the vesicles involved in the cleavage process.

In Oomycete species, which produce zoospores, sporangium development begins with an expansion of the sporangial initial hypha caused by an inflow of cytoplasm from neighbouring hyphae. The extent of this expansion varies in different species. As the sporangium develops secondary wall thickening also occurs, however, the biochemical changes associated with this are unknown. Other notable events in the formation of the sporangium are the differentiation of a basal plug which separates the sporangium from the support hypha and an apical papilla, which is the site of spore discharge.

Differentiation of zoospores has been examined in detail in four Oomycete species. In *Phytopthora*, *Pythium* and *Saprolegnia* cytoplasmic cleavage is achieved by vesicle fusion, but *Aphanomyces* is an exception and no *de novo* membrane synthesis is found. In *Phytopthora* and *Pythium* the cleavage vesicles arise from activity of golgi and dictyosomes, but their origin is unknown in the case of Saprolegnia. Fusion of the vesicles with themselves and with vacuolar membranes results in the establishment of furrows dividing the cytoplasm. Cleavage of spore units in the sporangium of *Aphanomyces* is achieved by activity of the plasma membrane and tonoplast of the hyphae present prior to sporogenesis.

Development of Sporangiospores

Our knowledge of the ultrastructure of sporangiospore development stems from some elegant studies on *Gilbertella persicaria*, a member of the Mucorales. The sporangium initial is very similar in ultrastructure to a normal hypha; it contains many nuclei but there is no evidence of nuclear division occurring before cytoplasmic cleavage. A unique feature is the interconnection of nuclei by endoplasmic reticulum with up seven nuclei in a group. As pre-cleavage development ensues, small peripheral vesicles develop from the cisternae. As the sporangium develops, the vesicles apparently disappear and become replaced by larger so-called cleavage vesicles. The larger vesicles are identified by the presence of granules around the inner surface of the vesicle membrane. The origin of the vesicles is unknown but they may represent a changed form of the peripheral vesicles.

Another important pre-cleavage change in the developing sporangium concerns the mitochondria which become associated with a double membrane which appears densely stained in electron micrographs. Eventually the mitochondria become totally engulfed by these membranes and later they are seen to disintegrate. Up to one third of the mitochondria may be lost at this time. The reasons for this mitochondrial degeneration are unknown and as yet there is no biochemical data to amplify the cytoplogical change.

The division of the cytoplasm into the individual spore units commences with the grouping and eventual fusion of the cleavage vesicles. The fusion events result in a system of membrane-bounded tubes that ramifies the cytoplasm of the sporangial protoplast. *Gilbertella persicaria* is typical of many Mucorales species having a columella as part of its sporangial structure. The columella is a cap-shaped membranous structure that delimits the spores from the sporangial stalk. Cleavage of the columella from the sporangium appears to be effected in a similar manner to the spores, i.e. by the activity of cleavage vesicles.

Following the cleavage the sporangium is filled with wall-less spore protoplasts with the membrane surrounding them originating from the membrane tubes derived from the cleavage vesicles. The granules first seen inside the vesicle are now apparent on the outer surface of the spore membrane. The granules eventually become fused into a network of irregular rods which then become part of one of the spore wall layers. As the spores continue to develop, the cytoplasm becomes denser, membrane contrast is lost and additional wall material is laid down.

Development of Conidia

Production of conidia is a feature of Ascomycete and Deutcromycete fungi. Conidia have three distinctive features, they are non-motile, not formed by cleavage and thus not enclosed within a spore-bearing structure. Three patterns of conidium development are recognised, fragmentation, fusion and extrusion, and these provide useful taxonomic criteria.

Arthrospores, blastospores and porospores are types of conidia formed by relatively simple mechanisms. During the formation of arthrospores the differentiating hyphal apex becomes expanded and meristematic, producing a chain of cells in basipetal sequence. Blatospores and porospores develop by extrusion methods from either somatic hyphae or from specialised spore-bearing hyphae called conidiophores. Blastospores are produced at apical or lateral sites of either hyphae or conidiophores by a budding process. Porospores develop as extrusions through pores formed in the outer wall layer of the conidiophore by enzymatic digestion. The inner wall layer is pushed out to form the spore wall. In many instances the mechanism of spore formation is not so clear cut, e.g. macroconidia formation in *Neurospora crassa.* The first stage of spore development is undoubtedly a blastic process or extrusion involving the enlargement of the conidium initial. Subsequent septa production and separation of the spores has been described as an arthic or fragmentation process.

Phialoconidia, typical of the genera *Aspergillus* and *Penicillium*, are characterised by their formation from a specialised conidiophore cell called the *phialide.* These are single cells produced as lateral branches of aerial conidiopores or in clusters at the apices of conidiophores. The spores arise as apical extrusions of these cells and in many fungi a chain of spores is produced in basipetal sequence.

Ultrastructural aspects of the development of the conidiophore and conidia have been investigated in various Aspergilli. Distinct stages in differentiation of the mature conidiophore can be recognised starting with the transformation of a hyphal segment into the foot cell of the conidiophore. The conidiophore stalk grows from the foot cell as an aerial branch and it can be distinguished from a somatic hypha by its greater diameter and thicker wall. In

Aspergillus niger as many as eight wall layers have been reported in the basal region of the conidiophore stalk. In other species the conidiophore wall is composed of two layers.

The intracellular structure of the young conidiophore is reminiscent of the vegetative hypha, the cytoplasm is rich in ribosomes and other organelles, with vesicles at the apex. As the stalk of the conidiophore lengthens the basal regions become vacuolated.

A characteristic feature of the *Aspergillus* conidiophore is the vesicle which represents the swollen apex of the stalk. As the vesicle forms so the stalk wall becomes thickened. Internally the spatial distribution of organelles, characteristic of the elongated hypha, becomes disrupted as the organelles become uniformly distributed throughout the vesicular cytoplasm. Another notable change is seen as the shape of nuclei and mitochondria change from an elongated form in the stalk region, to ellipsoidal and spherical forms.

The next stage of differentiation involves the development of numerous cells, called sterigmata, over the surface of the vesicle. They begin as spherical protrusions, forming in a fairly synchronous manner. Up to 500 sterigmata have been observed in *Aspergillus niger*. These cells become elongated and cylindrical in shape and eventually are separated from the vesicle by the formation of septa. In some species the sterigmata become phialides each producing a chain of conidia, but in others the sterigmata produce another elongated cell which functions as the phialide. The phialides are uninucleate, in most species, and contain other organelles.

The phialides give rise to spores in rapid succession by basipetal budding. This involve continuous synthesis of wall materials, repeated nuclear divisions and synthesis of RNA and protein to be packaged in the spore. The site of these latter syntheses in unknown and could be the vesicle with subsequent transport of the product to the spores.

Environmental Control of Sporulation

In fungi the genetical competence for asexual reproduction the switch from vegeatative growth to reproductive development is generally greatly influenced by environmental factors. The period of vegetative growth following the colonisation of a new substrate or medium, is variable but ultimately the exhaustion of a nutrient, particularly assimilable carbon and/or nitrogen, may act as a trigger to initiate reproduction. There are some exceptions to this general situation, however, e.g. *Aspergillus nidulans*, where surface colonies derived from single spores will produce conidiophores and conidia after only 20 hours at 36°C. Depletion of nutrients is unlikely in this short period of growth so an alternative explanation is necessary. One interesting view suggests that conidiophore induction depends on gene controlled initiation and that it takes up to 20 hours for this event to be brought about.

Our knowledge of the environmental conditions influencing asexual reproduction in fungi is now very extensive and although much of the earlier work was of a descriptive nature, it is possible to formulate some generalised statements about the interactions:

1. The environmental conditions that favour vegetative growth may not favour asexual reproduction.
2. The range of conditions that promote asexual reproduction is normally narrower than the conditions that promote vegetative growth.
3. Environmental conditions that influence the initiation of asexual sporulation may differ from the conditions that favour development and maturation of the spore-breeding structures.

In some fungi physical factors are important in the control of asexual sporulation and particular interest has been paid to the

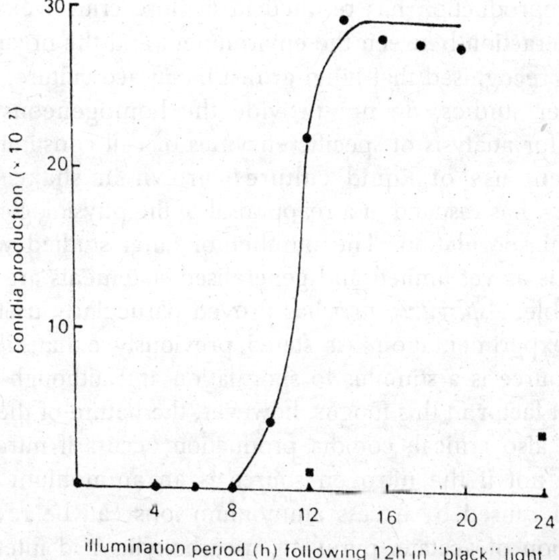

Fig. 7.4. The effect of light on conidiation in Botrytis cinerea.

role of light which may be either stimulatory or inhibitory. The major features of light controlled development are the wavelength and dose of light and the physiological status of the mycelium receiving the stimulus. In many of the light-sensitive fungi near-UV light is a common promoter and blue light also in other species, however, there are no reports of red photoresponse. Interestingly several of the species exhibiting near-UV controlled sporulation also show blue light inhibition but this effect is reversible following further exposure to near-UV. A photoreceptor system based on a pigment called *mycochrome* (M) has been proposed to explain this reversibility. The pigment exists in two forms, one absorbing near-UV (M_{NUV}) and the other, blue light (M_B).

$$M_{NUV} \underset{\text{blue}}{\overset{\text{near-UV}}{\rightleftharpoons}} M_B$$

Constituents of cell extracts having these light absorbing properties have been reported. The so-called M_{NUV} exists either as a soluble component in the cytoplasm or in a form loosely bound to the particulate fraction. The M_B is found tightly bound to the particulate fraction. The chemical nature of mycochrome is unknown as is the nature of its interaction with the metabolic processes associated with sporulation.

In recent years the shift in interest to the biochemical basis of asexual reproduction has resulted in a more critical examination of the interaction between the environment and the organism. It is generally recognised that fungi grown in surface culture, typical of the earlier studies, do not provide the homogeneous material required for analysis of specific enzymes or cell constitutents. The consequent use of liquid cultures, grown in shake flasks or fermenters, has resulted in a reappraisal of the physiological control of asexual sporulation. The number of fungi studied with these methods is as yet limited and generalised statements are therefore not possible. *Aspergillus niger* has proved particularly useful in this area of experimentation. As stated previously exhaustion of the carbon source is a stimulus to sporulation and although this is an important factor in this fungus, however, the nature of the nitrogen source is also critical; conidia production occurs if nitrate issued but does not if the nitrogen source is an ammonium salt. The inhibition caused by excess ammonium ions can be reversed by the addition of acetate, pyruvate, tricarboxylic acid intermediates and amino acids to the medium. Inhibition of conidiogenes is in

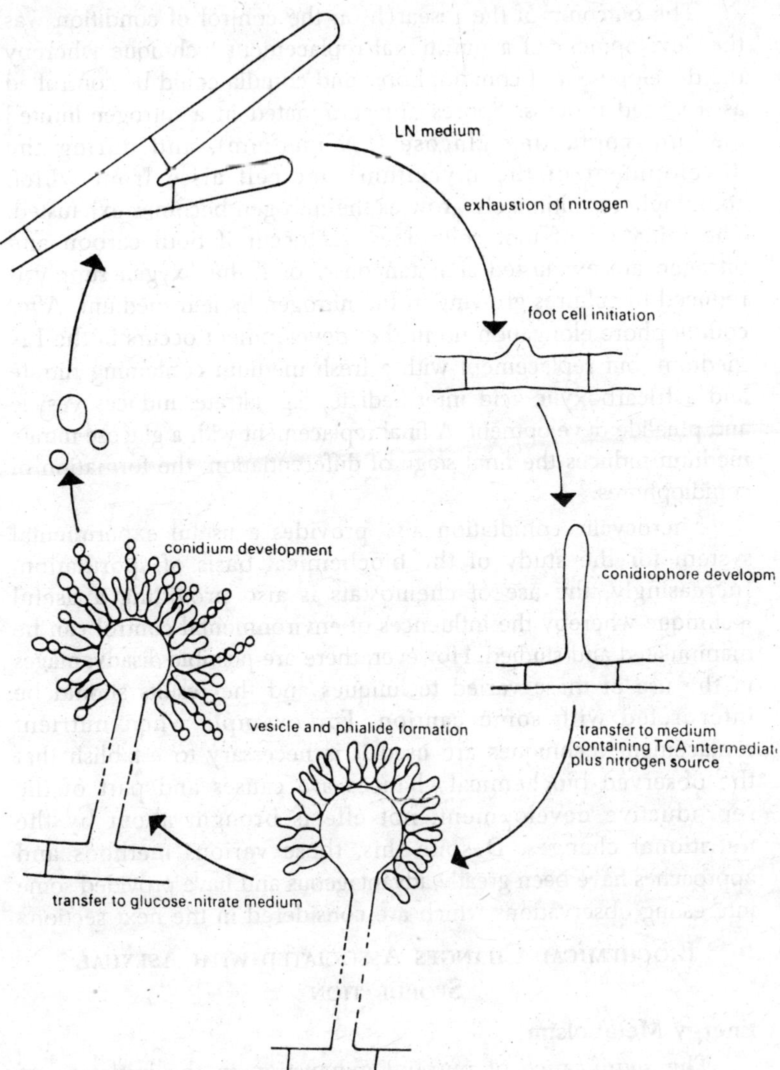

Fig. 7.5. Nutritional control of conidiogenesis in Aspergillus niger.

Neurospora crassa is also caused by the presence of ammonium nitrogen in a medium containing citrate as carbon source. Substitution of the ammonium salt with nitrate again stimulates spore formation. In this fungus the effect of ammonium ions can be overcome by aeration of the cultures.

The outcome of the research on the control of condition was the development of a nutritional replacement technique whereby the development of conidiophores and conidia could be controlled as a staged process. Spores are germinated in a nitrogen-limited medium containing glucose (LN medium), and during the development of the mycelium foot cell arise from which conidiophores ultimately grow as the nitrogen becomes exhausted. The initiation of foot cells does not occur if both carbon and nitrogen are exhausted simultaneously or if the oxygen supply is reduced to cultures growing in the nitrogen-limited medium. After conidiophore elongation no further development occurs in the LN medium, but replacement with a fresh medium containing nitrate and a tricarboxylic acid intermediate, e.g. citrate, induces vesicle and phialide development. A final replacement with a glucose-nitrate medium induces the final stage of differentiation, the formation of conidiophores.

Microcyclic conidiation also provides a useful experimental system for the study of the biochemical basis of sporulation. Increasingly, the use of chemostats is also providing a useful technique whereby the influences of environmental control can be manipulated and studied. However, there are possible disadvantages in the use of these varied techniques and the results should be interpreted with some caution. For example when nutrient replacement techniques are used it is necessary to establish that the observed biochemical chances are causes and part of the reproductive development, not effects brought about by the nutritional changes. Despite this, these various methods and approaches have been greatly advantageous and have provided some interesting observations which are considered in the next section.

Biochemical Changes Associated with Asexual Sporulation

Energy Metabolsim

The significance of nutrient exhaustion in the initiation of asexual reproduction was discussed in the previous section. Depletion of a key nutrient will result in a metabolic shift in the organism which in some way is probably associated with the new pattern of development. Observations on standing cultures of *Neurospora crassa* support this hypothesis. When plentiful supplies of carbon and nitrogen (glucose and nitrate) are available active hyphal growth occurs. Cytochemical studies have shown that the

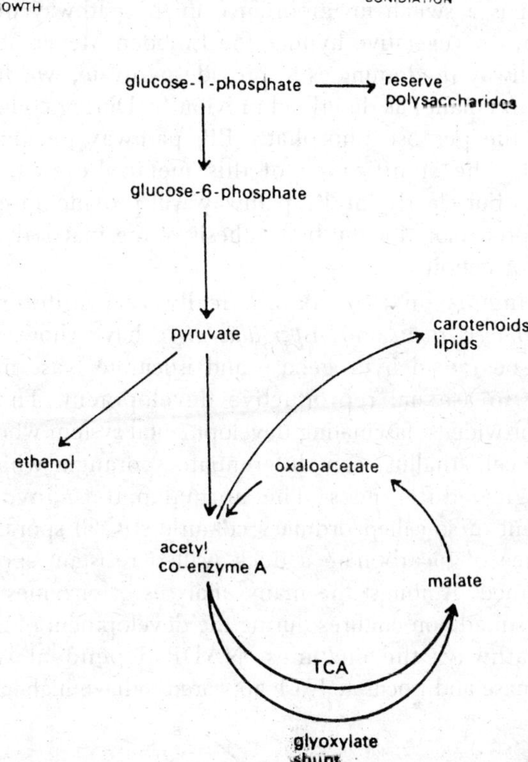

Fig. 7.6. Changes in carbon metabolism associated with conidiation in Neurospora crassa.

energy metabolism at the hyphal apices is fermentative and this is substantiated by the detection of ethanol in the culture medium. The high concentration of glucose causes the repression of oxidative metabolism and restricts mitochondrial development. As aerial branches grow up from the mycelium the glucose effect is gradually reduced, probably because of reduced translocation of the sugar, and oxidative metabolism can be demonstrated. Thus reduction in available glucose brings about depression and a metabolic shift in favour of the tricarboxylic acid (TCA) cycle. The aerial hyphae then go on to produce macro-conidia. The possible significance of this metabolic shift is a need for a great energy supply for macromolecule synthesis in the developing spore. Investigations on *Aspergillus niger* have also demonstrated the functioning of the TCA cycle in cultures undergoing conidiation.

Another detectable change recorded in several fungi undergoing conidiation is a switch in importance in the pathway of glucose metabolism. In vegetative hyphae the Embden Meyer hof Parnas (EMP) pathway predominates to provide pyruvate, which may be converted to ethanol as described previously. During conidiophore formation the pentose phosphate (PP) pathway becomes more significant. The significance of this metabolic switch is not understood, but clearly the PP pathway will provide an enhanced supply of precursors for the biosynthesis of the materials required for spore formation.

Experiments on two taxonomically very different fungi, *Blastocladiella emersonii* and *Aspergillus niger*, have shown that the enzymes isocitrate dehydrogenase and isocitrate lyase may have some role in asexual reproductive development. The former organism provides a fascinating developmental system whereby the vegetative cell (thallus) can differentiate sporangia which show morphological differences. The normal pattern involves the development of so-called ordinary colourless (OC) sporangia, but the presence of bicarbonate a thick walled resistant sporangium (RS) is formed. Amongst the many analyses of enzymes and cell constituents made on cultures during the development of these two distinct pathways the enzymes, NADP dependent isocitrate dehydrogenase and isocitrate lyase appeared to have higher activities

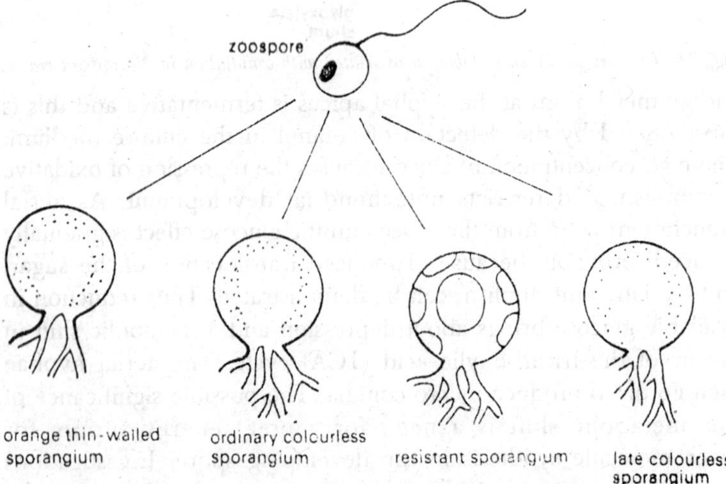

Fig. 7.7. Patterns of sporangium development in Blastocladiella emersonii.

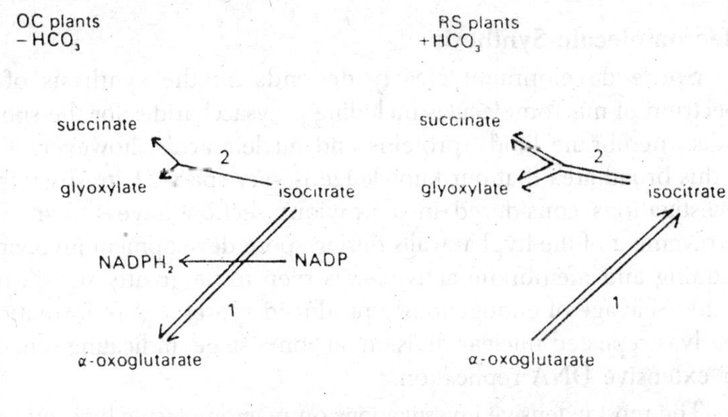

Fig. 7.8. The effect of bicarbonate on metabolism of Blastocladiella emersonii.

in the RS culture. The former enzyme was found to operate in reverse of its normal role in the TCA cycle yielding isocitrate from α-oxoglutarate by reductive carboxylation. Isocitrate lyase then brings about the cleavage of isocitrate to glyoxylate and succinate. The significance of this metabolic switch, it is claimed, is in the utilisation of glyoxylate in amino acid and subsequent RNA synthesis that occurs during differentiation of the RS sporangium. These conclusions were later challenged when comparisons were made between OC and RS developing cultures during the period of exponential growth.

In the OC cultures this lasts for the whole developmental period, but in the RS cultures exponential growth occurs for only part of the developmental cycle. Using, it was argued, more comparable material for these analyses, the differences described previously were not found in these later experiments. This controversy again raises the question as to whether the changes in enzyme activity were casual in the development process, or the effect of a changed environment.

Further support for a role for these two enzymes in asexual development can be found in experiments with *Aspergillus niger.* The induction of conidation in batch cultures, chemostate cultures or in cultures subjected to the nutrient replacement procedures, resulted in an increase in isocitrate dehydrogenase and isocitrate lyase in every case. Enhanced isocitrate lyase activity has also been described in conidiating cultures of *Neurospora crassa.*

Macromolecule Synthesis

Spore development clearly depends on the synthesis of a spectrum of macromolecules including polysaccharides for the spore walls, membrane lipids, proteins and nucleic acids, however, it is in this broad area that our knowledge is very sparse. Ultrastructural investigations considered in a previous section have shown the involvement of the hyphal walls during spore development involving budding and membrane activity was seen to be greatly significant in the cleavage of endogenously produced spores. Spore formation involves repeated nuclear division, at some stage, indicating a need for extensive DNA replication.

The most extensive investigations on macromolecule biosynthesis during asexual sporulation have been made on some of the more primitive fungi belonging to the Phycomycete group. In both *Achlya bisexualis* and *Allomyces arbuscula* active RNA and protein synthesis occur during spore development. From isotopic count determinations made of three parameters in studies on *Achlya*, i.e. rate of amino acid incorporation, specific activities of amino acid pools and the release of amino acids from prelabelled proteins, protein turnover values of 75 per cent were obtained. A large proportion of the amino acids released were re-utilised in *de novo* protein synthesis during zoospore differentiation.

The occurrence of macromolecule synthesis during zoospore formation in *Blastocladiella emersonii* has proved more difficult to demonstrate. The transition from vegetative growth to zoosporangium formation is characterised by a marked decline in the rate of synthesis of DNA, RNA and protein and under these conditions the determination of *de novo* synthesis is not easy. A requirement for mRNA and protein for this development step is a reasonable assumption but this requirement could be fulfilled by synthesis of the specific molecules at a time prior to spore differentiation, i.e. the later phase of vegetative growth or during the early period of zoosporangium development.

Experiments on *Trichoderma viride*, a fungus that requires a light stimulus for the induction of conidiation, have shown that RNA is produced continuously during critical period of the light induction period. However, DNA/RNA hybridisation failed to show differences in the RNA transcribed during photo-induction. Even so it was suggested that new species of RNA were being transcribed but could not be detected because the amounts produced were very

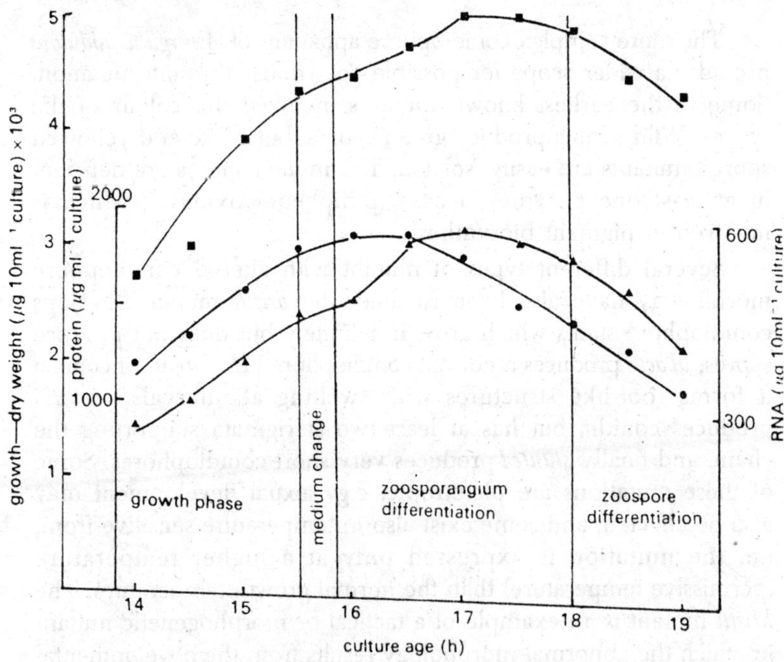

Fig. 7.9. Changes in dry weight and macromolecular composition of cells of Blastocladiella emersonii during growth and zoospore formation.

small with respect to the total RNA in the culture. Incorporation studies have been made with a few fungi and provide general evidence for RNA and protein synthesis during sporulation, but details regarding the specificity of the macromolecules are not yet available.

Genetic Control of Asexual Sporulation

Study of the genetical control of asexual reproduction has been restricted to two fungi, *Neurospora crassa* and *Aspergillus nidulans.* This is not unexpected in view of the broader genetical knowledge of these two organisms. In general many of the mutations which affect conidiation in these fungi also cause other unrelated defects, i.e. they are pleiotropic mutations. In *Neurospora*, however, mutants have recently been isoated in which only conidiation has been affected; these mutants affect the developmental steps. So far only mapping work and limited biochemical experiments have been carried out with these mutants (the gene loci involved are located on separate chromosomes), but they obviously provide interesting subject for detailed biochemical investigations.

The more complex conidiophore apparatus of *Aspergillus nidulans* provides a wider scope for possible deformities through mutation. Some of the earliest known mutants involved the colour of the spores. Wild strains produce green spores, but white and yellowed spored mutants are easily isolated. The mutant spores are deficient in at least one enzyme, laccase (p-diphenol oxidase), which is involved in pigment biosynthesis.

Several different types of mutant with altered conidiophore morphology have also been isolated: the *bristle* mutant develops conidiophore stalks which grow indefinitely but does not produce spores; *abacus* produces a normal conidiophore but instead of conidia it forms root-like structures with swelling at intervals; *medusa* produces conidia but has at least two sterigmata supporting the chain, and finally *stunted* produces very short conidiophores. Some of these mutations are pleiotropic, e.g. sexual development may also be affected, and some exist also in temperature sensitive from, i.e. the mutation is expressed only at a higher temperature (permissive temperature) than the normal growth temperature. The *bristle* mutant is an example of a tactical or morphogenetic mutant in which the abnormal morphology results from the developmental process failing to move from one stage to the next. There is repetition of the final stage, thus in this mutant very long conidiophore stalks are formed. The *abacus* mutant is similar, the abnormal structure of the conidiophore probably being the result of repeated phialide formation. The *medusa* and *stunted* types are mutations which affect auxillary loci, that is they affect functions that are supplementary to the conidiation process. Both these mutants produce conidia but in smaller numbers than the wild strain because of modifications to the conidiophore.

Both the *Neurospora* and *Aspergillus* mutants have been investigated for conidiation-specific enzymes, but none of those considered so far fulfils such a function. The occurrence of the enzymes, associated with pigmentation of conidiophores and conidia of *Aspergillus nidulans*, in the abnormal mutants has been investigated. Cresolase, the enzyme concerned in the synthesis of a pigment in the conidiophores and sterigmata, and laccase are both absent from *bristle* mutants, but *abacus* does produce cresolase. The interrelationship between the activities of these two genes still has to be resolved, however, it is likely that the *bristle* locus plays a key role in the regulation of conidiophore development.

8

Vegetative Growth of Filamentous Fungi

In the filamentous fungi the basic unit of structure is the hypha. When massed together the hyphae form a mycelium. The individual hyphae are tube-like structures which exhibit polarised growth and may produce branches at sub-apical sites. In effect hyphal growth involves two components, an increase in volume which results from extension of the tip and an increase in protoplasmic contents. These involve a complex interaction of biosynthesis of macromolecules and their subsequent organisation into recognisable physical structures. The latter aspect of growth occurs over long distances within an individual hypha. In recent years fungal hyphae have been the subject of broad investigation in regard to their ultrastructure, biochemistry and physiology and it is now possible to devise models relating these features to the growth process.

Hyphal Cytology

Light microscope studies, some of the earliest reports appearing in the latter part of last century, provided a lot of fundamental information about the fungal hypha which has stood the test of time and still remains central in our thinking on hyphal growth. Of particular interest were the reports an apical growth of hyphae and the discovery of the Spitzenkorper, a zone at the extreme apex of septate fungi that can be stained with iron-haemotoxylin. More sophisticated methods, e.g. autoradiography, fluorescence microscopy and phase contrast microscopy, have confirmed these early descriptions. The Spitzenkorper appears to be associated with

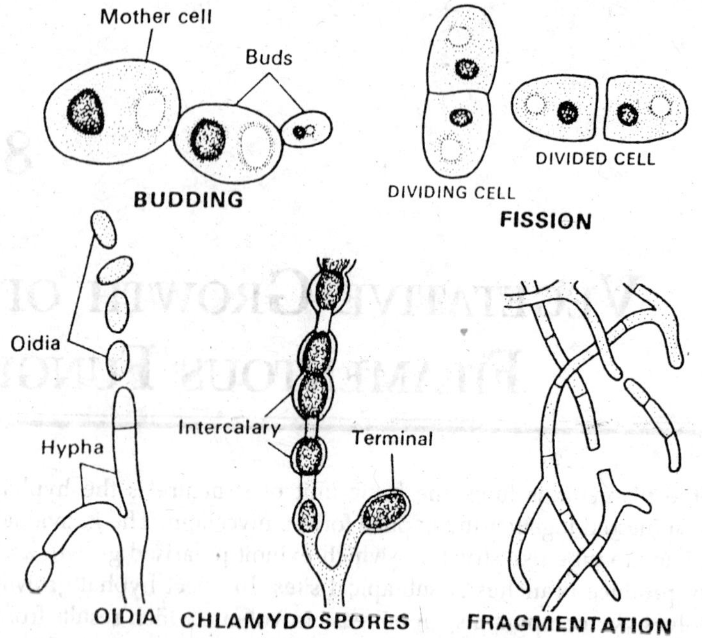

Fig. 8.1. Different types of vegetative reproduction in fungi.

active growth because it is not present in non-growing hyphpae. Another feature of growing hyphae, just resolvable by phase contrast microscopy in good preparations particularly of the large hyphae of Zygomycetes, is the presence of small particles which can be seen migrating towards the hyphal apex.

The tip region involved in increasing hyphal volume is known as the extension zone. The length of this zone has been determined in various ways including observations on the displacement of markers placed on hyphae and more recently by the use of optical brightener compounds, manufactured for incorporation into detergents, which bind to the extension zone of hypha of many fungi.

Hyphal Ultrastructure

Many of the early observations made by light microscopy, e.g. the Spitzenkorper and movement of particles, have been further resolved by the use of electron microscopy. However, since the proliferation of the fungal hypha involves the syntheses of new components of the rigid hyphal wall the following discussion will start with this component of the hypha.

Composition and Architecture of the Hyphal Wall

The boundary of the fungal hypha is the rigid cell wall. The analysis of the chemical composition of the wall involves a detailed procedure and preparation of the material prior to investigation. An unknown factor is the possibility of changes in composition during these preparative procedures. Once the wall material is prepared a protocol for fractionation is used and the result from an analysis are given in table 8.1.

Table 8.1. Wall composition of fungi representative of different taxonomic groups.

	% Composition		
	Phycomycete Phytophthora heveae	*Ascomycete Chaetomium globosum*	*Basidiomycete Agaricus bisporus*
Cellulose	36	–	–
Chitin	–	13.9	43
α-1,3-glucan	–	–	14
β-1,3-glucan	54	54.1	27
Lipid	2.5	3	1.5
Protein	4.6-6.7	29	16

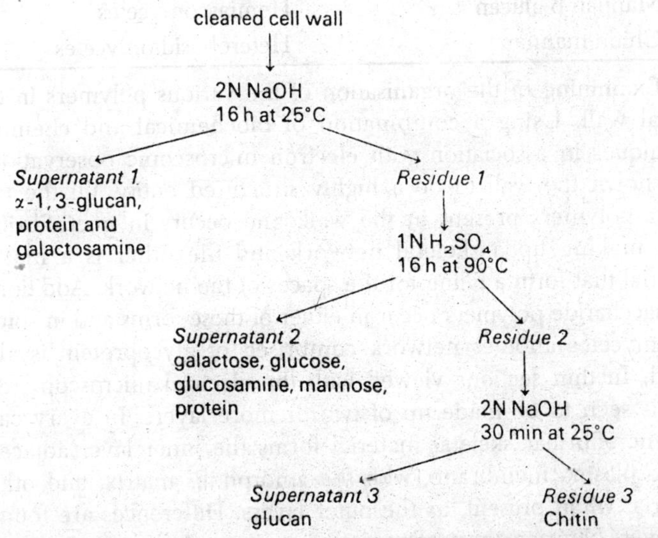

Fig. 8.2. Fractionation procedure for fungal cell walls containing chitin and glucans.

From the many reports on hyphal wall composition of different fungal species several basic features arises. Typically the walls are comprised of polysaccharides (75–80 per cent on a dry weight basis), protein, lipid, and melanin. A variety of polysaccharides occur–the homopolymers β-1,3-glucan, β-1,4-glucan, α-1,3, glucan, chitin, chitosan, and a variety of heteropolymers which are made up from mannose, galactose, fucose and xylose. Of these different polymers the β-1,3-glucan is found more broadly across the various taxonomic groups of fungi. The analyses of wall composition of the many fungi investigated has revealed the interesting situation shown in table 8.2 where there is a relationship between the nature of the two major polymers present and the different taxonomic groups.

Table 8.2. Cell wall composition and fungal taxonomy.

Major polymers	*Taxonomic group*
Cellulose-β-1,3-glucan	Oomycetes
Cellulose-chitin	Hyphochytridiomycetes
Chitin-chitosan	Zygomycetes
Chitin-β-glucan	Chytridiomycetes
	Euascomycetes
	Homobasidromycetes
Mannan-β-glucan	Hemiascomycetes
Chitin-mannan	Heterobasidiomycetes

Examining of the organisation of the various polymers in the fungal wall. Using a combination of biochemical and chemical techniques in association with electron microscopic observations, has shown the wall to be a highly structured entity. Of the two major polymers present in the wall, one occurs in microfibrillar form making up a skeletal network and the other is a matrix material that form a filling for the spaces of the network. Additional polysaccharide polymers occur in either of these forms and in some. Ascomycetes another network comprised of glycoprotein is also found. In thin sections viewed with the electron microscope, the wall is seen to be made up of two or more layers. In every case the microfibrillar skeletal material forms the inner layer adjacent to the plasma membrane, with the amorphous matrix, and other material when present, as the outer layers. Differences are found, however, between the wall structure in the apical extension zone

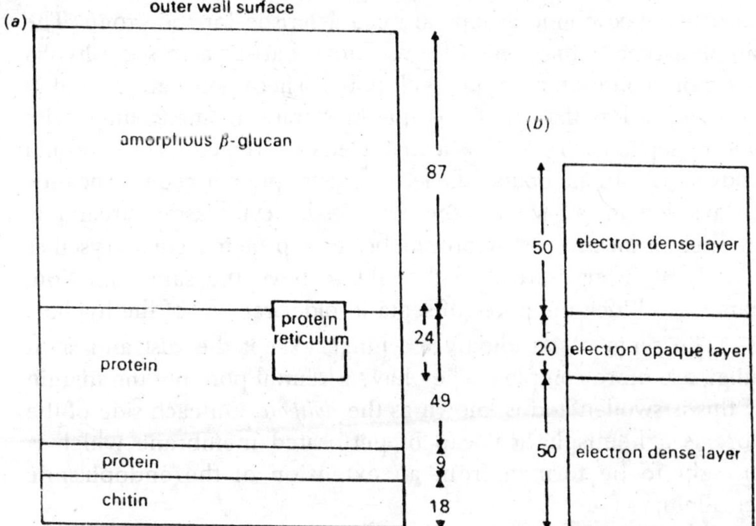

Fig. 8.3. Organization of polymers in the hyphal wall of Neurospora crassa, observations made on 5-day culture. (a) Model of the wall based on experiments in which wall material was subjected to specific enzyme treatment. (b) Structure of the wall as seen in thin section by transmission electron microscopy.

and in the sub-apical region. The wall of the extension zone is called the primary wall and is comprised almost entirely of the microfibrillar wall component but may have some amorphous component. In the sub-apical region the secondary wall is deposited on the outer surface of the primary wall and involves the thickening of existing microfibrils, the addition of new ones and the formation of the amorphous and other wall components.

The hyphal wall can be removed by digestion in a mixture of the enzymes specific for the major polysaccharide components. If this digestion is done in an osmotically stabilised medium the cytoplasmic contents of the hyphae are then released as discrete units termed protoplasts. Protoplasts have proved useful for studies on wall biogenesis in fungi because when incubated in an osmotically stabilised growth medium they regenerate a new wall and revert to the normal hyphal form.

Septa

Septa of crosswalls occurring irregularly along the length of a hypha are a feature of all fungal groups. The septa are produced in the apical region in what appears to be controlled manner. With certain exceptions, the septa found in representative species of particular taxonomic groups are characteristic for the group. The simplest type is that which forms a total barrier across the hypha but more common are septa with pores. These pores are central in location, unless the septum is multiperforate. In many fungi with porous septa, each pore has a characteristic structure. The Woronin body–a membrane bounded, electron-dense proteinaceous structure–always remains adjacent to the pore despite cytoplasmic streaming. In species lacking the Woronin body, a proteinaceous crystal is found. Both structures are thought to have the same function, namely to block the pores of septa in older regions of the hyphae.

The septa of Basidiomycete fungi, except the rust and smut fungi, are more complex. They have a central pore but the margin of this is swollen and is known as the *dolipore.* On each side of the pore is a hemispherical cap of perforated membrane which is thought to be formed from an extension of the endoplasmic reticulum.

The chemical composition of the septa is unknown except for two species, *Neurospora crassa* and *Schizophyllum commune.* In the former microfibrils of chitin are found in septa of young (one day old) mycelium but as time passes both sides of the septum are covered with an amorphous proteinaceous material. The β-1,3-glucan and glycoprotein components of the hyphal wall are not found in the septum. In *Schizophyllum commune* the septa contain chitin and a β-glucan, both of which are also found in the hyphal wall. The α-glucan found in the wall is absent from the septa. Dissolution of the septa in this and other Basidiomycetes is an important feature of reproductive development.

Protoplast

Ultrastructural observations on the fungal hypha have revealed an interesting picture of organisation. The apical region of a hypha may be conveniently sub-divided into two zones; the tip or apical zone, characterised by the presence of vesicles and ribosomes, and the sub-apical zone in which the organelles characteristic of a heterotrophic eukaryote are found.

The presence of vesicles at the tip zone is of particular interest and has led to the formulation of models explaning their involvement in wall formation and hence hyphal growth. The small particles seen by light microscopy have been interpreted as vesicles and the organisation of the latter at the apex provide an understanding of the nature of the Spitzenkorper. The vesicles fall into two groups on the basis of size. The smaller ones, termed microvesicles, have shown to contain chitin synthase in several fungi, and their involvement in wall synthesis must therefore be presumed.

The arrangement of vesicle in the tip zone is a variable feature with different patterns being characteristic for the different taxonomic groups. In the Oomycetes both sizes of vesicles are found distributed in a random fashion. In electron micrographs, however, the larger one is seen to contain electron opaque, granular material whereas the microvesicles have more uniform contents. The sub-apical zone of these fungi is characterised by large numbers of dictyosomes frequently seen in association with cytoplasmic vesicles. Some micrographs have shown the vesicles to occur in lines between the dictyosomes and the apex, strongly suggesting a relationship between the two organelles and events in the tip zone. In representatives of other taxonomic groups the organisation of the apical vesicles shows more order. In the Zygomycetes the larger vesicles form a distinct layer beneath the plasma membrane giving the appearance of a hemispherical cap. The microvesicles occur randomly in the region behind this cap. The microvesicles in the apical zone of Ascomycete fungi occur as a spherical cluster surrounded by the large vesicles and the complete cluster is located just behind the apical wall. The Basidiomycete arrangement is somewhat smaller with a spherical arrangement of large vesicles surrounding a vesicle free zone which includes a dense core of unidentifiable structure. The vesicular arrangements seen in the tip

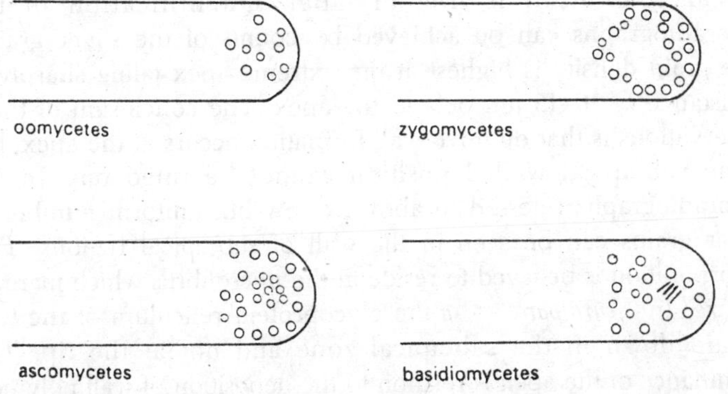

Fig. 8.4. Diagrammatic representation of the organization in hyphal apices of fungi from different taxonomic groups.

zones of these fungi are probably the Spitzenkorpes described from light micrographs.

Vesicles are also seen in the sub-apical zones of the hyphae of fungi. Regions of cytoplasm occur that are free or ribosomes and in these, membranous structures corresponding to the cisternae of golgi bodies are present, associated with vesicles. In the sub-apical region vesicles also occur intermingled with the other organelles, nuclei, mitochondria, endomembrane components and ribosomes. Involvement of the vesicles in events at the apex, including the synthesis of the hyphal wall, necessitate their transport through the mass of other organelles, but a mechanism for this process remains to be discovered.

Biosynthesis of Cell Wall Polymers

Location of Synthesis

As mentioned previously the existence of an apical region of hyphal extension was described many years ago following light microscope observations. Substantial experimental evidence confirming these observations has come from investigations in which hyphae were fed with radioactive sugars, e.g. ^{14}C glucose, ^{14}C N-acetylglucosamine, and then subjected to autoradiography. The sugars were shown to be rapidly incorporated into polymer such as chitin and glucan at the extreme apex. When a pulse feeding of isotopic sugar is followed by a so-called cold chase, i.e. replacement of the isotope by the normal sugar, autoradiographs reveal that the area of wall just behind the apex has become labelled, but the apical zone itself is free of label. Quantification of the autoradiographs can be achieved by counts of the silver grains. The grain density is highest at the extreme apex faling sharply at a distance of 10–15 μm behind the apex. The conclusion of these observations is that de novo wall formation occurs at the apex, but some sub-apical wall deposition cannot be ruled out. In the autoradiographs referred to above, a low but uniform number of silver grains can be seen in the wall of sub-apical regions. This incorporation is believed to reside in the microfibrils which increase in size. In *Neurospora crassa* the glycoprotein reticulum of the wall is laid down in the sub-apical zone and not at the tip. The dominance of the apex is relation to the deposition of wall polymers can be overcome experimentally, e.g. by growth in the presence of cycloheximide or by osmotic shock, and under these conditions radioactive substrates are incorporated uniformly throughout the wall. The effect of these treatments therefore is to overrule the mechanism that controls apical wall synthesis.

Biochemical aspects of Wall Synthesis

Of all the wall polymers described in the filamentous fungi, the biosynthesis of chitin is the most extensively documented. The enzyme involved in the final stage of polymer formation is chitin synthase which catalyses the reaction

$$\text{UDP–N acetylglucosamine} + (\text{N-acetylglucosamine})_n \xrightarrow{Mg^{2+}} \text{UDP} + \text{chitin}$$

The supply of UDP – N-acetylglucosamine, which is necessary for continued synthesis, is from a pathway starting with fructose-6-phosphate and glutamine as precursors. The complete pathway has not been followed in a single organism but there are descriptions of the four enzymes involved from different fungi.

The cytoplasmic location of chitin synthase has not been fully established although there is strong evidence to support a membrane site. In cell homogenates the highest levels of chitin synthase are

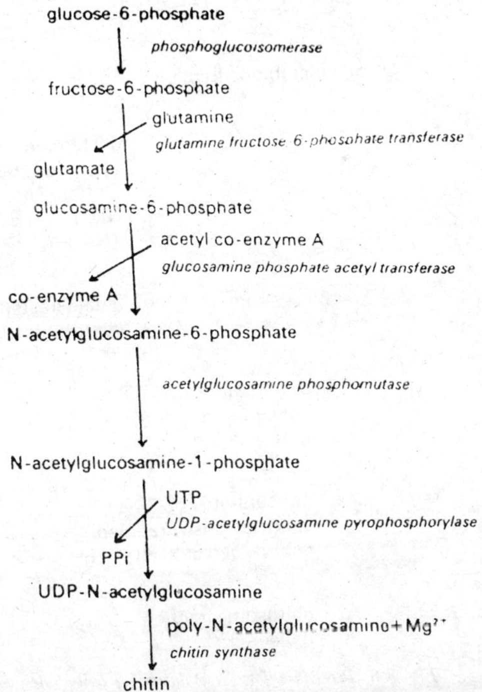

Fig. 8.5. Biosynthetic pathway leading to the formation of UDP-N-acetylglucosamine.

found in the microsomal fraction which is rich in membrane. In the intact hypha the functional site of enzyme is assumed to be on the outer surface of the plasma membrane. Evidence for this view stems from several experimental sources, in autoradiographs viewed with an electron microscope silver grains were found on the membrane surface and in the wall, and never in the cytoplasm, and protoplasts isolated from hyphae by digestion of the walls are able to synthesise new wall material and revert to the normal morphology. Recent studies on the yeast form of *Mucor rouxii* showed

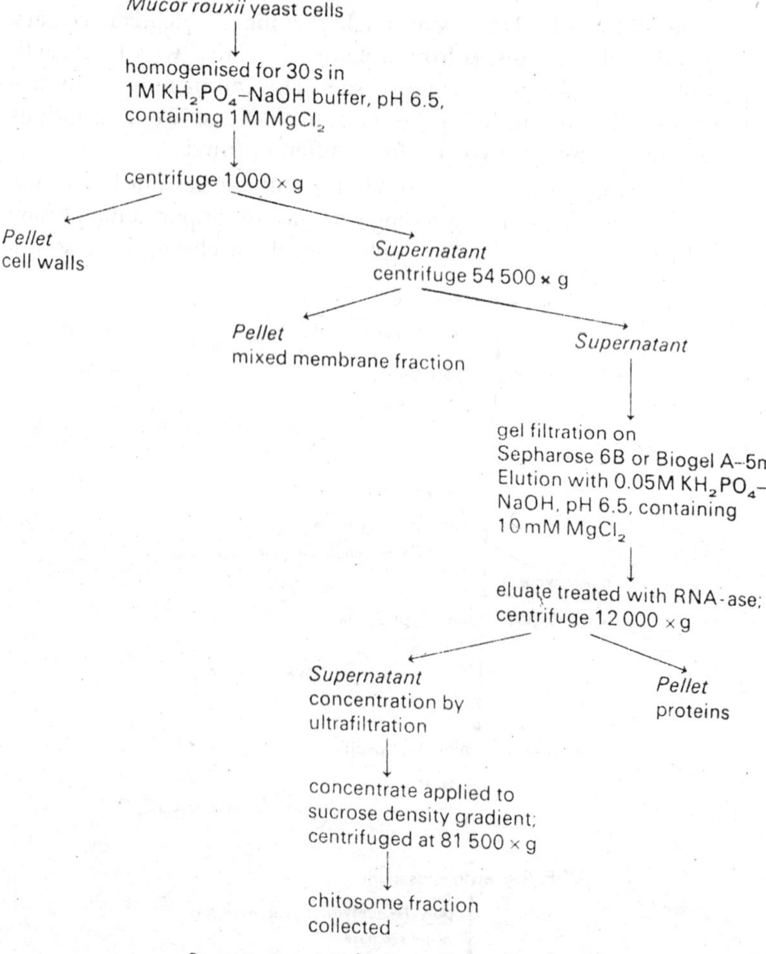

Fig. 8.6. Procedure for the isolation of chitosomes.

that microfibrils of chitin could be produced *in vitro.* The preparation of chitin synthase used took the form of particles obtained by a detailed purification procedure. The particles, called chitosomes, were subsequently shown to be microvesicular structures, 40–70 nm in diameter. During microfibril synthesis the chitosomes undergo a transformation such that when the microfibril is formed the shell of the membrane is shed. The *in vivo* function of the chitosome, which has since been described in several other fungi, is presumed to be that of conveyor of chitin synthase to the cell surface where the chitin polymer is normally synthesised.

Data on glucan synthesis in growing hyphae is more limited. In Neurospora crassa where a β-1,3-glucanis present as an amorphous component in the wall, a β-1,3-glucan synthase has been found in cell wall preparations. Also in *Phytopthora cinnamomi,* which has a β-1,3, β-1,6-glucan fibrillar wall component, a synthase was also found in the wall fraction, that catalysed the synthesis of the polymer from UDP-glucose. When the fungus was grown under conditions of high glucose concentration, thus reducing the level of protease in the hyphae, a membrane bound enzyme was also found that catalysed the formation of a polymer which contained only β-1,3-linkages. A synthase system associated with the formation of the wall β-1,4-glucan has been found in *Saprolegnia monoica,* an Oomycete fungus. The enzyme has been detected in fractions containing wall and microsomal material and more recently in vesicles believed to originate from the golgi system.

Model for Wall Growth

A proposed model for wall growth involves both wall lytic enzymes as well as synthases. The lytic enzymes are believed to function by cleaving the existing wall polymers to allow extension and insertion of new material. Evidence for the significance of these enzymes is claimed firstly from experiments on the bursting of hyphal tips and secondly from studies on the release of sugars from walls during incubation. Bursting of hyphal tips can be obtained by treatment with various chemicals which have been suggested increase the activity of those lytic enzymes located in the walls. This idea has been challenged by the suggestion that the treatments caused changes in the wall matrix materials, these changes in turn led to bursting. Evidence for wall lytic enzymes from a second group of experiments is more convincing. Up to 3 per cent of the polymers were found to be released from walls of

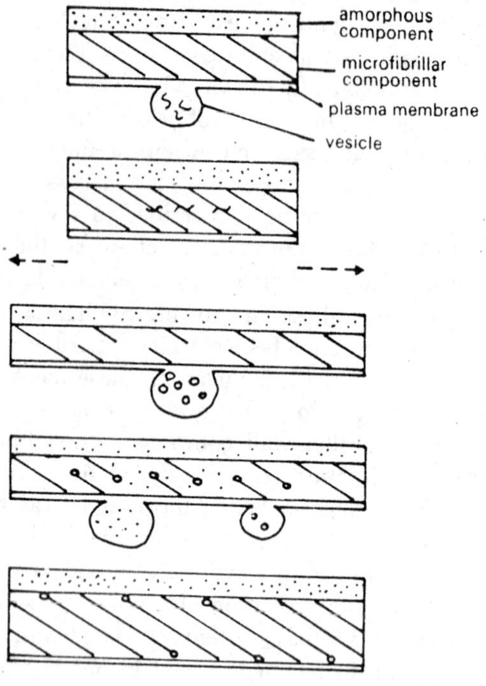

Fig. 8.7. Model of cell wall growth in fungi.

Asperegillus nidulans, in the form of glucose, N-acetylglucosamine, mannose and galactose. In further experiments using isotopically labelled walls it was shown that the release of sugars was preferentially from the newly formed wall at the apex rather than from sub-apical regions. More recent studies on the same fungus revealed the presence of six autolytic enzymes (autolysins) that hydrolysed chitin, laminarin, α-glucoside, β-glucoside, protein and β-N-acetylglucosaminide respectively. These enzymes were found bound to the wall fraction and inthesoluble protein of the cytoplasmic fraction of cell homogenates. The attachment of the wall bound enzymes was mediated by lipids. The cytoplasmic vesicles in the hyphae are assumed to function again as carriers of these enzymes but confirmation of this and the role of the autolysin in wall formation and hyphal growth must await further research.

Despite this evidence for the existence of wall-bound lytic enzymes direct information relating to their involvement in hyphal growth is lacking.

Whatever the precise details of the system, the synthesis of the hyphal wall clearly demands the secretion of a variety of materials at the surface of the plasma membrane in the region of the growing tip. Transport of these compounds is an obvious role for the vesicles described earlier in this chapter. Cytochemical data suggests that some of the vesicles may be carriers of carbohydrate material and recently there have been interesting reports that the microvesicles are carriers of chitin synthase as described earlier. Discharge of the vesicles contents will occur following fusion of the vesicle with the plasma membrane the latter event also provides a mechanism for membrane growth which is also a necessary part of hyphal extension.

The mechanism for vesicle transportation is not known, but two hypotheses have been proposed, both of which are based on reports of a gradient in membrane potential of some 100 mV in the apical zone, 7 nm long, in *Neurospora crassa* hyphae. The first hypothesis proposes that this gradient could generate enough current for the electro-phonetic movement of vesicles through the apical zone. The second view is based on the assumption that the gradient is a function of a decrease in the number of potassium pumps (ATP-ase dependent) in the membrane towards the hyphal tip. A consequence of this would be an osmotic flow forwards to the tip which could carry the vesicles.

Control of Localized Synthesis

The previous discussion has stressed that wall synthesis is localised at the hyphal apex, so how is this localisation controlled? In several fungi there is now evidence for the existence of an active and a zymogen form of chitin synthase (a similar situation to that described for the yeast *Saccharomyces cerevisiae*) and that the latter form can be activated by protease enzymes found in the hyphae. The significance of these two components in relation to the mechanism of control remains unknown. An attempt to understand their distribution within the hyphal organisation was made using the technique of protoplast isolation as a fractionation system. Protoplasts released from hyphae of *Aspergillus nidulans* during the first hour of wall digestion are predominantly from the apical tips. In the subsequent time period th ecytoplasm from sub-

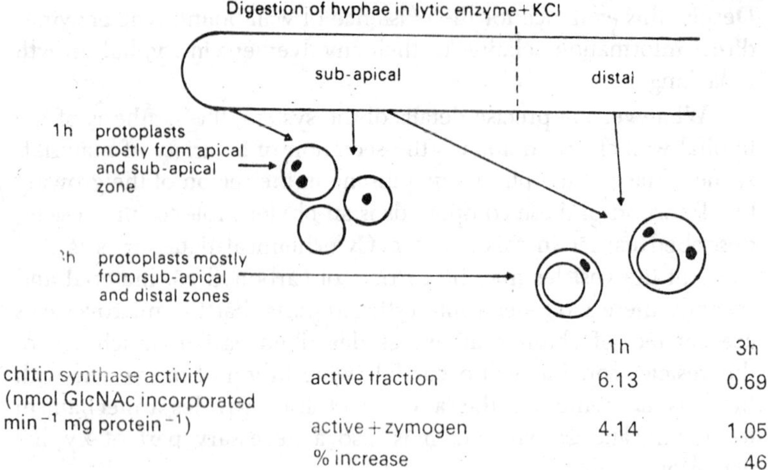

		1h	3h
chitin synthase activity (nmol GlcNAc incorporated min^{-1} mg protein^{-1})	active fraction	6.13	0.69
	active + zymogen	4.14	1.05
	% increase		46

Fig. 8.8. Chitin synthase activities in hyphal protoplasts from Aspergillus nidulans.

apical and distal zones is also released as protoplasts. Comparison of active and zymogen chitin synthase in the different preparations showed that protoplasts released in the first hour had a high level of active chitin synthase but no zymogen, and the protoplasts released later had a much reduced level of the active enzyme and higher amounts of the zymogen.

From the limited data available a possible working hypothesis would suggest that chitin synthase is released at the apex following fusion of the chitosomes with the plasma membrane. The chitin synthase, which is probably synthesised in zymogen form, is activated either in the chitosome prior to fusion, or on the membrane surface following fusion. As the hyphal tip advances the newly inserted membrane and the newly formed wall become lateral to it. In this position the majority of the enzyme molecules would become inactivated either reversibly or irreversibly.

The presence of inactivated enzyme in the lateral regions of the hyphae is one conclusion drawn from experiments on *Aspergillus nidulans* in which hyphae were either treated with cycloheximide or subjected to osmotic shock. In both instances the treated hyphae were pulse labelled with radioactive carbohydrate and examined by autoradiography. In this case the hyphae had a uniform distribution of silver grains over their whole length and not the restricted location as referred to previously. The changed pattern of incorporation and therefore presumed wall synthesis could be an expression of the reactivation of zymogen, however, an alternative explanation cannot be ruled out at this stage. This suggests that the cycloheximide treatment or osmotic shock disrupts the normal mechanism of vesicle transportation and discharged at the apex resulting in a random discharge over the whole hyphal surface, cannot be ruled out at this stage.

Branch initiation is also an important part of the model; in the case of a reversible inactivation situation the enzyme molecules would be reactivated. If the enzyme is inactivated irreversibly then the insertion of new active enzyme molecules would be required to initiate growth of a branch. Additional control of chitin synthase can be effected through N-acetylglucosamine, known to be an activator of the enzyme *in vitro*. In the hyphae this molecule would be released at localised sites following lysis of chitin by wall bound enzymes. If this lysis occurs only at the apex this would enhance the rate of synthase activity at this region.

Hyphal Growth and the Duplicatoin Cycle

In the unicellular bacteria and yeasts the events of growth and division are integrated, forming the cell cycle. It is now believed that similar integration of growth and nuclear division occurs in the apical compartment of hyphal fungi but owing to the absence of cell separation the term duplication cycle is used to describe these events. Some diversity exists in relation to the organisation of the apical compartment in different fungi. In some species the compartment is uninucleate, e.g. the *Phycomycete Basidiobolus ranarum*, and in the monokaryotic mycelium of many Basidiomycetes. In the dikaryotic mycelium of these latter fungi each hyphal compartment has two nuclei. In coenocytic species, e.g. *Aspergillus nidulans*, there are many nuclei present, in some instances up to 90. In all cases a regular sequence of events can be identified. Following septation the apical compartment elongates as a result of growth at the tip. In the species with uni- and binucleate compartments the nucleus(i) is seen to remain in a constant position relative to the tip and consequently must migrate at a rate equivalent to the rate of tip growth. When the apical compartment reaches a critical size nuclear division occurs followed by septation. The timing of these events is regular indicating a cyclic nature. The situation in coenocytic fungi is identical, with certain minor modifications. Nuclear division is not completely synchronous and where a large number of nuclei are involved the division event

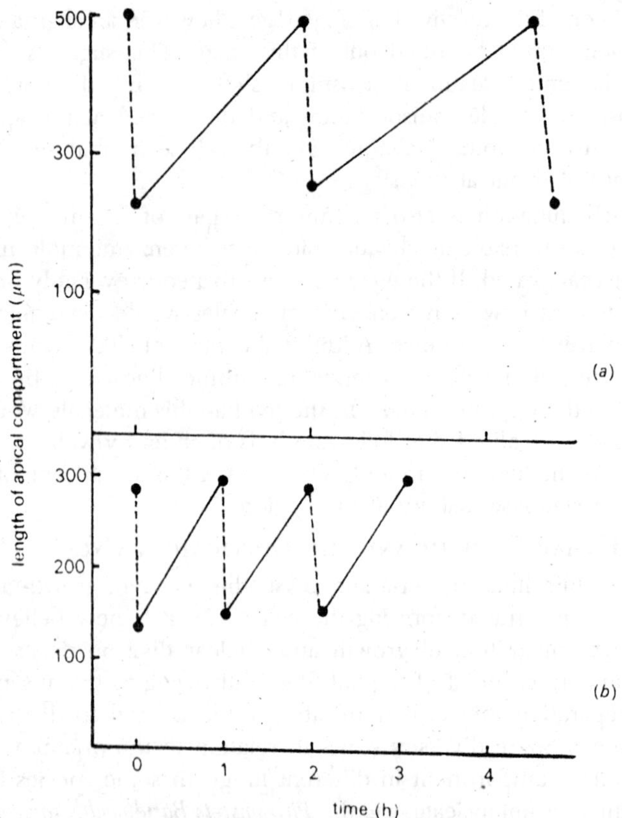

Fig. 8.9. The duplication cycle showing growth and septation of the apical compartment of fungal hypha in (a) Aspergillus nidulans and (b) the dikaryotic mycelium of Schizophyllum commune.

may last for up to 20 minutes, until each nucleus has divided. Septation is also different in coenocytes. The formation of several septa divides the compartment into an apical segment and a number of other segments each of variable size and nuclear number. In all cases the apical compartment is reduced to half its length at septation thus maintaining constancy in size.

An essential feature of the duplication cycle is the as yet unexplained relationship between the volume of cytoplasm and the DNA content of the apical compartment. As the cytoplasm attains a doubling in volume, division of the nucleus or nuclei and septation occur, thus restoring the nuclear-cytoplasmic ration. The observed events suggest that a positive control system is operating.

It has been proposed that a cytoplasmic factor is involved, similar to that suggested for bacteria. If such a factor is produced during the interphase period, at a rate proportional to the rate of increase of the cytoplasm, then mitosis would be induced when a critical level is attained. The apical cells of hyphae in a mycelium show no synchrony in respect of their duplication cycles, which suggests that if such a cytoplasmic factor is involved its effect is very restricted to individual apices. This would suggest that cytoplasmic mixing is very restricted even in those fungi in which the septa have pores.

Our understanding of the duplication cycle will hopefully improve in future years through the study of mutants. Several temperature sensitive mutants have been isolated in *Aspergillus nidulans* that have blocks in nuclear division and in septation. The nuclear division mutatns fall into several groups including those blocked at the stage of DNA synthesis and others blocked during mitosis. Investigations on these strains may tell us more about the control of nuclear division in the fungal hypha and the relationship between this event and septation.

Hyphal Branching

Branch formation is a feature of all fungal hyphae which allows the organism to rapidly colonise a particular substratum. A variety of patterns of branching are seen and the frequency of branch formation is variable–both aspects are subject to environmental and genetic control of apical or sub-apical hyphal compartments. In the former instance the septa are believed to be involved in branch formation, acting as a barrier to the flow of vesicles towards the hyphal apes. Fusion of the vesicles with the plasma membrane, as is also believed to occur at the apex, results in the discharge of their contents. This initiates changes in the hyphal wall and branch formation. Where branches arise independently of septa then the location is presumed to be random and is determined only by the sites at which vesicles accumulate. The stimulus for such accumulationis unknown. On occassions the hyphal apex is seen to branch dichotomously, this could alsobe an expression of vesicle accumulation. In this case a greater supply is required than is normally needed to maintain the normal polarised growth.

Dimorphism

As mentioned in the previous chapter some fungal species grow in two or more morphological forms, i.e. they exhibit dimorphism.

In the majority of cases the forms are distinct, yeast-like and mycelial. It is of interest that many of the species involved are pathogenic. The transformation from one form to the other is influenced by specific environmental conditions. These are diverse and may be nutritional, e.g. in *Candida albicans*, or physical factors, e.g. temperature in *Paracoccidiodes brasilieusis*, or the gaseous environment, e.g. in *Mucor rouxii.*

The morphological switch from a dividing or budding cell to a cell growing by elongation is presumed to be related to changes in the cell wall. However, examination of the cell wall composition of the three fungi mentioned above reveals major differences between each example and underlies the fact that a single mechanism cannot explain dimorphism in all fungi.

Research on *Paracoccidiodes brasiliensis* and *Candida albicans* has centred on the cell wall and in the former there is good evidence to suggest that a switch in wall biogenesis is the basis for the morphological change. Analysis of walls from the yeast and mycelial forms has shown significant differences in composition and structure. In the yeast form the major polymer is an α-glucan with chitin and in the mycelial form a β-glucan and a galacto mannan are the major components. What remains to be discovered is the nature of the underlying control mechanism which influences the pathway of wall synthesis and how this is affected by a temperature change. A much simpler change in the cell wall has been proposed for *Candida albicans* which involves the disulphide component of the wall protein. It is suggested that growth in the yeast form is maintained by high levels of sulphydryl residues in the protein, mediated by a $NADH_2$ dependent protein disulphide reductase.

The availability of the hydrogen donor stems from an excess generated during carbon metabolism. The experimental evidence for this proposal resides in the demonstration of the presence of the reductase enzyme, however, other causes for the dimorphic change cannot be ruled out and further study of this organism is necessary.

The biochemical basis of dimorphism in *Mucor rouxii* and related species appears to be different from the fungi described previously. The problem has been actively researched in recent years and whilst several mechanisms have been proposed one hypothesis implicated cAMP as the endogeneous initiator of the dimorphic switch from the yeast to hyphal form. Yeast cells growing in CO_2 and shifted to an aerobic environment switch to the mycelial form and a simultaneous decrease in the intracellular level of cAMP (about 3–4 fold) was found. Supportive evidence also came from the observation that the lipid derivative dibutyryl cAMP transformed hyphal cells growing in air, to the yeast environment of 100 per cent CO_2 have high levels of cAMP. Other evidence does not support an involvement of cAMP. In cultures in which the growth form is controlled by the flow rate of gaseous nitrogen into the culture both yeast and hyphal cells have similar levels of cAMP and the additon of dibutyryl cAMP to hyphal cultures failed to bring about the transformation as reported previously. It can only be concluded, at this time at least, that the dimorphic switch is not linked, in an obligatory way, to the intracellular concentration of cAMP.

Another explanation for the phenomenon was sought in the relationship between the yeat form and a high rate fermentative metabolism based on growth on hexose. In this respect the yeast form of *Mucor rouxii* contrasts with *Saccharomyces cerevisiae* because fermentative metabolism in the former does not repress mitochondrial development. Mucor species cultured aerobically in the presence of high concentrations of glucose are fermentative and take the yeast form.

It could be argued that the importance of hexose for the yeast pattern of development is an indication that the enzymes of carbon catabolsim are some significance in the morphogenetic switch, however, differences in metabolic patterns between the yeast and mycelial form under certain environmental conditions do not necessarily occur under another set of conditions. One case in point relates to the utilisation of the glycolytic and the pentose phosphate papthways.

In aerobically grown yeast cells 14 per cent of the glucose is metabolised through the pentose phosphate cycle compared to 28 per cent in the mycelial form. Comparison of yeast and mycelium under anaerobic conditions revealed that the balance of carbon metabolism through the two pathways was the same for both growth forms. Several other reports have been made of differences between enzyems produced in the two growth forms, e.g. isoenzymic forms of pyruvate kinase, but in no case so far described has a clear cut correlation between specific enzymes and a particular morphological type been shown.

The two patterns of development in dimorphic fungi are clearly the phenotypic expression of different parts of the genome. Attempts to unravel molecular differences that are associated with morphogenetic switches have been few, and mostly with the dimorphic species of Mucor. The RNA polymerases of the two growth forms have been found to be similar so differences might therefore be expected at the translational level. The switch from yeast to hyphal form is accopanied by a noticeable increase in the rate of protein synthesis but the nature of the newly formed proteins still requires elucidation.

9

Growth of Yeasts

The yeasts are a group of organisms that exhibit heterogeneity in virtually all aspects of their biology. Although the name yeast is widely used it has no taxonomic significance. Thus yeasts are typically unicellular fungi which, on the conventional criteria of mechanism of sexual reproduction, belong to one of the three major groups–the Ascomycetes, the Basidiomycetes and the Fungi Imperfecti. The processes of cell growth and division are closely integrated and, although cell division is in fact a reproductive process, it is clear that the event is very much influenced by the growth of the cell that proceeds it. For this reason these aspects are treated together in this chapter.

Cell Morphology

Yeast cells are diverse in shape but most common are types which are ellipsoidal and a smaller group where the cells are elongated. Dimensions of the cells are variable and subject to environmental factors but average sizes for ellipsoidal cells are 5 µm long axis by 4 µm short axis, and for the elongated cells are 3.5 µm diameter by 5 µm to 20 µm long. These two groups have different patterns of cell division. In the ellipsoidal forms division involves the formation of buds whereas in the elongated forms the process is a conventional fission of the cell into two equal parts. In the budding yeasts some variation is found in the relationship of mother and daughter cells. In some the daughter cell becomes detached when another bud starts to form and the position of the first bud is marked by a bud scar on the wall of the mother cell. In other species the daughters remain attached to the mother cell

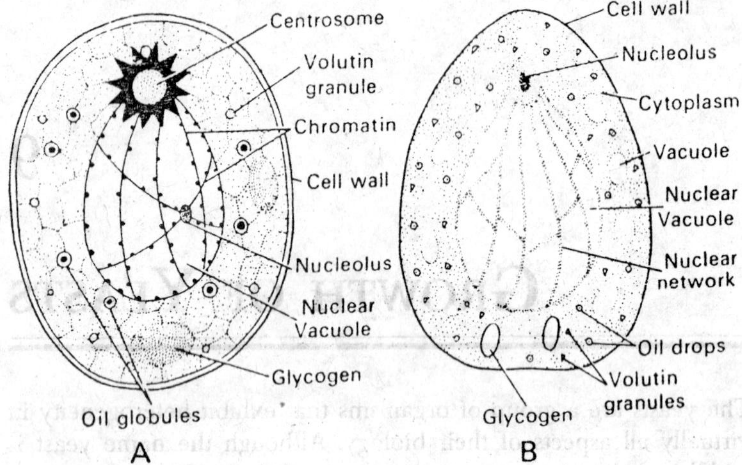

Fig. 9.1. Structure of yeast cell. A—According to Alexopoulos; B—According to Wager and Peniston.

and they in turn produce daughters as does the mother cell. This type of association of cells can take on the appearance of a pseudomycelium and in some yeasts this type of morphology is characterized by differentiation of the cells. Cells of the main chains become elongated and cylindrical producing clusters of buds at

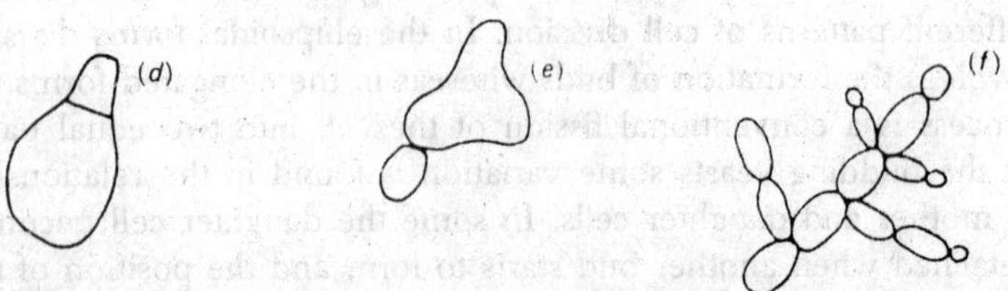

Fig. 9.2. Diversity in cell shape in yeasts; (a) spheroidal—Saccharomyces, (b) elongate—Schizosaccharomyces, (c) apiculate—Hanseniaspora, (d) flask-spahed—Pityrosporum, (e) triangular—Trigonopsis, (f) pseudomycelial—Candida.

their shoulders. A few yeasts, mainly those which are pathogenic on animals, have a dual morphology and are said to be dimorphic. These organisms grow in either the unicellular or filamentous form and the switch in morphology is environmentally controlled. In the Basidiomycete yeasts this dimorphism also occurs with the mycelial forms representing the dikaryon stage in the life cycle.

ULTRASTRUCTURE

Cell Wall

The wall surface of yeast cells is characterised by the presence of scars which are the sites of bud formation and fission. In the budding yeasts there are two types of scar; the birth scar which marks the site of attachment of the cell to its mother cell, and the bud scars which are the sites where buds were formed. The birth scar is normally sited at one end of the cell on the long axis, and the bud scars occur either randomly or at fixed sites depending on

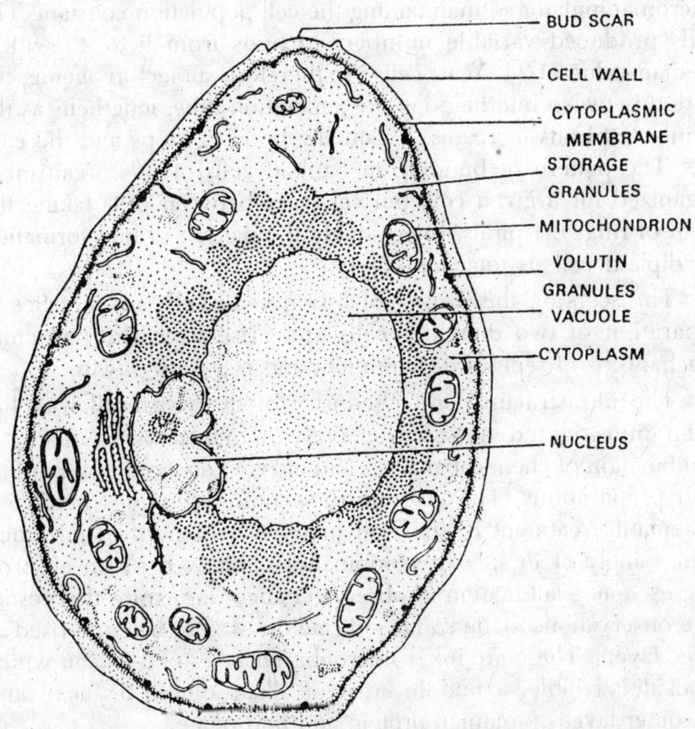

Fig. 9.3. Electron microscopic structure of Yeast.

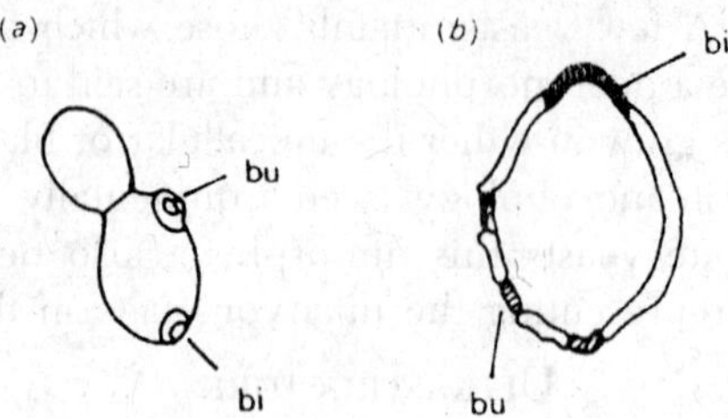

Fig. 9.4. Budding in Saccharomyces cerevisiae. (a) Mother and daughter cell showing birth scar (bi) and bud scars (bu).

the species. Is there a limit to the number of buds a cell can form?

It seems there is and some intricate experiments on *Saccharomyces cerevisiae* were done to prove it. To prevent nutrient exhaustion being a limiting factor in bud formation cells were subjected to microsurgery such that after a bud had formed it was removed by micromanipulator so maintaining the cell population constant. The cells produced variable numbers of buds from 9 to 43 with a median value of 24. Yeast cells are therefore subject to ageing, the period between budding, i.e. the generation time, lengthens as the number of buds increases. Eventually budding stops and the cells die. The pattern of budding in haploid cells of this organism is organized for a given cell, but can vary between cells taking the form of rings or spirals. Interestingly the pattern of bud formation by diploid cells is totally random.

The scars on the surface of fission yeast cells are the sites of separation of two daughter cells and arise from the particular mechanisms of septum development and cell wall growth.

The ultrastructure and chemistry of the yeast cell wall has been investigated most extensively in *Saccharomyces* using a combination of chemical analysis and enzyme dissection on purified wall preparations. The enzyme dissection techniques involve a systematic treatment of the wall material with purified enzymes, either singly or in specific combinations, followed by an electron microscopic examination to observe changes in texture that result. The observations made so far indicate the wall to be comprised of three layers. These are inner microfibrillar layer of β-glucan which is alkali-insoluble, a middle layer of alkali-soluble β-glucan and the outer layer of mannan-protein and phosphate.

The inner layer of glucan accounts for about 35 per cent of the wall, and is made up primary of a β-1,3-linked polymer with a small number of β-1,6-linkages that form branch points. Electron micrographs of this material reveal a dense microfibrillar network. A second (more highly branched) polymer is also present in small quantity and this has a greater number of β 1,6-linkages. The two glucans are separable using hot acetic acid in which the highly branched polymer solubilises. Alkali extraction of *Saccharomyces cerevisiae* wall removes the soluble glucan and the mannan-protein complex and the addition of Fehlings Solution separates the two components by precipitation of the mannan-protein. The glucan can be precipitated by neutralizing the alkaline extract with acetic acid. The alkali soluble glucan is amorphous in texture and comprises 20-22 per cent of the wall, it also differs from the inner layer of glucan in having β-1,6-linkages inserted in the β-,1-3 chains. Small quantities of mannose have been discovered in the alkali-soluble glucan fraction suggesting possible linkage of this middle layer to the outer mannan-protein-layer.

The outer mannan-protein complex makes up about 30 per cent of the wall. The polysaccharide component of this complex has a backbone of D-mannose units with α-1,6-linkages about 50 residues in length and side chains of variable length with α-1,3- and α-1,2-linkages.

Other types of yeasts show differences in the wall polysaccharides. The Basidiomycete budding yeasts have chitin as the skeletal component, replacing glucan. Chitin is also present in *Saccharomyces cerevisiae* walls, but in very small quantity and is highly localized in the septum which is formed during the budding process. The fission yeast *Schizosaccharomyces* pombe lacks both chitin and mannan in its walls. A polysaccharide found only in this latter species is a nigeran-like α-1,3-glucan.

The protein component of the wall has several roles. In bakers yeast (*Saccharomyces cerevisiae*) it may account for up to 10 per cent of the dry wall, the amount varying in response to environmental conditions, age of culture and species. Some of the protein is enzymatic occurring in the form of invertase and acid phosphates, and is found in the mannan-protein layer. An important feature of the structural protein in this layer is the high content of sulphur-containing amino acids which most probably form disulphide bridges. It has been proposed that these bridges have an important role in maintaining the structural integrity of the wall and are thus involved in the initiation of budding.

The other component of yeast walls is lipid and this may make up about 12-13 per cent of the wall.

Protoplast

The organization of the protoplast is typically eukaryotic, its limit is the plasma membrane. In *Saccharomyces cerevisiae* this is characterised by the presence of particles, 15 nm diameter, organized in parallel rows. The particles are of mannan and protein and are thought to be the units of construction of the outer wall layer. The particles have been found embedded at varying depths in the plasma membrane and this has been interpreted as an expression of their passage through the membrane. Another feature of the plasma membrane is the presence of many deep invaginations. In this sections these are seen to be up to 50 nm deep and about 30 nm wide.

An internal membrane system comprised of endoplasmic reticulum and nuclear membrane is readily distinguishable in electron micrographs of *Saccharomyces cerevisiae.* The nuclear membrane has typical pores and, as in other Ascomycete fungi, remains intact at division. In Basidiomycete yeasts the membrane breaks down at division. The presence of a golgi apparatus is a point of controversy. This organelle has been described in micrographs of freeze-etched preparations. However, the function of this organelle in wall synthesis may well be taken over by the endoplasmic reticulum. Alternatively the golgi may be an ephemeral structure arising prior to wall synthesis during bud growth and being consumed in the process. Vacuoles and mitochondria of typical eukaryotic form are present.

Associated with the retention of the nuclear membrane during division are certain other unique features, at least in *Saccharomyces cerevisiae.* In the life of a cell a distinctive region becomes apparent on the nuclear membrane called the spindle or centriolar plaque and associated with it are microtubules, some projecting into the body of the nucleus and the rest into the cytoplasm. Prior to nuclear division the plaque duplicates and the copies lie adjacent to each other. Later they rapidly migrate to opposite sides of nucleus and become connected by the intra-nuclear microtubules which form a spindle. These events occur simultaneously with bud development and immediately after the migration of the nucleus to the neck which connects mother and bud. Both the nucleus and the microtubular spindle subsequently increase in length very rapidly.

The nuclear membrane then pinches inwards at the neck region forming two nuclei. Genetic studies have indicated that *Saccharomyces cerevisiae* has at least seventeen chromosomes but these cannot normally be seen during nuclear division.

In the Basidiomycete types the nucleus becomes elongated, part in the mother cell and part in the bud. The chromosomal component moves into the region of the nucleus in the bud and the nucleolus stays in the mother cell. The nuclear membrane in the bud breaks down and the chromatin material divides. What happens to the nucleolus is not clear, it seems to break up but reappears again within the newly formed nuclei which arises through the formation of new membrane around the two chromatin components. When the nuclei are formed one moves back to the mother cell, the second remaining in the bud.

Cell Growth and Division in Budding Yeasts

Cell Cycle

Synchronisation of yeast cells for cell cycle investigations is normally achieved using centrifugation selection methods. In most budding yeasts the cell cycle follows the conventional eukaryotic pattern with four distinct phases, G_1, S, G_2 and M. S and M represent periods of DNA synthesis and mitosis, respectively and G_1 and G_2 are periods in the cycle separating the two events. Cell division follows mitosis, although cycle separating the two events. Cell division follows mitosis, although *Saccharomyces cerevisiae* deviates in this respect because they are temporally separated with cell division occurring the G_1 phase. Various other events associated with the budding process can also be used as markers in the cycle. After mitosis the mother and daughter cells start the G_1 phase of the next cycle, the daughter cells may increase in size during this time but this depends on environmental influences. The end of G_1 and the start of S are marked by the events of plaque duplication and bud initiation.

Budding

The development of a new bud begins with a localized weakening of the mother cell wall. The reduction of the sulphydryl bridges in the wall protein has been implicated in this event. Within the protoplast small vesicles accumulate adjacent to the budding site. Electron micrographs of freeze-fractured yeast cells show the plasmalemma to have small circular invaginations which could arise

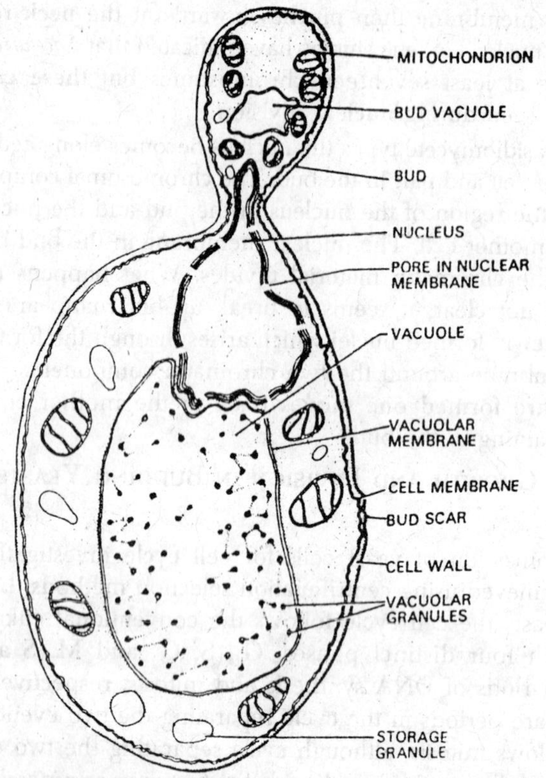

Fig. 9.5. Budding in S. cerevisiae showing division of nucleus.

from fusion of these vesicles with the plasmalemma. The bud continues to grow in size during the cell cycle and this involves the *de novo* synthesis of wall polymers. Investigations using autoradiographic and fluorescent staining procedures have shown that deposition of wall material occurs at a site distal to the parent cell.

The pattern of budding is a variable feature in different yeasts. In *Saccharomyces cerevisiae* and some other Ascomycete yeasts it is multi-polar and can occur at any point on the cell surface except tat a previous budding site. Other Ascomycete yeasts have bipolar budding which occurs repeatedly at the two ends of the cell. The Basidiomycete yeasts also produce bud at random sites but several buds can be formed at one site. The mechanism of orientation of the new bud in *Saccharomyces cerevisiae* is unknown. The spindle plaque has been implicated because at the stage of duplication it

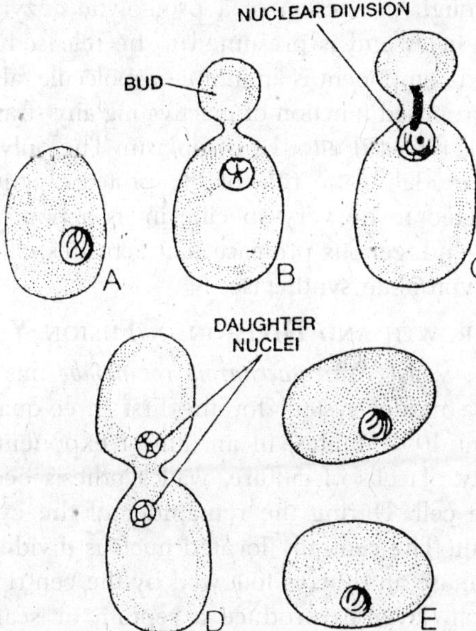

Fig. 9.6. Budding in Yeast.

becomes orientated in the direction of the new bud with the extranuclear microtubules extending into the bud.

The migration of the spindle plaque and bud growth are continuous through both the S and G_2 phases of the cell cycle. At the end of G_2 the nucleus has migrated to the neck connecting the bud and the mother cell. Mitosis then follows during which the spindle plaque exhibits the spectacular behaviour already described.

The budding process and the cell cycle, are completed with the formation of the septum. This apparently simple event in fact represents a highly significant morphogenetic process and has been the subject of intensive investigation. The interest in septum development resides in the fact that the first formed wall is composed of chitin and septa are the only site in which this polymer is found. The localized deposition of chitin, both spatial and temporal during the cell cycle raises fundamental questions relating to the mechanisms controlling such morphogenetic events. The model proposed is the result of many years of investigation and is based on the activities of three components. The enzyme concerned in synthesis of the polymer, chitin synthase, appears to exist in an inactive zymogen form bound to the plasma membrane. Activation is achieved through the action of a proteolytic enzyme which is contained in vesicles and is presumed to be released at the neck region. The third component is an inhibitor molecule, also a protein, which has the proposed function of inactivating any stray proteolytic enzyme which is released into the cytoplasm. The only weak point in this elegant model is the role of the protease activator which might be expected to be very specific in its action. It has been shown that the endogenous protease that activates chitin synthase also activates tryptophan synthetase.

Cell Growth and Division in Fusion Yeasts

The fission yeast *Schizosaccharomyces pombe* has been used extensively as a model system for the first three-quarters of the cell cycle (about 105 minutes) in an almost exponential manner. For the majority of cells of culture, wall synthesis occurs only at one end of the cell. During the remainder of the cycle the size remains constant. The centrally located nucleus divides at the end of the growth phase and this is followed by the centripetal growth of the inner wall layer to produce a septum or scar plug. The septum then splits transversely allowing the two daughter cells to separate. At the site of septum formation a scar remains where the

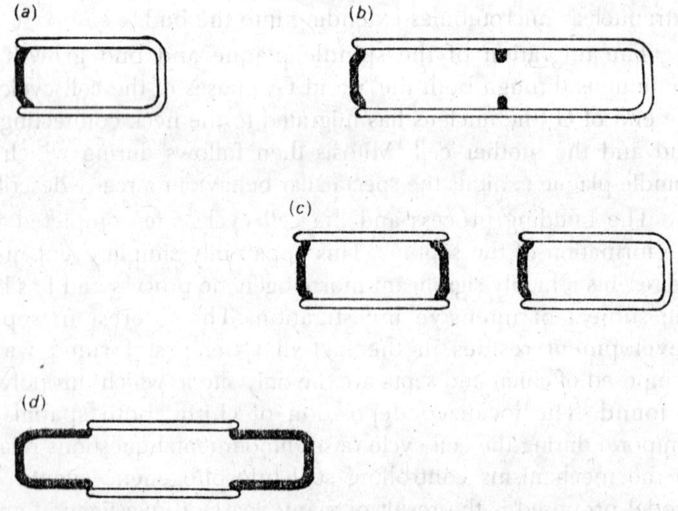

Fig. 9.7. Diagrammatic representation of cell division in Schizosaccharomyces pombe; (a) cell with single birth scar, (b) cell elongation and the initiation of septum development, (c) separation of daughter cells, one cell has a single birth scar and the other has two, (d) growth at both poles in the cell with two birth scars.

outer wall layers are severed. Where daughter cells have a scar plug at each end then new wall growth will occur at both ends of the cell.

DNA Synthesis during the Yeast Cell Cycle

Location of DNA

Haploid strains of *Saccharomyces cerevisiae* contain approximately 1.2×10^{10} daltons of DNA. Most of this is located in the nucleus and genetic studies suggest that it is organized into seventeen chromosomes. Mitochondrial DNA accounts for 5-20 per cent of the total and this occurs as circular molecular 25 µm long. The ratio of mitochondrial to nuclear DNA remains fairly constant under a variety of growth conditions. A few years ago an additional DNA molecule was discovered in the form of closed circles 2 µm in circumference. These small molecules have since been shown to have properties similar to bacterial plasmids, and the term plasmid is now used to describe them. Yeast plasmid DNA accounts for 1-5 per cent of the total. It can be separated as a distinct component independent of nuclear and mitochondrial DNA but its location is as yet unknown.

DNA Synthesis

A constant time relationship between growth rate and DNA synthesis has also been suggested for yeasts and there is some experimental support for the idea. When *Saccharomyces cerevisiae* was grown in a chemostat at the different growth rates induced by glucose limitation it was found that the time period between bud emergence and cell separation was similar over a wide range. At the slower growth rates the phase that did vary was the period prior to bold emergence. A corollary of this finding concerns bud emergence and DNA synthesis which are generally co-incident, in which the time from the initiation of DNA synthesis to cell division is also constant. This observation led to the proposal by Carter (1978) of two phases to the growth cycle of the yeast cell, an expandable phase which will vary according to growth rate and a constant phase.

Further support for this hypothesis was derived from experiments in which a numerical analysis technique devised to determine the duration of cell cycle phases in asynchronous populations of cells was used with temperature sensitive cell cycle (*cdc*) mutants which have blocks at different stages of the cycle.

The experiment involves the culture of the mutant at the permissive temperature at which the cycle is normal and then switching to the non-permissive temperature. The cells in the culture that have passed the stage at which the block occurs will continue their cycles but the remainder will be subject to arrest when they reached the blocked stage. The percentage of cells in the population that have passed the block is determined and this information is used to calculate the timing in the cycle that the block occurs using the equation

$$X_1 = \frac{1 - \ln(N/N_0)}{n2}$$

Where X_1 = stage of cell cycle block

N_0 = number of cells at the time of the temperature shift

N = final number of cells

Carter (1978) applied this approach to a mutant of *Saccharomyces cerevisiae* cdc 7, which is defective for the initiation of DNA

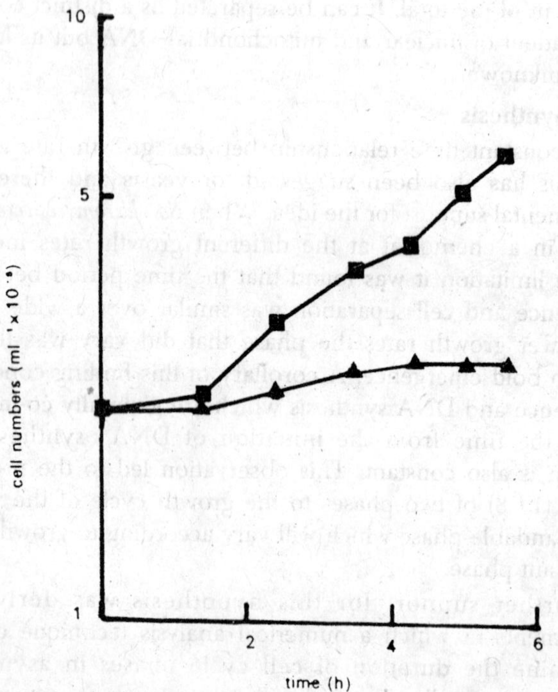

Fig. 9.8. Effect of a temperature shift on cell division in Saccharomyces cerevisiae cdc 7.

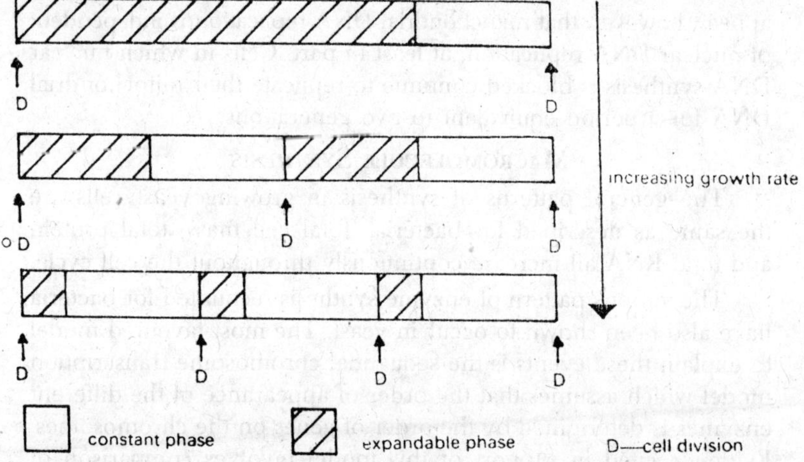

Fig. 9.9. Schematic model for the cell-division cycle in Saccharomyces cerevisiae.

synthesis. Cultures of the mutant were grown at different glucose-limiting rates in a chemostat and cells then transferred to the non-permissive temperature (38°C). The cells that continued to divide were measured. As the growth rate declined the block on DNA synthesis was apparent later and later in the cell cycle, but with the exception of very low growth rates, the time from initiation of DNA synthesis to cell division remained constant over a wide range of growth rates.

The initiation of DNA synthesis in the yeast nucleus is dependent upon active protein synthesis but completion of a round of synthesis will occur, if protein synthesis is stopped. In experiments involving the use of the inhibitor cycloheximide it was shown that exposure up to 10 minutes prior to the start of S had no effect on DNA synthesis, clearly the synthesis of the requisite proteins can occur in a relatively short time. The significance of protein synthesis to the mechanism of initiation of DNA synthesis remains obscure, however, it is hoped that this might be resolved as more detailed studies are made on the *cdc* mutants.

The pattern of mitochondrial DNA synthesis is variable in different yeasts, in some it occurs continuously throughout the cell cycle in others the timing is specific. Details on the control of the process are restricted. Thus it is not known whether the nuclear genome or the mitochondrial genome carries the message for synthesis of the components of the replication system. It does appear, however, that mitochondrial DNA replication is independent of nuclear DNA replication, at least in part. Cells in which nuclear DNA synthesis is blocked continue to replicate their mitochondrial DNA for a period equivalent to two generations.

Macromolecule Synthesis

The general patterns of synthesis in growing yeast cells are the same as described for bacteria. Total cell mass, total protein and total RNA all increase continuously throughout the cell cycle.

The various pattern of enzyme synthesis recounted for bacteria have also been shown to occur in yeast. The most favoured model to explain these events is the sequential chromosome transcription model which assumes that the order of appearance of the different enzymes is determined by the order of genes on the chromosomes. Evidence cited in support of this model involves comparison of the sequence of enzyme production which is coded by chromosome II. The genes for lysine requirement and the utilization of galactose are closely linked and this is reflected in the close timing of their translation. In a particular mutant strain this region of chromosome II has been altered due to a translocation event, and the two genes are a greater distance apart. This genetical difference correlates with the biochemical situation in which the timing of synthesis of two enzymes is altered.

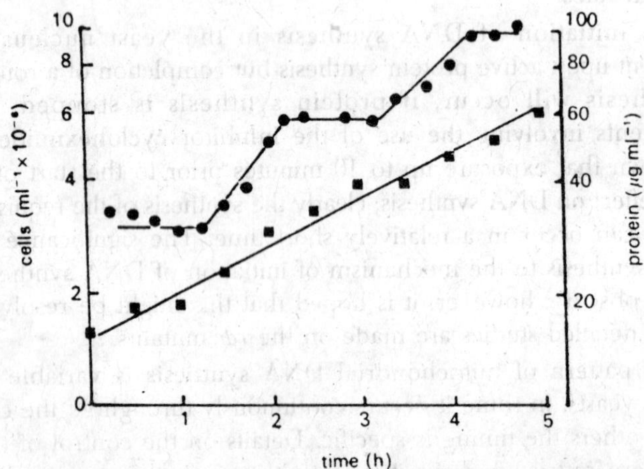

Fig. 9.10. Growth and protein synthesis in a synchronous culture of Schizosaccharomyces prombe.

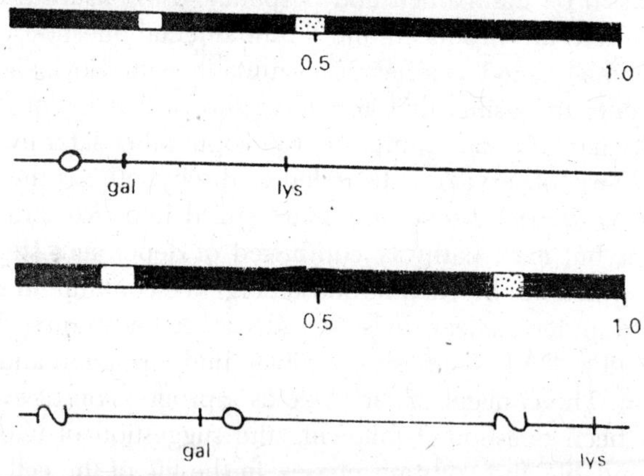

Fig. 9.11. Cell cycle and genetic maps showing the relationship between timing of enzyme synthesis and gene location.

Synthesis of Cell Wall Polymers

Information on the biosynthesis of the major cell wall polymers is limited and relates primarily to the mannan component. Preparation of intra-cellular membranes have been isolated that utilize guanosine diphosphate-mannose as the mannosyl donor for the synthesis of the mannan component of the mannan-protein complex. Mannosyl transferases, highly specific for the synthesis of α-1,6-, α-1,3- and α-1-2-linkages, transfer the mannosyl residues for the polymerization of the chains. Attachment of the mannan chains to the protein involves a lipid intermediate, dolichol monophosphate.

A β-glucan synthase system has yet to be isolated from yeast cells, and a lipid intermediate has not been identified to date. The evidence for β-glucan synthesis is therefore rather limited and rests with studies in which cells, treated with lipid solvents and made permeable, were found to incorporate UDP glucose into the β-,1,3-linked glucan.

The pattern of cell growth during the cell cycle suggests that wall synthesis proceeds continuously.

Co-ordination of Growth and Cell Division

An obvious but still fascinating feature of growth and division of yeast cells is the order associated with the various events. This order has to be maintained and perpetuated to ensure that viable daughter cells are produced and a fundamental question relates to how it is maintained. Studies using mutants with blocks at specific cycle events and using inhibitors have shown that a block early in the cycle may also cause inhibition in some other later events, but not all. These observations have led to the hypothesis that the cell cycle of *Saccharomyces cerevisiae* is organized into two independent pathways, but each pathway comprised of dependent steps. Thus mutants which are blocked at bud emergence continue to synthesis DNA and under nuclear division and conversely strains which can be blocked at DNA synthesis do initiate bud formation and nuclear migration. The concept of the cycle as a progression of events has recently been questioned following the suggestion of two phases, the expandable and constant phases, in the life of the cell with the constant phase being equivalent to the cell cycle.

There is, as yet, little sound information on the mechanism of integration of growth and division. One proposal suggests that increments of growth might be controlled by the occurrence of particular cell cycle events, alternatively the requirements for one

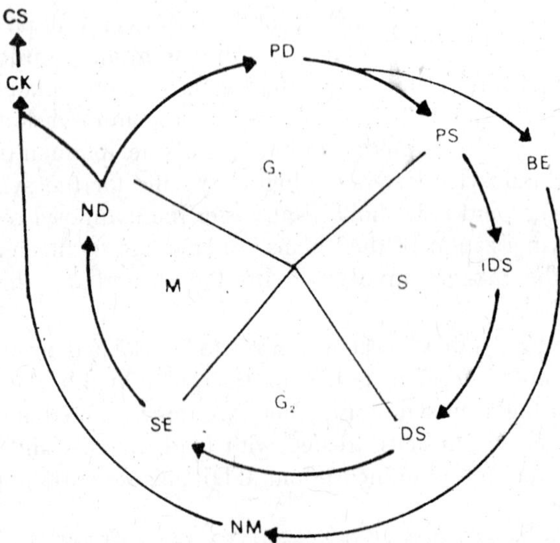

Fig. 9.12. Dependent and independent events in the cell cycle of Saccharomyces cerevisiae. PD, plaque duplication; PS, plaque separation; iDS, initiation of DNA synthesis; DS, DNA synthesis; SE, spindle elongation; ND, nuclear division; BE, bud emergence; NM, nuclear migration; CK, cytokinesis; CS, cell separation.

of these particular cell cycle events may be satisfied only when a specific size is reached.

The positive involvement of cell size as a controlling factor in the cell cycle raises interesting implications in relation to the newly formed buds. After a cycle of budding the mother cell remains at a size about the same as its pre-division size. This cell might then be expected to commence another cycle immediately. In contrast the bud is considerably smaller and a period of growth will be required before bud emergence and the initiation of DNA synthesis commences. In cultures of fast growing cells this period will be short but in slow growing cultures the cells will attain the required status after a much longer period. Consequently the population will be comprised of two groups of cells with different generation times. This period of growth to attain the critical size prior to the commencement of the next round of division, could be interpreted as the expandable phase.

In the fission yeast *Schizosaccharomyces pombe* a slightly different view of the integration of cell cycle events has been put forward. There is evidence for at least two sequences, the DNA-division cycle which involves DNA synthesis, motosis and cell division, and the growth cycle which includes the processes of macromolecule synthesis, including enzymes. The two sequences can be dissociated by a variety of experimental procedures. Inhibiting DNA synthesis or nuclear division was found to have no effect on the growth cycle markers but the cells grow to unusual lengths. Accleration of the DNA-division cycle caused the formation of small cells even though the rate of events in the growth cycle were not affected.

The growth of the yeast cell provides a simple yet elegant system for the study of "biochemical differentiation". The ordered sequence of events occurring during growth and division are an expression of the genome of the organism under a particular set of environmental conditions. Our knowledge of these important organisms has expanded greatly but the challenge of some of the fundamental questions still remains.

10

Cellular Life Cycle

Mitosis may be divided into two major activities: (1) a series of morphological and biochemical changes in the nucleus that lead to nuclear division, or *karyokinesis*, followed by (2) a cytoplasmic cleavage, or *cytokinesis.* As stated previously, the end result of these two activities will be two identical daughter cells. The time element for the entire process is variable, being completed in a few minutes for some bacteria under optimal conditions, or extending over a period of months for some cells, such as those of the mammalian liver.

The life cycle of a cell may be divided into a number of distinct periods, each characterized by its own particular morphological and biochemical events. Every cell has a particular period in its life cycle during which the genetic material (the DNA of the chromosomes) replicates. That is, the DNA must produce an exact copy of itself if the cell is to divide. This period of duplication of the genetic material is called the *synthetic (S) period.* Preceding the S period is a period of growth during which the cell carries on its normal metabolic activity in preparation for DNA synthesis. The period is called the *G_1 period* or *presynthetic gap.* Following the period of DNA synthesis there is a second period of cell growth called the G_2 or *postsynthetic period.* At the end of the G_2 stage, *mitosis* occurs: the chromosome material separates into two equal parts and the cytoplasm divides, forming two cells. Mitosis constitutes the shortest period in the life cycle of the cell. Once cell division is completed, the life cycle of the two new cells may begin again.

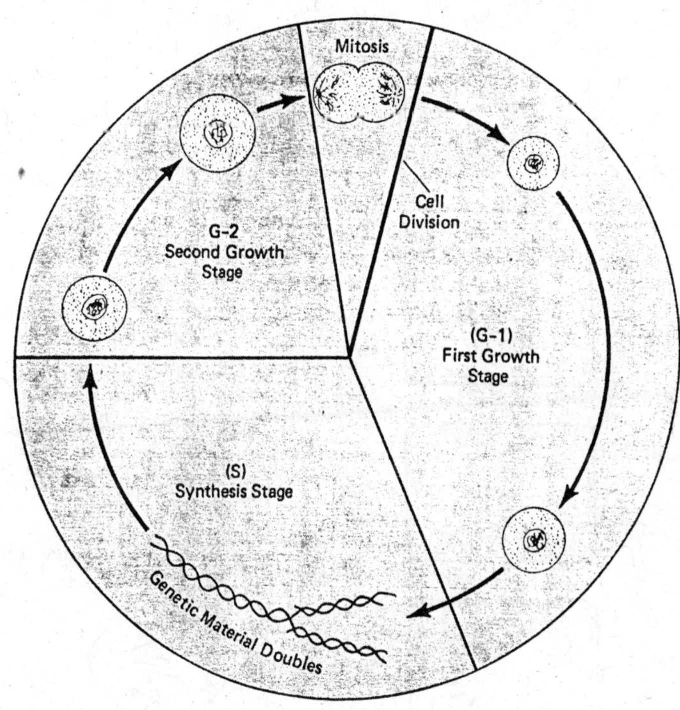

CELL CYCLE

Fig. 10.1. Life cycle of a cell.

Interphase

That portion of the cell cycle represents by the G_1, S and G_2 period constitutes *interphase*, sometimes referred to as the "resting" phase. The latter term is widely used but is so inappropriate that it should be abandoned. The only resting cells is a dead cell, and the interphase cell is far from dead. It is during this period that the cell carries out its normal metabolic and physiological activities as well as making all the preparations necessary for the mitotic process. These include the synthesis of genetic material, both DNA and RNA, an activity that normally occurs within the first few hours following the previous mitotic division. It may be advantageous at this point to indicate that although the majority of cells continue to undergo mitotic division during the life of the organism, there are exceptions to this. For example, nerve and muscle cells–cells that achieve a high degree of cellular

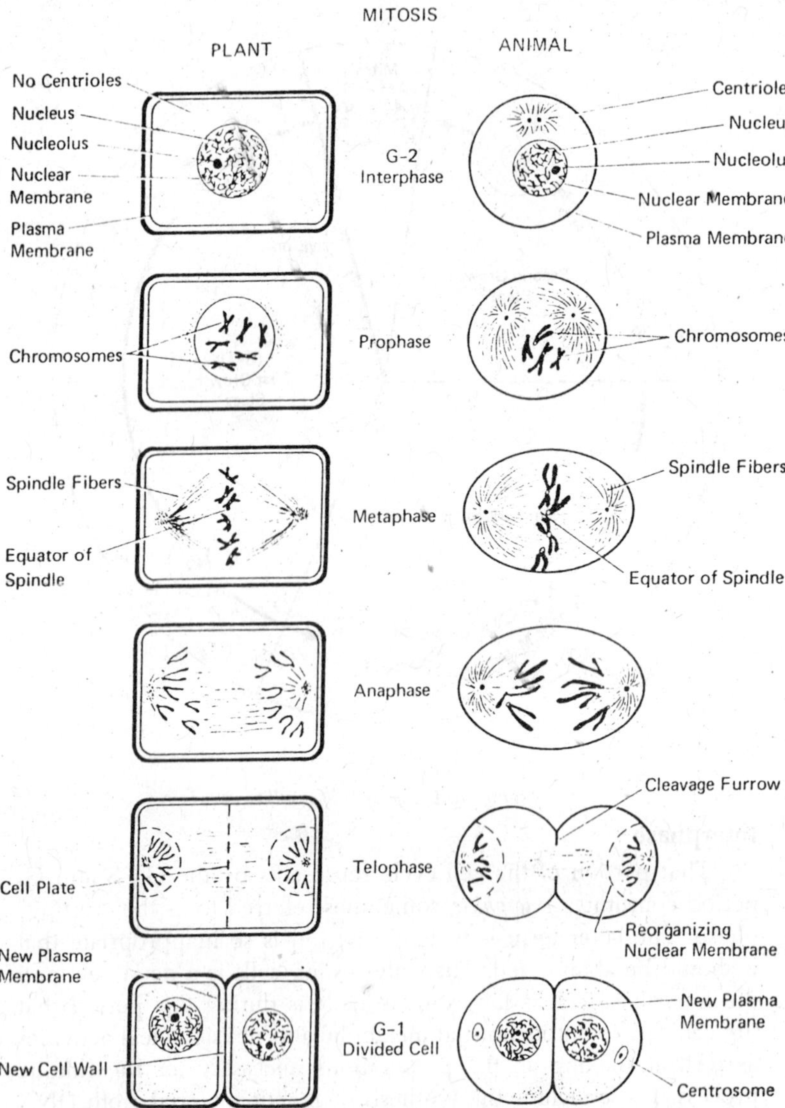

Fig. 10.2. Summary of mitotic cell division in plants and animals.

specialization–do not divide in the adult. Apparently, the "decision" to divide again is made within the first few hours following a mitotic division. When our understanding of the processes controlling such "decisions" becomes greater, we will have made a tremendous advance in explaining such phenomena as organismal development and pathological conditions such as cancers.

The classic picture of the interphase cells is one in which the nucleus and nucleolus are evident, chromatin material is seen only as readily stainable material within the nucleus with little apparent structuring, and the cellular organelles are seen in their "normal" states within the cytoplasm. Individual chromosomes are not evident in the nucleus and will appear only in later stages of cellular division. The material that will constitute the chromosomes is what is called the chromatin material. During interphase the centrioles of the cell usually replicate. With the initiation of the actual process of mitosis, these organelles will move away from each other to lie at opposite sides or poles of the cell and will serve to establish and organize the mitotic apparatus (= mitotic spindle). This is true for animal cells; most plant cells appear to lack centrioles.

The replication of the genetic material–the DNA and RNA of cells–that occurs at this time is a most interesting event and will be discussed in some detail. The replication of this chromosomal material is also considered, from the viewpoint of its organization.

Karyokinesis

Before entering into a discussion of the mitotic process, it must be emphasized again that the stage considered are not always clearly defined and separated from one another. They are simply selected periods during the entire continuous process of mitosis when certain morphological changes are evident. Each stage blends into the next in such a manner that it is sometimes difficult to state at which point one stage ends and the next begins. The mitotic process in the living cell is both continuous and dynamic. The stages used are for the convenience of description only, although they do point out certain morphological and biochemical events that take place.

The stages to be considered are: (1) prophase, (2) metaphase, (3) and (4) telophase.

Prophase

During prophase the first stage of mitosis, a number of events take place, all apparently associated with nuclear changes prior to division of the cell. Coupled with the movement of the centrioles to opposite sides of the nucleus is a gradual shortening and condensing of the chromatin material so that the bodies known as chromosomes become visible as attenuated threads separating from one another. By the end of prophase the chromatin material has become extremely shortened and condensed and stains heavily with basic dyes (dyes containing a positively charged chemical group with the ability to combine chemically with the negatively charged groups of other molecules, such as in this case of the nucleic acids and proteins of the chromosomes). The shortening and condensing process is a result of two factors. First, it appears that water is lost from the chromatin material. Second, and probably more important, there is a progressive coiling of the chromosomes to form a compact helical structure which then coils on itself again.

It should be emphasized that what is called the chromatin material of the nucleus represents the nature of the chromosomes in the interphase period. The chromosomes are extremely attenuated in form and become visible as such only when the chromatin material condenses. Just prior to prophase the chromosome material replicated, and during prophase it is these doubled structures that become visible as distinct bodies. At this stage each daughter chromosome is usually called a *chromatid.* The two chromatids of each pair are connected by a specialized region of the chromosomes known as the *centromere.*

Just prior to or at the beginning of prophase the centrioles divide and move to opposite sides of the nucleus. From each centriole a cluster of filaments extends outward in a structure known as the *aster.* A *spindle*, consisting of *spindle fibers*, develops between the centroles. Each of the double-stranded chromosomes is attached at the centromere to a spindle fiber. Chromosomes move to align themselves on an *equatorial (central) place* of the spindle apparatus. During this time the nuclear envelope has been undergoing a progressive disintegration and finally disappears, as do the nucleoli.

Metaphase

Once the chromosomes have moved and become aligned on a central plane, the cell is said to be in metaphase. During this, the shortest period of mitosis, the chromosomes appear in a linear arrangement when viewed laterally and in a ring or disk when viewed laterally and in a ring or disk when viewed from one of the poles of the cell. At this time, or sometimes earlier, the single centromere that attaches each pair of chromatids to a spindle fiber divides, At this time each chromatid becomes a separate single-stranded chromosome.

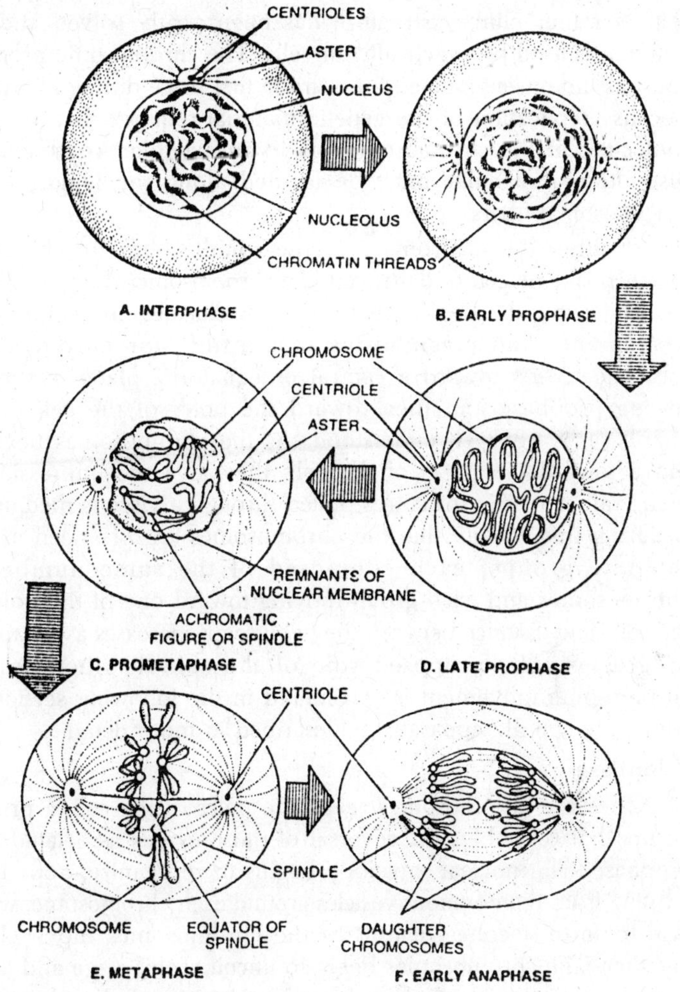

Fig. 10.3. Mitotic cell division.

Anaphase

Once the chromatids of each chromosomal pair have separated, the daughter chromosomes begin a migration toward opposite poles of the cell. This is the initiation of anaphase. During this period the functional significance of chromosomal coiling becomes evident, because it is physically easier for a compact mass of material to move through the cytoplasm than it is for an elongate fibril such as an interphase chromosome to do so. The problem of what to do with a rather clumsy structure has been neatly solved and this solution is used by practically all cells. The chromosome probably must assume its attenuated form in the first place during interphase because replication of the genetic material requires that the giant molecules of the chromosomes be in an extended condition and also other molecules must be formed along the length of the chromosomal fibers.

If either the centromere or the spindle fiber to which it is attached is damaged or destroyed, the chromosomes do not separate, thus indicating the need for both of these structures in chromosomal movement. The mechanisms responsible for chromosomal movement, first toward a central or equatorial plane of the cell during prophase and then toward the poles of the cell during anaphase, is not well understood. As the chromosomes begin to move toward the poles of the cell, they give the impression of being rather plastic structures pulled through a viscous medium. It is during this activity that the chromosomes are first seen as two separate groups, each composed of the same number of chromosomes and each group moving toward one of the poles of the cell. The distance between the two groups increases as anaphase progresses. The proposed role of the spindle fibers in this chromosome movement is considered in the following section, in which the spindle apparatus is described in more detail.

Telophase

Once the chromosomes reach the poles of the cell, a process occurs that resembles the reversal of those events associated with prophase. The nuclear envelope begins to reorganize, apparently forming from membranous vesicles around each chromosome, which fuse to form a cohesive unit as the chromosomes move closer together. The chromosomes begin to uncoil at this stage and again resume the attenuated threadlike appearance of the chromatin material. The nucleoli reappear, seemingly reorganized from a specific site on one particular chromosome, the nucleolar reorganizer. The spindle apparatus disappears. During telophase the centrioles undergo replication so that each daughter cell will be supplied with a pair of these organelles. At the end of telophase the nuclei have assumed all the characteristic of the interphase nucleus.

Summary of Karyokinesis

The mitotic process began with a cell in which a number of chromosomes, each consisting of two strands of chromatids, were present. As mitosis progressed the double-stranded chromosomes became aligned on an equatorial plane during metaphase. The chromatids of each pair separated into single-stranded chromosomes and moved toward opposite poles of the cell during anaphases. During telophase the single-stranded chromosomes resumed the attenuated threadlike form characteristic of the interphase nucleus. When the chromosomes appear at the next mitotic prophase they will again be double-stranded structures, indicating that they have replicated during interphase. The actual process of chromosomal replication is the only phase of mitosis that cannot be observed directly.

The function of karyokinesis is to separate identical chromosomes from each other and move them to opposite sides of the cell, thus ensuring that each daughter cell that is to be formed will receive a complete complement of chromosomes identical to that of the parent cell. In this way each daughter cell receives an identical set of genetic information and the genetic continuity of the cell line is assured.

Although the above discussion of karyokinesis holds true for a majority of cells, there are some exceptions. Notably, cellular division in procaryotes appears to be a simpler process. In such cells there are no specific events concerned with karyokinesis to be seen. In most instances a centrally located ring of new cell-wall material begins to form and grows inward, separating the cell into two equal halves, a process known as *binary fission.* Such a process might seem to result in an unequal distribution of genetic material, but such is not the case. Investigation of the genetics of procaryotic cells show that each daughter cell so formed is endowed with a complete set of genetic information, thus indicating a duplication of genetic material and its equal division between daughter cells. In the procaryotic cell, remember, the genetic material is not associated with structures such as chromosomes but is a mixture of protein, DNA, and RNA. Nor is a structure such as the nucleus present. Mitosis is not the only method of cellular division; rather, it is a special type of cell division found in the majority of cells.

Mitotic Apparatus

No discussion of mitosis would be complete without a description of the mitotic apparatus. This structure, easily seen in the light microscope and recognized as structure formed from microtubules when viewed in the electron microscope, has presented the cytologist with one of the most fascinating and perplexing problems in the analysis of mitosis.

The spindle consists of protein fibers. These fibers were shown to be actual structures and not preparation artifacts by experiments in which the spindle apparatus was moved about in the cell by micromanipulation and also by the fact that the spindle apparatus can be isolated in a purified form by cell homogenization and cell fractionation. From such pure preparations it has been shown that the spindle contains 3 to 5 percent RNA, with the remainder of the spindle material being a single type of protein, similar in some respects to the contractile protein, actomyosin, of muscle. The

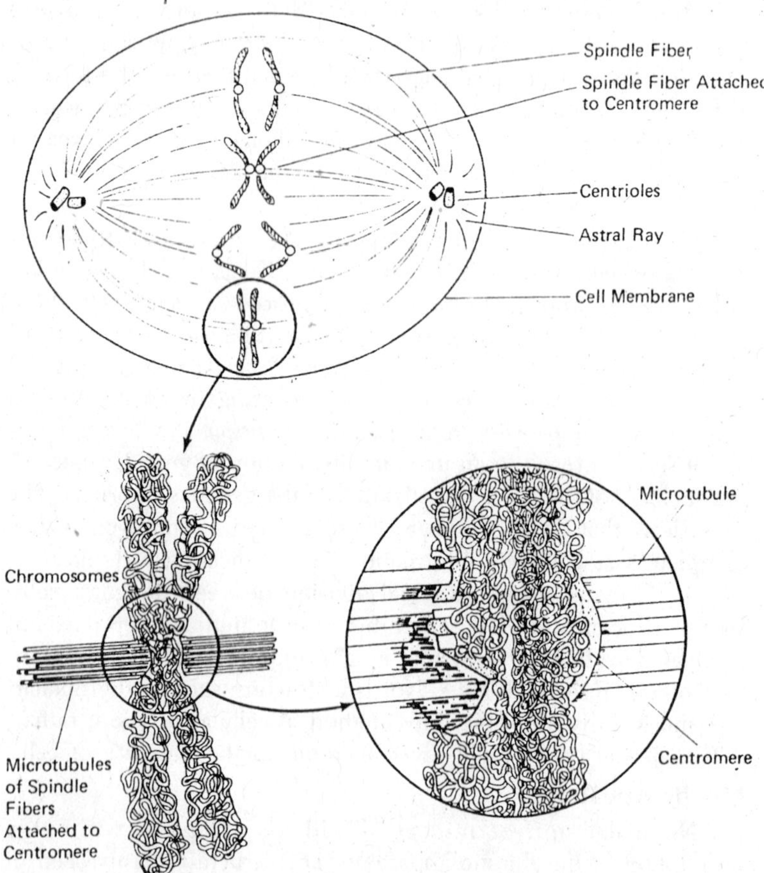

Fig. 10.4. Relationship of spindle fibers and chromosomes.

material of the spindle fibers are formed into microtubules, and there are two types present. Some lead from the centriole at one pole of the cell; these are pole-to-pole or continuous spindle fibers. Others extend from a centromere of a chromosome pair. The formation of the microtubules that make up the spindle fibers is associated in some way with the centriole in animal cells, but this is not necessary requirement for their organization because plant cells lack centrioles and can still form spindle fibers. Spindle-fiber formation may be dependent on the formation of disulfide bonds, –S–S, between cysteine residues of protein molecules. Development of the spindle-fiber of system is a prerequisite for normal cellular division to occur, and any agent that interferes with spindle-fiber formation acts as a deterrent to mitosis. For example, if one subjects a cell to the chemical colchicine, normal spindle-fiber development is arrested and mitosis cannot take place. Colchicine is a reagent that inhibits the formation of –S–S–bonds, thus indicating that such bonding is a requirement for spindle-fiber formation and organization.

Once spindle fibers are formed, there remains the problem of explaining their role with respect to centriole and chromosome movements. It is thought that both pulling and pushing actions are accomplished by the fibers. In prophase, the migration of the centrioles to the poles is perhaps due to pushing action of the fibers. In anaphase, spindle-fiber contraction and shortening is observed in the early stages. Other evidence indicates that the two sets of chromosomes are pulled apart by an elongation of the spindle fibers between the poles. Chromosomal and continuous spindle fibers might well differ in their actions. Movements might result from different proportions of the two types of microtubules being active at any given time. The most widely accepted hypothesis for explaining spindle-fiber activity is based on the addition or subtraction of protein monomeric units to the microtubules that compose the spindle fibers. Such a mechanism would mean that shortening of a spindle fiber resulted from the removal of protein subunits from the fiber. The seeming pushing action of spindle fibers on chromosomes would result from a lengthening of the fibers due to the addition of protein subunits to them. In either case the observed contractions and lengthenings differ from the shortening or relaxation of muscle cells, especially because during spindle-fiber contraction there is no change in the diameter of the fibers as is seen in the muscle. However, as with muscle contractile proteins, ATP is required as an energy source for the contraction.

Cytokinesis

The process of cytoplasmic division may begin either during late anaphase or in telophase and proceeds in conjunction with the final stages of karyokinesis. Although this is the general rule, there are exceptions. For example, in certain algae and fungi, cytokinesis does not take place at all. The result of this type of cellular division is a multinucleate cytoplasmic structure in which few, if any, partitions separating the nuclei are present. Such organisms are called *coenocytic* and include not only those forms mentioned above but also the cells of some lower invertebrate animals. The process is also found in the reproductive stages in some higher plants and is responsible for the formation of both cardiac and skeletal muscle cells, which are multinucleate structures in the adult. There are also instances in which multinucleate structures are formed by continuous karyokinesis and the cytoplasm only later is partitioned into separate cells by the onset of a rapid cytokinesis. This is found in the development of some insects. It seems, however, that the selective advantage of the two processes occurring simultaneously has been so great that the majority of organisms have retained this mode of mitosis.

Animal Cells

The first sign of cytokinesis in a dividing animal cells is the appearance of *cleavage furrow* running circularly around the cell at the equator. From experimental data it has been concluded that the orientation of the spindle determines the location of the cleavage furrow. If the spindle is displaced from its normal position in the cell by micromanipulation, the cleavage furrow will form in the area of the new equatorial plane. The time factor appears to be critical, because if one waits too long in displacing the spindle, the cleavage furrow will form in the old position of the equatorial plate. The cleavage furrow becomes progressively deeper and deeper, constricting the cytoplasm of the cell and eventually cutting completely through the cell to produce two daughter cells.

Although the position of the cleavage furrow is dependent on the orientation of the spindle, very little is known as to the nature or mechanisms of the actual formation of the material that makes up this structure. It has been postulated that the cytoplasm in the areas of the cleavage furrow gains some type of contractile ability, thus making it capable of pinching the cells in two. Perhaps new cell membrane is formed at these points and by a process of growing inward gradually separates the cytoplasm into two halves. The spindle fibers have also been implicated in the process, because some investigators believe their experiments show that some of the fibers become attached to the cell membrane at the area of the cleavage furrow, and they hypothesize that these fibers actually pull the membrane inward. Unfortunately, this would not explain those cases in which cytokinesis does not occur until quite some time after the spindle has vanished. It would also fail to explain why a sea urchin egg (a cell that has been widely used in mitotic studies) is capable of cytokinesis even if the entire spindle apparatus is removed by microsurgical methods.

The final result of cytokinesis, then, is the formation of two daughter cells, each surrounded by a plasma membrane and each containing a nucleus and the genetic material that was evenly distributed during karyokinesis.

Plant Cells

In the previous discussion of plant and animal cells it was pointed out that one of the obvious differences between the two is the presence of a rigid cell wall in plants. As might be expected, the plant cell cannot undergo cytokinesis by the formation of a central invaginating cleavage furrow. Cytokinesis in plant cells is indeed different from that in animal cells and relies upon the formation of a *cell plate* in the central region of the cell, halfway between the two telophase nuclei. Organization of the cell plate results from the alignment of membranous vesicles that form in the central region of the cell. The vesicles then fuse to form the middle lamella of the plant cell. Cellulose is deposited upon the middle lamella to form the cell wall. Thus the cell wall is neither a membranous structure, in the sense that a cell plasma membrane is, nor a simple localized deposition of cellulose. The cell plate continues to grow from the central portion of the cell toward the periphery and cytokinesis is complete when the cell plate has fused at the periphery of the cell to effectively form two daughter cells. Again, cytokinesis in animal cells begins at the periphery and proceeds inward, whereas in plant cells cytokinesis is initiated centrally and proceeds toward the cell periphery.

Meiosis

In the discussion of mitosis it was pointed out that the end product of cell division of this type is a pair of daughter cells, each identical to the parent. This means that each daughter cell has exactly the same number of chromosomes (and hence the same amount of the genetic material) as the parent cell. If, however, one examines the chromosomes either in a living cell or in a fixed and stained preparation, it can be found that not all the chromosomes of the cell are different. By either photographing or drawing chromosomal preparations, one can construct an idiogram that indicates the differences in chromosome size, arm length, and centromere location. In so doing it becomes evident that the eucaryotic cell contains pairs of morphologically identical chromosomes. An organism with 12 chromosomes contains six pairs of chromosomes rather than 12 different chromosomes. This holds true for the majority of plant and animal cells. Any cell falling into this category is called *diploid.* In humans, for example, any somatic (body) cell contains 46 individual chromosomes. And there are 23 pairs of chromosomes, that is, 23 distinct morphological types of chromosomes. During a mitotic division of any somatic cell in the human body, the end result will be two daughter cells, each containing 23 pais of chromosomes.

But when sexual reproduction is considered, there is a problem to be faced concerning the number of chromosomes that end up in the zygote. During sexual reproduction a male sex cell and a female sex cell unite to form the *zygote.* If each sex cell (=*gamete*) contained a normal diploid number of chromosomes, the resulting zygote would contain twice as many chromosomes, that is, twice the diploid number. Thus, if the production of gametes depends on mitosis as described above (each daughter cell containing a diploid number of chromosomes), the number of chromosomes in the first-generation offspring of the organism would double as gametes combined; and each succeeding generation would double its chromosome content. Obviously, such a situation does not exist and some other mechanism of cellular division must be responsible for the formation of gametes to be used in sexual reproduction. The mechanism must operate in such a manner that only half the diploid number of chromosomes is present in each gamete. Then when two gametes combine, the diploid number will be restored. The process by which the number of chromosomes in gametes is halved is a special type of cellular division called meiosis or *reductional division.* The cells resulting from meiosis are *haploid*– they contain only half the number of chromosomes (n) present in adult somatic cells (2n). In a haploid cell one finds a complete set of each type of morphologically distinct chromosome. The diploid pairs have been redistributed so that during meiosis each gamete receives a full complement of each of the various chromosomes.

Meiotic Process

Meiosis involves two cellular divisions accompanied by only one replication of the chromosomes. The result is usually four cells each of which is haploid. The two divisions are known as meiosis I and meiosis II. The stages of meiosis are similar to those of mitosis but often several additional stages or substages are added to more adequately describe the complex chromosomal movements that take place in meiosis.

Meiosis I-Prophase I

The majority of events occuring in prophase I resemble those found in mitotic prophase. Although prophase I is often divided into six substages, here it is convenient to treat prophase I as a single unit.

During prophase I the nuclear membrane breaks down and it and the nucleoli gradually disappear. During the early stages of prophase I the centrioles move to establish the poles of the cell and the spindle apparatus is formed. In early prophase I the chromosomes are present as long, thin, attenuated threads, and they thicken and condense to form chromosomes of recognizable nature as happened in mitotic prophase.

The major difference between mitotic prophase and prophase I lies in the relationship established between chromosomes. Each chromosome consists of a pair of identical strands, the chromatids, because as in mitosis the chromosomes have replicated during the preceding interphase. But in prophase I, unlike in mitotic prophase, homologous pairs of chromosomes pair with each other in process known as *synapsis.* The pairing of homologous chromosomes is exact. Each point on one member of a pair comes to lie close to the identical point on the other member of the pair. In addition, the centromeres of the two chromosomes come to lie adjacent to one another. By the middle of prophase the pairing of homologous pairs makes it appear that the chromosomal number has been halved.

It should be noted that each chromosomal unit in this case is a *tetrad* or *bivalent*, made up of four chromatids arranged in pairs.

Toward the end of prophase I, the paired homologous chromosomes begin to move away from each other. The separation is not complete because homologous chromosomes remain united at points of interchange known as *chiasmata*. Chiasmata are thought to represent points of crossing over where chromosomal segments can be exchanged between member of a pair of homologous chromosomes. Crossing over is an important mechanism for rearranging the genetic material.

Meiosis I-Metaphase I

This is the step that distinguishes meiosis from mitosis. Whereas in mitosis the chromosomes become individually aligned on the equatorial plate during metaphase with no pairing of homologous chromosomes, such is not the case during metaphase I. At this stage we are dealing with synaptic or homologous pairs of chromosomes. It appears that only half the diploid number of chromosomes are present homologous chromosomes have joined together in pairs. Again each member of a pair consists of two strands or chromatids. These chromatids are not the same as the pairs of homologous chromosomes.

Using human cells as an example, the activities of chromosomes in mitosis and meiosis may be described as follows. During a mitotic division one can see 46 double-stranded chromosomes aligned on the equatorial plate, but in meiotic metaphase I there are 23 pairs of chromosomes arranged on the equatorial plate. Only one half the number of spindle fibers will be used in meiotic metaphase I as are used in mitotic metaphase. Forty-six spindle fibers are used during the latter process with the centromere of each pair of chromosomes being attached to a spindle fiber. In metaphase I the centromere of each pair of chromosomes occupies the same spindle fiber, thus using only 23 spindle fibers. The centromere of each synaptic pair will lie opposite the other on the equatorial plate, equidistant from each other and occupying the same spindle fiber.

Meiosis I-Anaphase I

During anaphase I homologous pairs of chromosomes separate and move to opposite poles of the cell. The important point here is that the centromeres of the synaptic pairs have not divided but

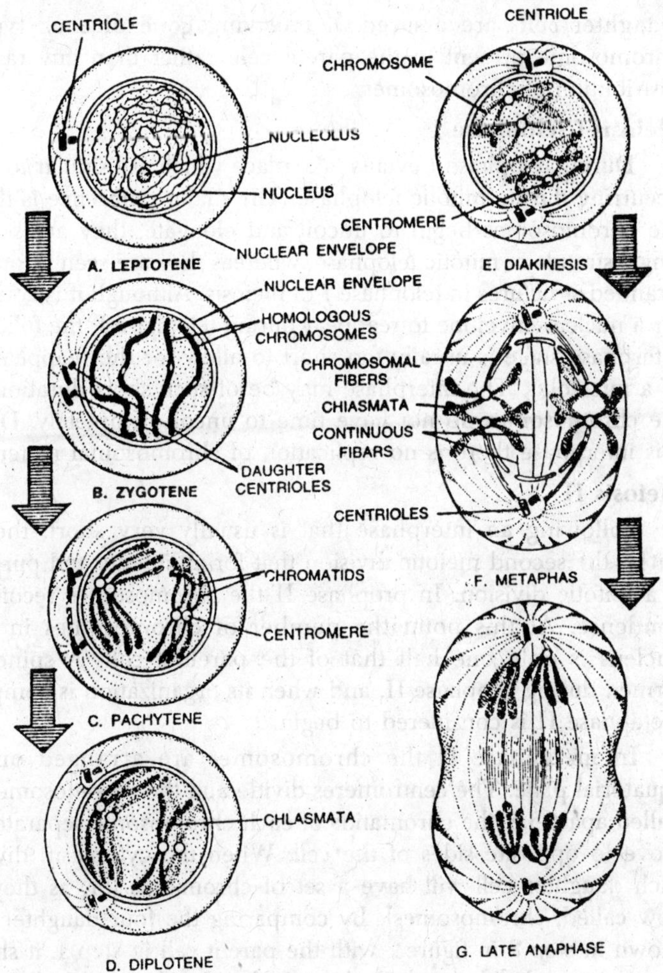

Fig. 10.5. Meiosis I.

remain single structures. In mitosis the onset of anaphase is marked by a replication of each centromere of a chromatid pairs so that chromatid separation occurs as the centromeres move toward the poles of the cell. Such is not the case in meiotic anaphase I. Rather, one is presented with a picture of complete chromosomes, each composed of a pair of chromatids united by an as-yet-unduplicated centromere, moving toward the poles. It is also important to realize that each of these groups of chromosomes possesses an identical chromosomal complement because homologous pairs are involved.

Daughter cells are assured of receiving gone of each type of chromosome present in the parent cell, rather than any random distribution of chromosomes.

Meiosis-I-Telophase I

During telophase I events take place which are similar to those occurring during mitotic telophase. The major difference is that as the chromosomes begin to uncoil and elongate, they are seen as single strands in mitotic telophase, whereas they are seen as double-stranded structures in telophase I of meiosis. Although it is common for a nuclear envelope to reappear during telophase I, the following interphase stage is usually too short to allow for the reappearance of a nucleolus. The interphase may be of such short duration that the chromosomes do not have time to uncoil completely. During this interphase there is no replication of chromosomal material.

Meiosis II

Following an interphase that is usually very short, the cell enters the second meiotic division that for all intents and purposes is a mitotic division. In prophase II the chromosomes recoil and condense. At this point the number of chromosomes in each nucleus is only one half that of the parental cell. A spindle is formed during prophase II, and when its organization is complete, metaphase II is considered to begin.

In metaphase II the chromosomes are arranged on the equatorial plate. The centromeres divide and the chromosomes are pulled apart–i.e. the chromatids of each chromosome separate and move to opposite sides of the cell. When the cytoplasm divides, each daughter cell will have a set of chromatids (or, as they are now called, chromosomes). By comparing the four daughter cells shown in step 8 in figure... with the parent cell in step 1, it should be obvious what happened: the initial eight chromatids (grouped into four chromosomes of only two types) have been redistributed into four cells; each cell ends up with two different chromatids.

Meiosis in Animals

Adult animals exist as organisms in which each cell of the body, with two notable exceptions, possesses the diploid number of chromosomes and mitosis as the normal form of cellular division. Mitosis allows for repair of damaged tissue, replacement of aging dying tissues, and tissue growth. The two exceptions to this general rule are those organs of the body responsible for production of

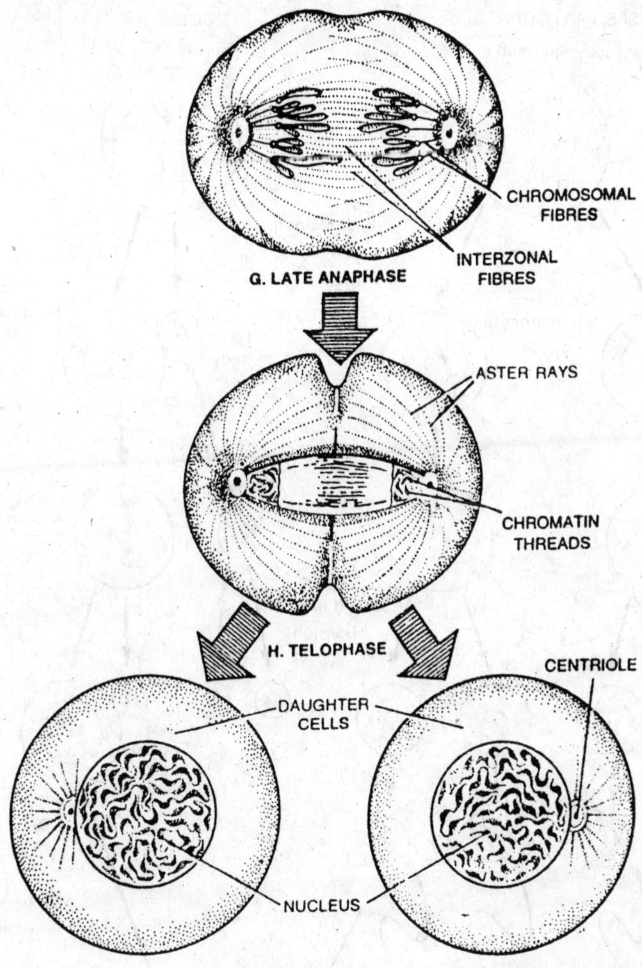

Fig. 10.6. Meiosis II.

sex cells, or gametes. The ovaries in the female and the testes in the male are organs responsible for gamete production. The ovaries produce eggs and the testes produce sperm. During sexual reproduction the union of these two types of haploid cells produces a diploid zygote, which will then undergo mitotic division to produce the sexually mature adult organism. The gametes represent a haploid stage in the life cycle of animals, whereas the sexually mature organism represents the diploid stage. The process of sperm production in the sexually mature organism is termed *spermatogenesis*

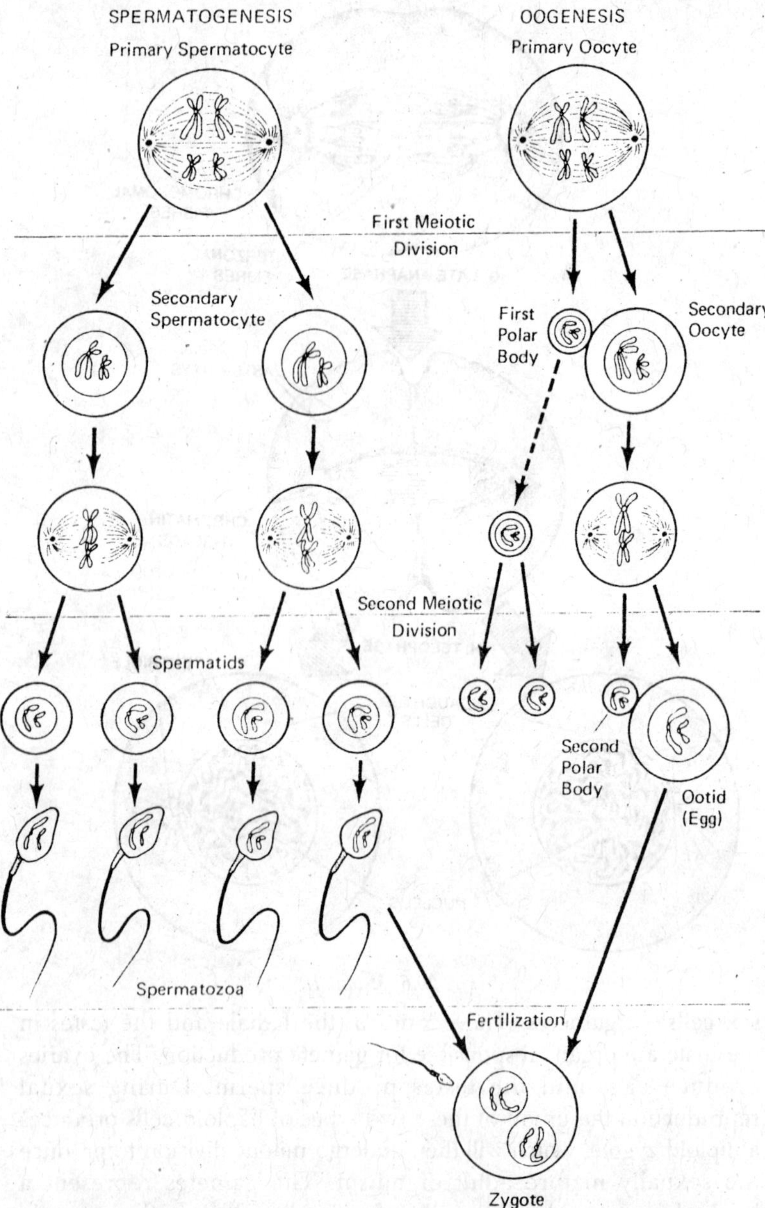

Fig. 10.7. Formation and development of sperm and eggs by meiosis.

and takes place in the testis. Meiotic division of a single cell results in the formation of four small haploid cells: the spermatids.

Following the second meiotic division the spermatids undergo a process of differentiation during which they develop the characteristics of the spermatozoa associated with each species. This usually involves the development of flagella, imparting a certain degree of mobility to the sperm. The anterior region of the spermatozoa, the head, contains little more than the nucleus and its genetic material, with only a very small amount of cytoplasm.

The process of egg formation in the female is *oogenesis.* During this process a follicle cell of the ovary undergoes meiosis. Unlike spermatogenesis, oogenesis results in a single egg rather than four gametes. If one examines a dividing follicle cell, one observes that an unequal distribution of cytoplasm occurs between daughter cells during both stages of meiosis. The end products of meiosis I are two haploid cells, but one is considerably larger than the other. The larger of the two cells undergoes the second meiotic division, again giving rise to two cells of unequal size. The smaller cells are called *polar bodies* and the first of these may occasionally undergo a second meiotic division to give rise to two polar bodies. The polar bodies are nonviable cells, and although one may occasionally be fertilized by a spermatozoan, the resultant zygote soon dies.

The unequal distribution of cytoplasm during oogenesis is of immense value when one considers that the egg must initially supply all the nutritive material for the developing embryo. It is obvious that the size of the spermatozoan is much too small for it to contribute little more than nuclear materials to the zygote. The cytoplasm of the ovum contains all the nutritive materials necessary for the early stages of embryogenesis in mammals and other organisms. Later stages of embryogenesis in mammals are supplied with nutrients by a fetal-maternal circulation established between the mother and the zygote implanted in the wall of the uterus. In all other animal groups the embryo must be nourished by the materials that originally compose the ovum.

Meiosis in Plants

The meiotic process in plants proceeds in the same manner as in animals, but the end products are not always gametes of the type described above. While all flowering plants are diploid as adults, other plants are often haploid as adults. In the latter the reproductive cells formed in meiosis give rise to spores which then undergo a mitosis and a cellular differentiation to give rise to an adult haploid organism. In plants we have a group of organisms in which the haploid and diploid stages of the life cycle are variable; the haploid is dominant in some plants and the diploid is dominant in other plants. In still other plants the haploid and diploid phases may be approximately equal.

The life cycle of a plant may be separated into a maximum of five phases: (1) a diploid multicellular phase, (2) a spore phase, (3) a multicellular haploid phase, (4) gametes, and (5) the zygote. In primitive plants the multicellular diploid stage is not present (e.g., in most algal forms). The zygote gives rise to spores which then undergo mitosis with a following differentiation to form the haploid multicellular unit. The *spore* is an asexual reproductive cell. The haploid multicellular adult mitotically produces gametes, the union of which forms a diploid zygote. The zygote then undergoes meiosis to give rise to four haploid cells, each of which is capable of developing into the multicellular haploid organism. In this type of life cycle we are dealing with almost the complete reverse of the type present in animals, because the haploid stage constitutes by far the greater part of the cycle, whereas in animals the haploid stage of the organism is confined to a very short gametic period. Thus the haploid stage is dominant in the life cycle of many primitive plants.

In those plants in which all five phases are present in the life cycle, the haploid and diploid stages are approximately equivalent. In this type of life cycle the union of the gametes produced mitotically by the haploid stage results in the formation of a zygote, just as in primitive plants. But this zygote undergoes mitotic rather than meiotic division, to form a multicellular diploid adult. The diploid adult will then produce spores by meiosis, which then develop into the multicellular haploid phase of the life cycle. This type of life cycle is present in the flowering plants, which spend the majority of their lives as multicellular diploid organisms.

Some generalities can be drawn from examination of life cycles in animals and plants:

1. Meiosis does not necessarily produce gametes but may form either gametes or spores.
2. If gametes are formed, their union will produce a zygote that will divide mitotically to give rise to a diploid adult organism.

3. If spores are the end product of meiosis, they will then undergo mitosis to form a multicellular haploid organism.
4. Although both haploid and diploid stages are generally present in all organisms, the haploid stage is much reduced in animals, represented only by *gametes*. In plants the relative importance of haploid and diploid stages is quite variable and in lower plants the haploid is dominant. In higher plants the diploid is dominant.

11

GROWTH IN PLANTS

In the previous chapter, the structure and ecology of tropical forest communities have been considered as somewhat static structures. Of course these ecosystems are continually changing through growth and reproduction, death and replenishment, and these more dynamic aspects will be covered in the next chapter. Here the physiology of growth and development, especially of the trees, the major component of those communities, will be discussed against the background of the climatic and other factors described. The emphasis is firmly on whole-plant physiology, subjects such as metabolism, nutrition, translocation and water relations being mentioned only where they have special relevance to the central questions concerning the responses of particular organs. How does temperature, for instance, influence the production or growth of leaves? What are the facctros which control and integrate the allocation of carbon resources to the formation of new organs, against growth or storage in existing parts of the tree?

Such topicss are likely to be of the greatest interest both to biologists and to managers seeking to understand and use tropical forest land, for in general terms they appear to be more specific to the tropics than the underlying processes such as respiration,uptake of mineral ions or protein synthesis. Since the term growth implies the self-multiplication of living material, it includes in its widest sense all the increases and changes that occur in the life of an individual. However, a convenient distinction is usually made between quantitative increases in size or weight–growth in its narrower sense–and development, which covers qualitative changes

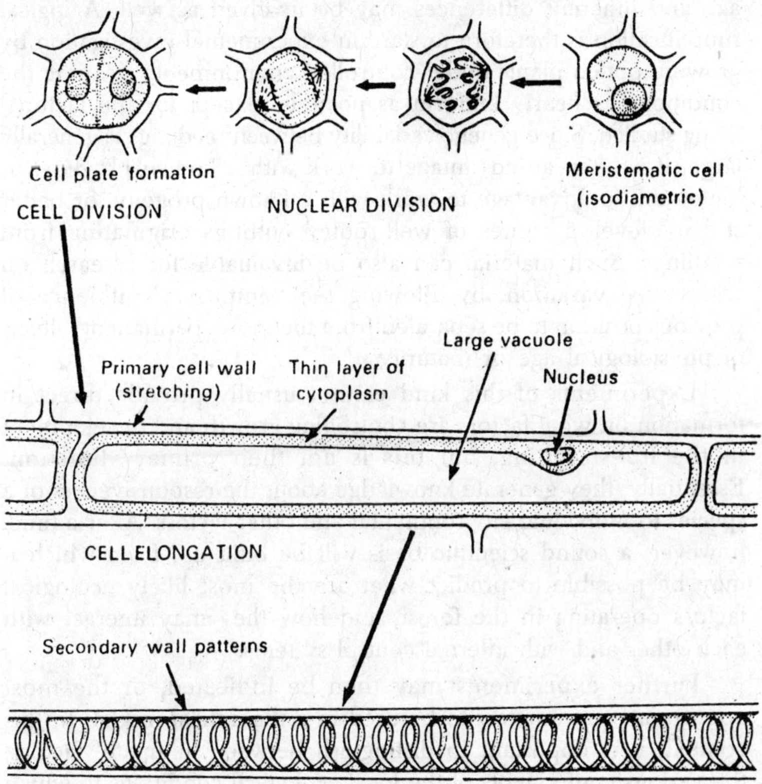

Fig. 11.1. Figure showing Cell formation, Cell elongation and Cell differentiation.

such as the intitiation of new organs, the starting and stopping of growth, and other relatively abrupt alterations to the existing physiological state. For example, measurements of the rate of expansion of a leaf blade or elongation of its petiole come into the first category; while assessments or observations on the emergence of new leaves, the duration of their period of growth and their ultimate loss from the tree involve the second. Thus the size and shape of fully expanded leaves are a function of developmental as well as growth processes.

On a given tree, they will vary because the same genetic potential has been expressed in differing environments, and also because of within plant competition and internal control systems. When comparing with a neighbouring tree, inter-plant competition, age and inherent differences may be involved as well. A logical simplification is therefore to start an experimental investigation by growing potted plants under controlled environments, keeping the conditions as nearly uniform as possible, except for the factor(s) being studied. Since genetic variability between seedlings is generally large, it is often an advantage to work with a known progeny, or better still to advantage to work with a known progeny, or better still to develop clones of well-rooted cuttings originating from seedlings. Such material can also be invaluable for research on within-tree variation, by allowing the 'temporary' influence of previous position to be separated from the more 'permanent' effects of physiological age or maturity.

Experiments of this kind cannot usually provide direct in formation on what factors are controlling growth and development in the forest setting, but this is not their primary function. Essentially, they generate knowledge about the responsiveness of a species to individual environmental and other factors. After a time, however, a sound scientific basis will be built up, from which it may be possible to predict what are the most likely ecological factors operating in the forest, and how they may interact with each other and with internal control systems.

Further experiments may then be indicated, or the most appropriate prarameters to be used in a theoretical model. In this context, it is important to distinguish between 'ultimate' factors, the various components of the local ecological conditions to which an indigenous species is normally adapted through evolution; and 'proximate' factors,some of which act as 'triggers' or signals that set in train a particular developmental change. It is often imagined that all the trees in the tropical forest are producing new leaves and stems continuously throughout the year. The fact that their shoot growth is typically periodic, with longer or shorter intervals, of activity and rest alternating with each other has been stressed since the classical studies of Schimper (1898) and Coster (1923). Nevertheless, reference to 'continuous growth in the tropics' still persists, particularly in the more journalistic or articles.

On the other hand, Kramer and Kozlowski (1960) go too far in stating that 'no matter how favourable the environment, woody plants do not grow continuously but alternate between flushes or periods of activity or dormancy'. The true position is that seedlings of many species may grow without a pause for a few months or

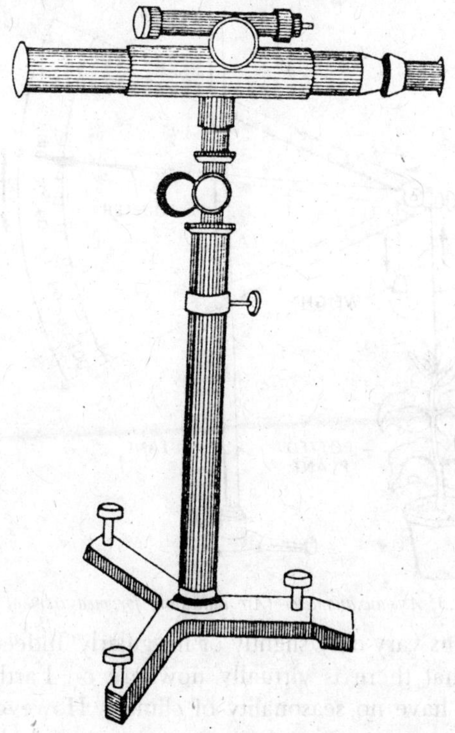

Fig. 11.2. Horizontal microscope for measurement of growth.

years, but the proportion of woody species which continue to do so afterwards appears to be less than 20 percent in Java and West Malaysia. Many of these were shrubs or small trees, and it may well be that the only groups in which continuous growth of large specimens is common in the natural habitat are the tree-ferns, conifers and palms. Once it is accepted that periodicity of shoot growth is typical of the dominant members of the forest community, the important question follows : are their periods of growth and rest appreciably synchronized or seasonal, or do they occur more or less at random?

The latter view was in fact taken by the pioneer Danish ecologist Eugen Warming (1895; 1909), who considered that 'there is no periodicity in the life of the forest as a whole'. However, subsequent work shows that it is quite usual to find seasonal peaks and troughs in flushing, leaf-fall, flowering and fruiting, not just in climates with a pronounced dry season, but in rain forests where

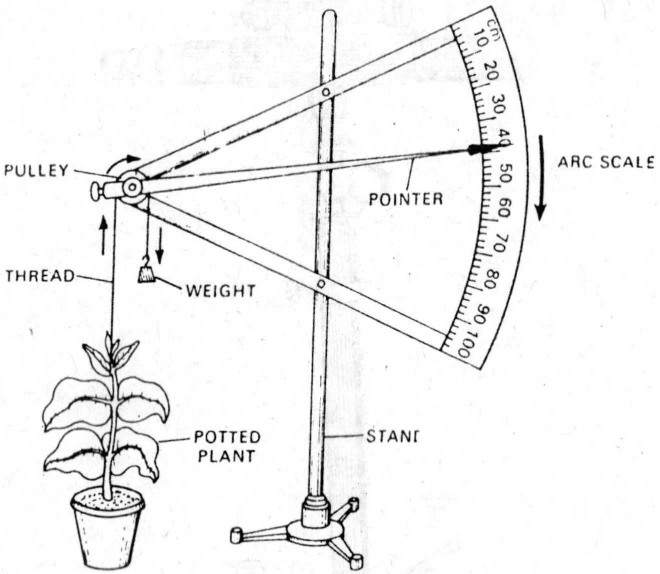

Fig. 11.3. Arc auxanometer (Arc indicator) for measurement of growth.

the conditions vary only slightly or irregularly. Indeed, Lieth (1974) considers that there is virtually nowhere on Earth that can be assumed to have no seasonality of climate. However, in order to show whether trees respond to seasons, repeated observations on tagged shoots over several years is needed, because of the presence of so many species, and considerable variation between trees of different ages, individual specimens and even parts of a single tree. Some of the seasonal, non-seasonal and irregular patterns that have been detected are described in the following sections, which deal with the growth and development of the main organs of tropical trees.

Bud-break

The first sign that shoot growth is about to start after an inactive period is often an elongation of minute leaves in terminal and some lateral positions. This may be preceded by the enlargement of the bud-scales where these are present. The subsequent, usually rapid phase of leaf expansion and internode elongation is generally referred to as flushing, and in some species is quite a striking feature of the forest which is frequently described in the literature. However, it is clear that the important physiological changes leading to renewed shoot extension must occur some time before any visible signs of outgrowth can be detected. In an early study, Simon (1914) concluded that there was probably no month without any flushing in the tropical forests of west Java, where the climate is fairly constant. The same may well be true also of some more seasonal environments, but it probably reflects the variability of the trees in the forest, rather than their uniformity. Many authors have shown that usually there is increased flushing at certain times of year, and less at others.

Some species, particularly in rather uniform climates such as that of Singapore, seem to flush regularly but to be 'out of step' with the calendar months. Others show irregular periods of bud-break, often with a great deal of variability both between and within individual specimens. Nevertheless, a general tendency towards seasonal flushing appears to be a characteristic feature of the majority of tropical forests. An example of a single-peak, annual flushing cycle, which is based on the classical studies of Coster (1928) in Java. Around twice as much flushing was found in November as in June and July, and furthermore bud-break appears to be triggered before and not at the time of the heaviest rains.

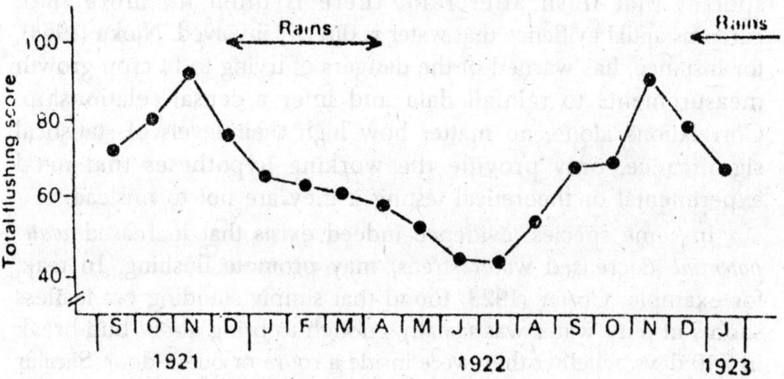

Fig. 11.4. Seasonal variation in the frequency of flushing in a tropical forest a Toeban, W. Java. 52 tree species scored on a 0-3 flushing scale.

Alvim (1964) has drawn attention to the frequency with which peak flushing coincides roughly with the equinoxes, in both Northern and Southern Hemispheres and even very close to the Equator. For example, when cocoa plants were grown under partial forest cover, they showed major flushing peaks around the time of both equinoctial periods. Other trees showing similar behaviour include *Terminalia catappa*, *Peltophorum peterocarpum* and *Cola acuminata*. Some trees show even more frequent recommencement of shoot growth, including cocoa when cultivated in full sunlight where the conditions are less comparable with the natural habitat of this small tree of the forest undergrowth.

Bud-break can therefore occur at any time of year, but is more likely in certain months. The pre-rains flushing already referred to is frequently though not invariably found, and flushing peaks just after the wettest season also occur. At first sight, this seems surprising, since the rains are so obviously the growing season for many crop plants, grasses and other herbaceous species. However, it is evident that for tropical forest trees as a whole this assumption may not be correct. The 'ultimate' factors which may perhaps have influenced the evolution of such seasonal replacement of the majority of the forest's assimilating surfaces will be further considered, in section 5.3, but attention will now be focused on the question of what 'proximate' factors may tigger bud-break.

It is often tacitly assumed that renewed flushing is caused by rainfall, but this clearly cannot be correct in the common situation when outgrowth starts in dry weather. Indeed, even for those tree species that flush after rain, there is often no more than curcumstantial evidence that water is directly involved. Njoku (1964), for instance, has warned of the dangers of trying to fit crop growth measurements to rainfall data and infer a causal relationship. Correlations alone, no matter how high their levels of statistical significance, only provide the working hypotheses that need experimental or theoretical testing if they are not to mislead.

In some species, evidence indeed exists that increased *water potential* (decreased water strees) may promote flushing. In teak, for example, Coster (1923) found that simply standing cut leafless shoots in pure water was usually enough to bring about bud-break in 7-10 days, whether they were inside a room or out-of-door. Similar effects were found in preliminary studies by Njoku (1963) with *Terminalia superba*, *Bosqueia angolensis* and *Millettia thonningii*. The most likely explanation for these results is that the buds were in a state of quiescence or 'post-dormancy' requiring only an increase in water content to flush, rather as a seed may lack just water for germination.

However, the same authors found that cut twigs of *Bombax malabaricum* and *Sterculia tragacantha* could only be forced in water when taken either early or late in the rest period. Thus these species may resemble many woody plants of the temperate zone in showing a period of 'true dormancy' during which the terminal buds will not flush even though conditions are favourable for short extension growth. (Lateral buds can be subject to internal control by terminals): The early and late portions of the rest period, in which outgrowth of new shoots can occur under favourable conditions, are referred to respectively as 'pre-dormancy' and 'post-dormancy'. Much more research is needs before the role played by water can be properly assessed, but two points may nevertheless be made. Firstly, many observations and experiments with well-watered potted seedlings, cuttings and mature grafts indicate that intermittent growth and rest can still occur, suggesting that other controlling factors are involved. Secondly, rainfull could have effects other than relieving water stress,such as the leaching of water-soluble inhibitors from the buds, or indirectly through the sudden temperature drops associated with tropical rain storms.

In many ways *temperature* is the most likely environmental factor to be involved in bud-break. It is perceived by all living cells, and affects virtually every chemical reaction, as well as the physical background for life processes. It is also known to have overall developmental effects: in many temperate trees, for example, chilling during the winter at temperatures just above freezing point breaks the true dormancy of their buds, while rising temperatures in the spring control their flushing in the post-dormant condition. It is not yet clear whether there are analogous systems in tropical trees, although in *Acer rubrum*, which has a very wide range from eastern Canada to South Florida, the tropical ecotype has no chilling requirement for bud burst. In cocoa (which produces repeated short cycles of shoot growth) mature cuttings flushed much more frequently in growth rooms at 30° or 31°C than at 23°C. Day temperatures appeared to be rather more important than those at night, but there was no suggestion that diurnal fluctuations as such were involved. Earlier correlations of flushing frequency with field temperature measurements had suggested that bud-break might be promoted when maximum temperatures exceeded about 28°C, and by diurnal fluctuations greater than 9°C.

A more unexpected factor requiring consideration in the tropics is *photoperiod.* It has often been assumed that because the changes in day-length occurring near the Equator are relatively small, plants might well be insensitive to them. However, it is equally possible that they might be more sensitive to small changes than species of the temperature zone, and indeed some rice varieties have been claimed to detect differences of 5 minutes. It is also known that some tropical animals are sensitive to photoperiod. Regarding woody plants, Alvim (1964) suggested that cocoa at 10-15°S showed a period without flushing because the days were shorter at that time. However, increasing day-lengths in a plantation to 15-16 h by means of bright lights did not stimulate flushing. In unpruned tea bushes growing about 27°N, on the other hand, promotion of bud-break did occur when the short natural day-lengths of November-March were extended to 13 h with weak supplementary illumination.

Experiments in controlled environments with various tropical forest tree species have shown clear sensitivity to photoperiod. Leafy seedlings of *Hildegardia barteri* remained dormant and did not flush under short-days (9 h 10 min). However, bud-break did occur when most of the leaves had become senescent and/or had been abscised, or where the plants had been leafless at the beginning of the experiments. Under long-days (17 h 10 min.), flushing took place whether leaves were present or not. Bud-break inmature cuttings of *Cedrela odorata* was also prevented by short-days unless the mature leaves had been shed. Interestingly, *Bauhinia acuminata* plants that showed earlier leaf senescence under short-days also resumed shoot elongation more quickly than the still leafy seedlings growing under long-days.

It would appear that the buds in these experiments were predormant; prevented from flushing through inhibition by the mature leaves. They could be released either by transfer to a favourable photoperiod or by loss of the leaves. In this connection, it is interesting to note that out-of-season flushing can be stimulated when *Brachystegia laurentii* is defoliated by caterpillars and that experimental defoliation induced bud-break in *Couroupita guianensis* grown in a glass-house at Heidelberg, Germany.

Besides these environmental factors and the inhibitory influence of mature leaves, it is likely that in some species endogenous rhythms are involved, though it is not necessary to invoke general hypotheses of long-term, yearly rhythmic cycles to account for the regular annual occurrence of bud-break. The internal control systems of cocoa have been particularly studied, and it has been found that periods of rapid root growth alternate with shoot flushes. When portions of the root system were removed, budbreak was delayed, suggesting that the close relationship between root and shoot activity might involve growth substances. Moreover, application of abscisic acid (ABA) to the leaves of the recent or previous flush also delayed bud-break.

When ^{14}C-ABA was applied to the leaves, it accumulated at the shoot apex, particularly during the dormant period. Conversely, applying the gibberellin GA_3 or the cytokinin zeatin promoted flushing, the radioistopes of these growth promoters accumulating at the apex most markedly during bud-burst and leaf-expansion. Further research is needed on internal factors, which might for instance control bud-break via the leaf-exchanging habit or prevent flushing in species where the vegetative buds do not grow out while a branch is flowering, such as *Ceiba pentandra* and *Hildegardia barteri*. Carbohydrate levels may be important as well as growth substances, for Fink (1982) has shown that the starch levels in stemwood of ten Venezuelan tropical tree species declined during flushing and increased again afterwards.

Rate of Stem Elongation

Two processes are involved in the production of new stems: the formation of additional nodes and the elongation of the internodes between them. Sometimes the new nodes (together with foliage leaf and bud-scale primordia) remain as a resting bud after their initiation by the shoot apical meristem, and internode elongation occurs later on. The stem units produced in a flush shoot may all have been pre-formed, or the apex may continue to form additional nodes after bud-break, so that 'fixed growth' is followed without a break by 'free-growth'. When the shoot is growing continuously for an extended period, new nodes are being formed and internodes extending without a break. This section is concerned with the many factors affecting the rate at which these processes occur, including the production of new leaves. Those influencing leaf expansion, which usually takes place at the same time, are considered. The rate of stem elongation in some tropical woody plants can be extraordinarily high.

Certain species of bamboos may even elongate pre-formed tissue at almost 1 m per day, and vines and lianes can also grow very fast. On a longer time scale, early growth rates of young trees under natural conditions in Zaire, the three climax species increasing in height at an average rate of 0.3-1.2 m per year, and the three pioneers at 1.2-3.7 m per year. It is also noticeable that the former

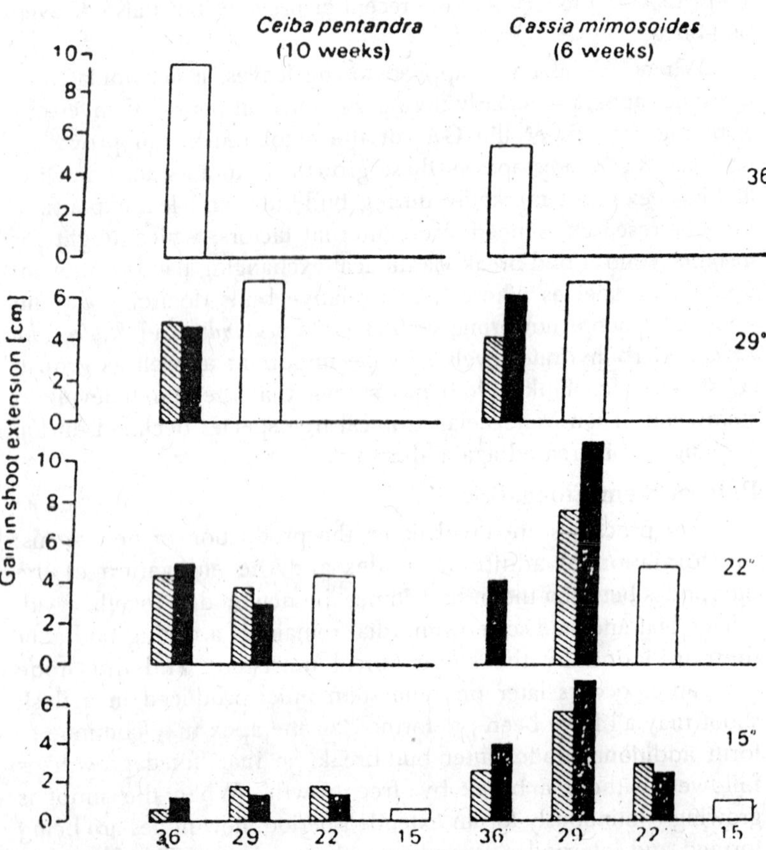

Fig. 11.5. Height growth of young trees under natural conditions in Zaire. Full line–canopy species: (1) Scorodophloeus zenkeri; (2) Oxystigma oxyphyllum; (3) Gilbertiodendron dewevrei. Dotted line–secondary species in clearings; (4) Calancoba welwitschii; (5) Terminalia superba; (6) Musanga cecropioides.

group increased their height increment as they got older, whereas rates steadily declined in the latter. In their study of over 1,500 trees during early succession of cleared forest in Ghana, Swaine and Hall (1983) found that after 5 years the majority of the late seral stage trees had grown at less than 0.5 m per year. Conversely, there were 8 species of pioneers that had exceeded an average rate of 2 m per year. Rates of height growth in the first year can be as much as 3-4 m for such heliophilous species, when thriving as seedlings or coppice shoots in large clearings or plantations. Exceptionally, 8-9 m in the first year has been recorded in *Albizia falcataria. Sesbania grandiflora, Eucalyptus deglupta* and *Trema micrantha,* the latter exceeding 30 m in 8 years. Although height growth in tropical trees can clearly be very rapid indeed, it is easy to exaggerate the general growth rates in the forest by concentrating attention on the performance of the leading shoots of a few species in particularly favourable conditions.

Most of the other shoots on a tree grow more slowly, and by the time the middle and upper tree levels are reached, stem extension, even of the most vigorous shoots, may well be reckoned only in centimetres per year. At the lower tree level, woody plants often survive with minimal increase in height for many years, while growth in small gaps is ususlly much less than in the open. The rate of stem extension can be affected by all the environmental factors discussed as well as by others such as the force of gravity and the presence of atmospheric pollutants. Not surprisingly, *water availability* is important, rates often being affected by quite small deficits. Currently expanding internodes can be inhibited, and reduced numbers of nodes may be formed during rest periods.

Cell expansion in plants clearly depends directly on turgor, while photosynthesis, metabolism and growth often slow down because water deficits stimulate an abscisic acid-induced closing of stomata that may persist for some days, even if the stress has been relieved. Daily fluctuations in growth rate of stems, as well as longer term effects, can result from alteration in plant water potential. Sixty years ago, Coster (1927) used an automatic recording technique to follow in detail the elongation of a young tropical bamboo shoot. During the night it grew at 13 mm h^{-1}, but there was a very pronounced check to extension growth during a hot day. The following day was also hot and sunny, and by 11.30 h the rate had fallen to 5 mm h^{-1}. At this point, he removed the four tall mature canes growing from the same clump, whereupon before noon the growth rate increased to 16 mm h^{-1}. Since the change was so large and so rapid, the conclusion is almost in escapable that elongation had been retarded by temporary water stress, and that it increased because the main transpiring surfaces had been removed.

In contrast, Coster found that the vine *Aristolochia gigas* grew at a rather steady rate throughout a night and a hot day, while in the liane *Congea villosa* day-time growth exceeded that made in the night. In view of such diversity of response, it would be unwise to make any general assumptions that shoot growth is typically greater at night, as is sometimes done.

Diurnal fluctuations of *air temperature* in the tropical forest can sometimes be considerable, particularly in the upper canopy. Seasonal changes are often less pronounced or minimal, and comparison with temperate zone conditions has led some to conclude that temperature might not be a particularly important factor. However, experiments with well watered potted plants in controlled environments indicate quite clearly that many tropical tree species are particuarly sensitive to small temperature differences. Thus *Ceiba pentandra* seedlings kept at 36°C grew 23 times faster than others from the same batch places at 15°C. In several experiments, increasing only the night temperature by 5-10°C promoted shoot elongation rates substantially. Cocoa appears to be a particularly sensitive species, for in one trial raising only the day temperature by 3.5°C stimulated shoot growth of 250 percent.

The full range of temperatures in which shoot growth occurs has not been thoroughly explored. Minimum temperatures may often lie in the range 12-15°C, below which many tropical plants in any case show chilling injury *Guarea trichilioides* and *Avicennia marina* appear to have a high minimum temperature of 21°C for shoot growth. Maximum temperatures for shoot elongation (and for survival) often seem to lie between 40°C and 50°C. As a general rule, the optimum temperature for shoot growth lies nearer the maximum than the minimum end of the range. A value of around 30°C was suggested for young seedlings of *Ceiba pentandra* and *Leucaena leucocephala*, which is perhaps rarely likely to be experienced in nature. Another feature of the shoot elongation of *C. pentandra* is that it is more rapid with constant rather than fluctuating temperatures.

Moreover, if the pairs of shaded and solid histograms are compared, it is clear that interchanging the day and night conditions does not make a great deal of difference. However, in the small woody coastal plant *Cassia mimosoides*, the warm day/cool night regimes were generally more favourable, and alternating temperatures of the led to more growth than constant conditions. The optimum regime appeared to be 29°C day/22°C night, close to conditions that might often be experienced in nature. Similarly, young coffee plants made most shoot growth with the combination 30°C day/23°C night.

Table 11.1 Effect of night temperature on shoot elongation of seedlings of tropical forest trees.

Species	*Night temperature (°C)*			
	Lower	*Higher*	*Rise in temperature*	*% Increase in shoot growth*
Cedrela odorota	20	30	+ 10	+ 94
Triplochiton scleroxylon	20	30	+ 10	+ 65
Ceiba pentandra	26	31	+ 5	+ 74
Ceiba pentandra	31	36	+ 5	+ 6
Bombax buonopozense	26	31	+ 5	+ 72
Bombax buonopozense	31	36	+ 5	+ 21
Gmelina arborea	26	31	+ 5	+ 140
Gmelina arborea	31	36	+ 5	– 14

Although cooler nights may well conserve assimilates, it should not be assumed that most species will necessarily be similar to the two just described. Even in *C. mimosoides*, when the day temperature is kept at 15°C or 22°C, shoot growth is promoted when nights are 7-14°C warmer than the days. *Gmelina arborea* apparently has a night temperature optimum for leaf production near to 36°C, while it is probably around 31°C for internode elongation of *Ceiba pentandra.* Some species actually have a higher night temperature optimum than that for the day, such as the tropical herb *Saintpaulia ionantha* and some provenances of the subtropical conifer *Pinus ponderosa.*

One way in which warm or hot nights might perhaps stimulate shoot growth would be if they promoted leaf production and/or expansion, and thereby increased the plant's subsequent capacity for photosynthesis. The rate of elongation by shoots may also be affected by the photoperiod under which the tree is grown. Notonly are at least 14 tree species sensitive to this factor, but the effects can be quite large. For instance, shoot growth in *Terminalia superba* seedlings was trebled by increasing the day-length from just over 9 h to the value found in June near the coast of West Africa. This species is so sensitive that statistically significant photoperiodic effect could be detected after as little as three days treatment. The effect involved both internode elongation and the number of nodes and leaves produced.

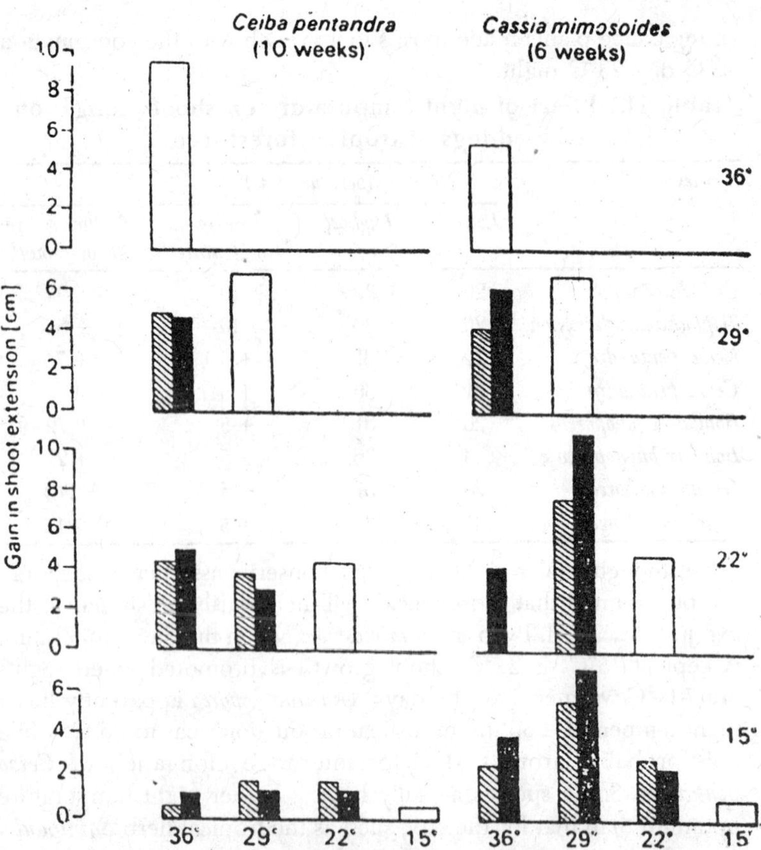

Fig. 11.6. Effect of constant and fluctuating air temperatures on the rate of shoot elongation in Ceiba pentandra and Cassia mimosoides seedlings. Open column–constant temperatures; solid column–fluctuating temperatures, with cooler night; shaded column–fluctuating temperature, with warmer night. Both day and night regimes lasted 12 h; day-length=13.2 h. Almost all Cassia plants in 22°C day/36°C night died before week 6.

The latter shoot growth parameter, and also short and total dry weights, were especially stimulated in *Chlorophora excelsa* when day-lengths were lengthened to about 17 h. All the treatments in these experiments recieved similar photon flux densities of shaded natural sunlight in glasshouses and fluorescent light in growth rooms, the day-lengths being extended as necessary by supplementary incandescent illumination too dim to influence photosynthesis appreciably. Thus it is clearly established that many tropical trees are sensitive to photoperiod, influencing either the rate of leaf production or internode extension or both. Other species in which long-days promote shoot growth include *Pinus caribaea* var. *hondurensis*, *Rauvolfia vomitoria*, *Hymenaea courbaril*, *Bauhinia acuminata*, *coffee* and *cocoa*; while extension of *Gmelina arborea* shoots appears to be unaffected by changes in day-length.

Table 11.2. Effect of day-length on shoot elongation of seedlings of tropical forest trees.

Species	*Photoperiod (h)*			
	Shorter	*Longer*	*Increase in day-length*	*% increase in shoot growth*
Cedrela odorota	11.0	14.5	+ 3½	+ 51
Theobroma cacao	12.5	14.5	+ 2	+ 31
Terminalia superba	9.2	13.2	+ 4	+ 195
Terminalia superba	13.2	17.2	+ 4	+ 59
Chlorophora excelsa	9.2	13.2	+ 4	+ 22
Chlorophora excelsa	13.2	17.2	+ 4	+ 355
Ceiba pentandra	9.2	13.2	+ 4	+ 130
Ceiba pentandra	13.2	17.2	+ 4	+ 1
Bombax buonopozense	9.2	13.2	+ 4	+ 82
Bombaxbuonoponzense	13.2	17.2	+ 4	– 7
Hildegardia barteri	9.2	17.2	+ 8	+ 200

Synergistic effects between day-length and night temperatures have been found in *Terminalia ivorensis*, and in *Ceiba pentandra* where 50 percent more leaves were formed when both promotive factors were applied, but neither had a significant effect when given alone. More complex interactiosn appeared to be involved in the control of internode length and total shoot elongation,which were both strongly influenced by day-length and night temperature. Thus the question as to whether trees under natural conditions will be influenced by photoperiod is difficult to answer. However, it is clear that shoot growth of a number of species can be substantially modified, increases of up to 75 percent per hour having been demonstrated (Table 11.2). Some evidence exists of sensitivity to alterations in the range experienced in the humid tropics, where natural day lengths vary by just over 30 minutes at a latitude of 5°C, and by more than an hour at 10°C. Besides possible photoperiodic effects, the shoot growth of forest trees may be affected by the light quality received, especially by the red/far-red ratio. The photon flux density is clearly also an important consideration, both indirectly through carbon assimilation and perhaps also directly. Effects of shading on growth are often profound, as has been shown for instance with seedlings of several species of Dipterocarpaceae.

It should be remembered that differences may not always be due to light itself, since treatments may also produce modifications of light quality, leaf temperatures and water potential. Clearly, many other factors will affect the rate of shoot growth. For instance, there was a very high correlation (r= 0.99) between the level of phosphorus fertilizers added and the height growth of *Terminalia ivorensis* seedlings. Besides external conditions, it is important to bear in mind that other processes which are proceeding simultaneously in th tree, such as root growth, may affect the shoots, as pointed out more than 25 years ago by Kursanov (1960) regarding the aerial roots of an epiphytic *Ficus* sp. Flowering or fruiting may divert carbohydrates away from shoot growth,or perhaps modify it through the mediation of growth substance.

It is also possible that sizeable differences may arise through the presence or absence of specific mycorrhizal fungi associated with the root system. Genetic differences between individuals in a population or species can be surprisingly great, as is demonstrated by the very substantial clonal variation in height growth that is typically found between clones of tropical forest trees, which in *Triplochiton scleroxylon* may be related in part ot different net assimilation rates. Higher photosynthetic rates per unit leaf area have also been reported for *Ceiba pentandra* than for *Musanga cecropiodes*, *Terminalia ivorensis* or *Chlorophora excelsa*.

Onset of Bud Dormancy

It is often possible to detect that a flush or a longer period of extension growth is coming to an end by observing that there are few or no expanding leaf primordia at the shoot tips. Once the terminal bud has entered some form of dormancy, stem elongation is confined to the completion of the extension of the last few internodes. The shoot apex itself is not necessarily in active, and often continues to initiate new leaves and 'stem units' during the dormant period, at least for some time. Besides foliage leaves, the apex may produce but scale primordia; and there are some species in which dormancy is signalled by the abscission of the whole of the terminal shoot tip, leaving lateral buds to continue axial growth. Such shoot tip abortion is a common feature of the shoot growth of *Cedrela odorata*, *Xanthophyllum curtisii* and *Citrus*. Thus the height of a tree may depend on three additional considerations, besides the factors influencing the rates of node production and internode elongation discussed in the previous section. These are the duration of the period of shoot elongation, the number of preformed nodes within resting buds and the duration of the period of bud dormancy.

Fewer ecological studies have covered the time of cessation of shoot elongation than the more obvious phenomenon of flushing. What information there is suggests that in plantation crops that grow by repeated flushes, the duration of elongation may be 6-7 weeks, for example in cocoa, and as little as 1½-2 weeks in rubber and mango. In the forest, it seems likely that periods of shoot extension of 1-2 months may become fairly genera after the seedling and sapling stages. The onset of bud dormancy will therefore occur at different times of year in trees which flush intermittently or irregularly, but in more regularly flushing species it will tend to be seasonal.

In Ghanaian forests, for example, the commonest month for the ending of shoot elongation appears to be April which is in the transitional period between the main dry and rainy seasons. Even in a species such as cocoa, which may flush several times a year it was observed several decades ago that most of the growth of mature cocoa occurs in the drier months ... while during the main wet season when conditions of water suply and humidity are most stable,little or no growth occurs.

In a tropical dry forest in Costa Rica, production of new leaves and shoots declined as the rainy reason advanced, and in a wet evergreen forest it decreased as the drier part of the year passed.

Although there are exceptions, it is remarkable that many tree species should be ceasing shoot elongation at a time when the 'growing season' is just starting. There is a parallel here, however, with the situation in the north temperate zone, where many species cease producing new leaves in May, June or July, when the environment for growth seems ideal. As was pointed out a century ago, trees often re-start shoot growth under less favourable temperature and other conditions than those obtaining when they stop elongation. However, perennial plants might well gain an important selective advantage by producing new foliage in the latter part of the dry season. Unlike annual grasses and many short-lived crops,their recently expanded leaves, at the peak of their photosynthetic capability, will then be present at the time of the greatest daily totals of photon cloudiness in the rainy season, and sometimes by dust or haze at the height of the dryest season, giving sunlight maxima around the equinoctial periods. An additional 'ultimate' factor that could be operating is that a leaf expanding in the latter part of the dry season may be subject to less overall water stress than one formed at any other time.

If the dormant period in many mature trees is as long as 10-11 months, this will curtail the yearly height increment even when the rate of elongation is rapid. In seedlings too, growth of *Triplochiton scleroxylon* and *Hildegardia barteri* fell behind that of *Musanga cecropioides* mainly because they became dormant for a while, when the *Musanga* grew continuously. This may also be one reason for the decline in height growth often found with increasing age : for example 3-year-old *Bombax buonopozense* stop height growth for only 3-4 months, whereas older trees are dormant for 9-10 months.

The physiological signals which induce a change from active shoot growth to terminal bud dormancy appear to include both external and internal factors. For instance, it has been demonstrated using controlled environments that shoot extension can be halted in a number of tropical tree species by reducing the *photoperiod.* Seedlings of *H. barteri* stopped extension growth quickly under 11½ h days, and more slowly in 12 h days. When given 12½ day-lengths, however, they grew continuously throught out the 26 week experimental period. Very rapid onset of dormancy also occurred inleafy *Cedrela odorata* plants exposed to short days, but theprocess was slower in *Terminalia superba* seedlings, about 50 percent of the treated trees stopping temporarily, while the other half continued to grow slowly. A proportion of *Ceiba pentandra* seedlings grown under short photoperiods also showed temporary dormancy. The length of the shoot extension period was decreased by about 6 weeks in *Bauhinia acuminata* when day-lengths were shortened, but a reduced daily light energy supply or temperature differences may have been involved as well as photoperiod in this experiment using a thick tarpaulin screen to apply the short days.

Cooler night *temperatures* tended to induce bud dormancy in *Gmelina arborea*, some plants stopping height growth with nights of 26°C, but none doing so at 31° or 36°C. Cool nights caninteract with short-days, as in *Bombax buonopozense.* Under natural conditions in the forest, a reasonable hypothesis might be that warmer nights and longer days might promote faster and more prolonged shoot extension, while cooler nights and shorter days could lead to reduced shoot growth and to cessation of height growth. However, it is very probable that other external factors will hasten the onset of terminal bud dormancy,for example reduced light intensity, a low red/far-red ratio, water stress and shortage of mineral nutrients.

Internal conditions within the tree also appear to be important incontrolling the onset of dormancy. Species which show repeated flushed of growth, such a cocoa, tea, mango and rubber, are good examples of this, although it should not be assumed that they will be unaffected by their external environment. One possibility is that a rapidly growing shoot might stop simply because new leaves were not being initiated quickly enough by the apes, which may happen when tea is grown in the tropics. Alternatively, each shoot flush may be confined to those leaves and stem units pre-formed in thebud. In this case, dormancy is as it was predetermined, perhaps withan 'articulate' shoot morphology reflecting the cyclic production of bud scales and foliage leaves.

A series of detailed studies on the photosynthetic capacities of different leaves of cocoa, and the translocation of assimilates within the plants, has shown that first year plants are flushing rapidly and expanding many new leaves, they are using much more carbohydrate than they are producing. Thus, it is possible that lack of assimilates might lead to cessation of shoot elongation; for example, successive removal of expanding leaves can prolong shoot elongation periods, for instance in rubber and in *Couroupita guianensis.* In temperate species of the tropicallly originating genus *Quercus*, it is even possible to turn an intermittently flushing tree into one which grows continuously. Some instances are also known in which individual specimens of intermittent tropical species exhibit continuous growth for long periods, as invigorous leading shoots of unplucked tea. Examples such as 'lamp-brush' rubber and 'foxtail' pine may be due to genetic differences, perhaps becoming expressed when the plant is grown out of its natural environment.

It is likely that there are often negative feed-back control systems that change or maintian the phases of shoot elongation or terminal bud dormancy. The interactions between roots, leaves and buds may well involve changes in the endogenous levels and distribution of both growth promoters and inhibitors, and changes in levels of these have been detected at different times in the flushing and dormant cycles of cocoa. In this species, internal water stress was apparently not involved, but it has been implicated in studies with other tropical trees.

Growth of Leaves

Young, growing leaves quite often attract attention in tropical forests because their colour or appearance is different from that of the mature foliage. They are often a lighter shade of green,because the formation of chloroplasts and the synthesis of chlorpohyll *a* and *b* has only just started. Sometimes they are very pale– for example in the half-expanded flush shoot of cocoa grown under shade–and they can even appear white, yellowish or blue in some species. The commonest colouration is red, generally caused by anthocyanins in shade to a bright colour which masks any green present, as seen for example in *Cynometra ananta* and *Carapa procera.* In some species, such as *Amherstia nobilis,* the new flush of pale-coloured young leaves all hang downwards for some time, which may be an adaptation that reduces waters tress until the leaves are capable of active photosynthesis.

A feature of tropical trees which is fairly common is the capacity for active leaf movements. The orientation of leaves and leaflets can be altered, not just by growth curvature of the petioles, but at swollen *pulvini* or 'leaf-joints' at the base of the organ in question. The turgor changes which are responsible for the movement can be relatively rapid, so that the plant reacts quite promptly to a change in environment. The best-known example is the 'sensitive plant', *Mimosa pudica,* which is a cosmopolitan weed of open spaces in the tropics. here the reaction time to the stimulus of a leaflet being touched is of the order of a second or less, and the leaflets also close (more slowly) in darkness. Light at wavelengths of 403 nm (violet) or 726 nm (far-red), but not at 585 nm (yellow) or 656 nm (red), has been found to cause opening of leaflets, but this was only effective when given between 0600 h and 1600 h, indicating the existence of an endogenous diurnal rhythm. Applied indole acetic acid also led ot opening, even in darkness, and autoradiography showed that it could spread from the rachis throughout the pinna in 4 minutes. In *Albizia,* a red/far-red seversible system has been shown to control opening and closing of the leaflets, the turgor changes in the pulvinus apparently being triggered by a massive influx or efflux, respectively, of pottassium ions.

In many leguminous trees, the leaflets close at night, and sometimes also in response to other stimuli–for instance those of *Brachystegia spicaeformis* have been shown to close during the day when leaf surface temperatures surpass 32-34°C while dark closure was promoted and the duration of the closed phase increased in *Cassia fosciculata* when temperatures were reduced form 35° to 15°C. Movements are not confined tothe Leguminosae, being noticeable at night in *Triplochiton scleroxylon*, for example, while an unusal example is the formation of 'night buds' in *Espeletia sahuttzii* at 3600 m in the Vanexuelan Andes. The rosettee of leaves close around the apical buds at night, which has been shown by measurement with thernoccuples to protect the young leaves from freezing temperatures, and from rapid heating up just after sunrise, either of which can be lethal to them, but not to the older leaves.

The physiological significance of changing leaf orientation probably lies in the achievement of a successful balance between maximal photosynthesis and minimial water stress. It addition, as noted, it has important ecological effects on other leaves, by altering the proportion and the quality of light that reaches them at different times. Similar consideration apply to the regular occurrence of upper canopy leaves that are more or less permanently oriented at a steeply inclined angle, rather than horizontally. The leaf mosaic can also change rather abruptly under a previously leafless tree crown that suddenly produces a new flush of leaves, and there are continual, slighter modifications resulting from the expansion and loss of individual leaves,or grazing by herbivores.

Seasonal changes which affect the forest as a whole may also occur. Peaks of new leaf growth generally occur shortly after those for bud-burst, because the expansion of leaves typically occurs at the same time as internode elongation. Sometimes the two processes appear to be closely linked; thus, for example, in tea a single deciduous bud scale amongst expanding foliage leaves will have a correspondingly short internode proximal to its insertion. Often the first foliage leaves on a new shoot are smaller and may be similarlyl associated with shorter stem units. On the other hand, the first internode of a sylleptic branch is typically much longer thanteh subsequent ones, but the leaves are usually of a similar size. In *Terminalia ivorensis*, for example, experimentally reducing the are of young expanding leaves did not significantly affect the internode lenghts.

Growth rates of leaves are often rapid, especially at the time when they are about half their half size. Leaflets of *Amherstia nobilis* can increase in length by 18mm per day, and petioles by 41mm per day. Leaf growth in this rapidly flushing species is sometimes completed in less than two weeks, although it takes another fortnight for the leaves to spread out horizontally. The rachis of the large fern *Angiopteris evecta* can even extend at 90mm per day, and Coster (1927) found that leaf night when the weather was sunny, but not on a cloudy day. In the great majority of species, leaf growth is determinate, so that in some species with large,compound leaves each one may take several weeks to complete its growth. In a few instances, such as *Guarea guidonia* and *Chisocheton* spp., the leaf-tips remain meristematic, producing new leaflets continuously or periodically over a longer period of time.

Leaf growth usually involves both cell division and cell enlargement, and the final size reached, and sometimes also the leaf shape, can be affected by a range of external, internal and genetic factors. A striking example of unequal growth of adjacent leaves is provided by plants showing strong *anisophylly*, for example the tropical American *Columnea sanguinea*, the Malaysian *Anisophyllea trapezoidalis* and the West Africal *A. laurina*. The branches have two rows of large leaves below, and two of very small leaves above, while on vertically growing main stems all leaves are large. A possible hypothesis might be that the horizontal position of the branch in relation to the gravitational field imposes a *gravimorphic* effect, so that either inhibition of growth occurs inleaves inserted on the upper side, or the two types of leaf arise from primordia that contain very different numbers of cells. Less pronounced anisophylly is found in other genera, for example, *Alstonia* and *Gmelina.*

Water stress is one of the factors likely to reduce leaf growth, possibly through the production of large amounts of ABA which may close stomata. In cocoa, for instance,when the available water content of the soil had decreased by a third, leaf growth had slowed considerably. By the time the moisture supply had decreased by a further third, expansion had stopped entirely, so that leaves only reached a length of 25-60 mm, while the contrly leaves on plants in moist soil were twice as long and still growing.

Shading commonly influences leaf development, generally increasing the rates of growth and final sizes, and sometimes modifying leaf shape. In cocoa and coffee, for instance, which are small trees of the forest understory, the leaves that expand in full sunlight are often smaller and yellowish-green, while those growing in shade tend to become larger and darker green. Increasingly, it has been found that cocoa leaves in full light grow considerably larger if the lower mature leaves on the same shoot have been removed (Jan Krekule, unpublished data). The implication is that the effects of light intensity and quality may be indirect perhaps operating via water stress. Even in rigorously controlled environments, it is quite difficult to distinguish between genuine control by photon flux density and secondary effects such as water stress or leaf temperature changes, or tertiary responses such as the closure of stomates. In shading trails in the field, differing air and soil temperatures also have to be taken into account, together with variation in atmospheric saturation,vapour pressure deficits and any root competition from shade trees.

Temperature and photoperiod can affect leaf growth; thus seedlings of *Triplochiton scleroxylon* grown under 11 h days + 20°C nights produced leaves that were 35-40 percent shorter than on plants receiving either 14½ days, or 30°C nights, or both. *Ceiba pentandra* and *Gmelina arborea* also showed an interaction between these two factors, while in *Chlorophora excelsa* leaves were 50 percent longer when grown under long rather than short-days. The effects on leaf area were likely to have been even greater than is suggested by these assessment of linear dimensions. Leaves of a paler, yellowish-green colour tended to occur when *Triplochiton scleroxylon, Terminalia ivorensis* and *T. superba* plants were grown in cool day/hot night regimes, in comparison with other combinations of fluctuating or constant temperatures.

There are likely to be many other influences upon the development of a leaf, including such factors as the availability of nitrogen and mineral nutrients, and the influence of herbivore damage to it (and elsewhere in the same or neighbouring trees). Changes in leaves which occur with the increasing physiological age of the tree are considered, while the longevity of individual leaves is discussed in the section which now follows.

Leaf Senescence and Abscission

A leaf typically reaches a peak of photosynthetic capacity around the time that expansion ceases, thereafter showing a gradually decline. Sooner or later a point is reached when it rather suddenly begins to turn yellow, brown, or less commonly red, and is then actively shed from the tree by completion of a separation layer in the abscission zone at its base. This change of colour is termed senescence, and it appears to be mediated by alterations to the balance of several endogenous growth substances. It is the terminal irreversible phase in the functioning of the leaf, which signals the whole sale breakdown of chlorophyll, ribonucleic acid and protein, and the rapid translocation out into the stem of some but not all of its organic and inorganic nutrients.

Abscission normally follows quickly, and the protective layer of the leaf-scar becomes an effective barries to invasion by pathogens, except for instance in cocoa where it may become a centre for *Phytopthora* infection. Lack of leaf abscission, with shrivelled leaves still retained on the stem, is often a sign that sudden stress has prevented the usual sequence of changes; retention of dead leaf basesis, however, a normal feature of some plants, including palms and tree-ferns. In *Cupressus* and *Pinus*, whole leafy branchlets areshed, rather than individual leaves, while in *Taxodium* leaves are shed from long-shootss, but the entire short-shoot is abscised.

A very wide range of factors appear to promote leaf-fall in trees,including lowered light intensity,changed temperature and photoperiod, mineral nutrient deficiency and water stress. Older leaves are much more likely to be abscised than those which have recently expanded,and it is probable that there are also interactive effects between different organs. For instance, container-grown cocoa plants subjected to drier conditions abscised a higher proportion of their leaves, and there was also evidence for an influence of the shoot/root ratio. Correlative effects were also suggested in an experiment with *Hildegardia barteri* seedings, in which, unexpectedly, more leaves were abscised under long-days than in short-days.

In the former treatment, new leaves were being produced and expanded throughout the experiment, whereas the terminal buds on plants in short photoperiods were dormant for a long time. Table 6.3 shows what appears to be a ,direct, additive effect of short-days and hot nights in enhancing leaf abscission in *Bombax buonopozense.* Two stages of leaf loss were detected in *Plumeria acuminata*, both of which were inhibited by light interruptions during the dark period. An interesting example is that of a disease of coffee caused by the fungus *Omphalia flavida*, which leads to premature leaf-fall and greatly reduced yields. Abscission appears to be stimulated not so much by leaf damage *per se* as by disturbance of the balance of hormones. One the flow of auxin from leaf to stem is reduced, the separation layer at the base of the petiole completes its development, and the leaf is abscised within week.

Table 11.3 Interaction of day-length and night temperature on leaf abscission in *Bombax buonopozense* seedlings. Mean number of leaves lost from each plant in a period of 4 months.

	Long-days	*Short-days*
26°C nights	0.6	1.5
36°C nights	3.0	6.0

Leaf-fall occurs all the year round in tropical forests, but peaks and troughs are usually found. The classical studies of Bray and Gorham (1964) on seasonal production of litter (about two-thrids of which was foliage) showed that in the rain forests of Colombia and Ghana, for example, most litter accumulated in March and least in July. In evergreen seasonal forest, it appears that leaf-fall may peak in the first half of the dry season, as some of the trees become completely leafless; evergreen specimens may also lose a proportion of their leaves at this time. Fluctuations in leaf-fall have many ecological implications, for instance regarding that the nutrient and water status of the soil, the photon flux densities reaching the lower tree layer, and the amount of food available to herbivores.

Every species does not behave in the same fashion of course, individual specimens in a population are often not synchronised, and conditions may vary from year to year. Nevertheless, regular annual leaf-fall has been recorded, for instance, around February for *Instia palembanica* in Malaysia. In the life of an individual tree, the partial or complete loss of its old foliage represents an important change in photosynthetic, transpiring and respiring tissue, except when leaves are produced and lost steadily throughout the year. Rather little attention has been paid until recently to the important topic of the longevity of tropical leaves.

There is no difficulty in knowing the age of a leaf on the majority of temperate zone trees: if they are deciduous species,then any leaves must be of the current year; while in many, though not all evergreens, there is distinctly 'articulate growth', such that the leaves produced in successive years are in separate 'cohorts'. In the tropics, on the other hand, it is usually impossible to know the age of a leaf unless trees are observed leaves. The minimum life-span for undamaged leaves is likely to prove to be about three months, as recorded for the woody climber *Grewia carpinifolia* in Ghana. Although the majority of leaves probably do not last more than about fifteen months, there are some well-documented cases, such as the understorey trees *Drypetes parvifolia, Vepris heterophylla,* and canopy species *Diospyros abyssinica* and *D. mespiliformis,* and conifers of montane forests, in which the life-span is at least 2-3 years.

There is a considerable amount of confusion in the literature over the distinction between the evergreen and deciduous habits in tropical forests. Basically, the question of whether a forest, tree or branch is leafy or leafless is a function of longevity and the relative timing of bud-break and leaf abscission, perhaps modified by herbivory and other damage. Because all shoots on a tree, all individuals of a species,and firm 'leafiness' categories becomes increasingly difficult the larger the unit. Indeed, even within a single shoot, the older leaves may be shed before those more recently expanded,while in compound leaves some of the leaflets may be abscised and other retained. Taking these points into consideration, the following four broad categories are proposed as useful groupings that are based on recognisably different growth strategies, and can be applied at any level:

A. Periodic growth (deciduous)

Leaf-fall occurs well before bud-break; life-span about 4-11 months. In this case, the branch, whole tree or forest is leafless (or nearly so) for a definite interval, varying from a few weeks to several months. Even in rain forest receiving a high rainfall evenly spread throughout the year, canopy trees of *Cordia alliodora* may for instance remain leafless for weeks in the middle of the rainy season in Costa Rica. In seasonal forests in West Africa, some specimens of *Terminalia ivorensis* can remain leafless for more than 6 months, mainly during the drier months. In this type, leaf-fall and bud-break are apparently not directly connected with each other.

B. Periodic growth (leaf-exchanging)

Leaf-fall associated with bud-break; life-span often about 12 (or 6) months. Here the flushing of new leaves starts around the time at which the majority of the old ones are absicised, with in about a week either way. This habit is clearly seen in *Terminalia catappa*, where the old leaves turn red and fall and the simultaneously replaced by new. In this species, the process often occurs twice a year, while in *Entandrophragma angolense*, *Dillenia indica*, *Ficus variegata* and *Parkia roxburghii*, for example, it happens once. In upland forest in Costa Rica, *Quercus* spp. can be seen to shed all the old leaves, and then within a few days to be expanding many new red leaves. It is likely that it is the senescence of the old leaves which provides the physiological signal for bud-break, placing type B responses in a different category from either deciduous or evergreen. The term 'leaf-exchanging' should be distinguished from the somewhat confusing expression 'leaf change' (*Laubwechsel*), which tends to incorporate elements of flushing, leaf growth and abscission. 'Semi-deciduous' is unsuitable because it is often used to imply that a substantial number of trees in a forest are deciduous.

C. Periodic growth (evergreen)

Leaf-fall completed well after bud-break; life-span about 7- 15 months or more. This is a truly evergreen category,which is characteristic of woody plants of the lower and middle tree layers, and some of the upper tree layer. Many dipterocarps are periodic/evergreen, and other examples include *Clusia rosea*, *Fagraea fragrans*, *Mangifera indica* and *Pinus* spp. As in stype A, there may be no direct connection between leaf-fall and bud-break,but it is possible that the expansion of new leaves might lead to senescence of the old through a hormonal control system, by competition for nutrients, or because of shading.

D. Continuous growth (evergreen)

Continual formation and loss of leaves; life-span variable from about 3-15 months. This is a second evergreen class,and is characteristic of some palms, conifers and tree-ferns, and is found in seedlings of many dicotyledonous trees. Examples of older trees include *Dillenia suffruticosa*, *Trema guineensis*, *T. micranthum*, and the shrub *Hamelia patens*. In this type there is, of course, no bud-break, but the rates of leaf initiation, enlargement and abscission may each vary considerably, according to chages in the environment, or to external and internal competitive effects. The 'leafiness' may thus fluctuate, although this was not found to be the case in the red mangrove, *Rhizophora mangle*.

The presence of type A trees even in the most uniformly moist of rain forests is an important point that is often overlooked. It provides a reminder that the microclimate in which leaves in the euphotic layer develop, particularly on the exposed crowns of emergents, is quite unlike the ever constant, equable conditions of journalistic fables, and can be quite stressful. In evergreen seasonal forests, one-third of the tree species may show the deciduous habit, and the timing of the leafless period in Ghananian trees closely followed the rainfall patterns. There were some leafless trees to be found in every months except June, but the proportion of trees without leaves increased to a peak at the height of the dry season in December and January. The percentage declines as flushing occurs in the majority of species in February and March, and falls to a very low level before the onset of the rainy season.

In evergreen seasonal forests, as in other drier formations, there seems little doubt that the chief adaptive value of the deciduous habit lies in the avoidance of severe water stress in the middly part of the dry season. Yet it should not be forgotten that many other trees operating on type B and C strategies endure the same dry seasons in a leafy state, just as various evergreen trees survive the cold winters at higher latitudes, while other species do so leafless. Not just one, but several kinds of developmental strategy commonly evolve in response to environmental or biotic problems. Moreover, drought is clearly not the only stress involved because some species lose their leaves in the rainy reason, perhaps because light intensities are lower than in the summer dry season. It should also be borne in mild that, as Janzen and Wilson (1974) have pointed out, there may be an extra 'cost' attached to being leafless in the tropics, since stored carbohydrate can be reduced by as much as 50 percent because of high respiration rates in the twigs, boles and roots.

Although the usefulness of the underlying feature of 'leafness' is beyond doubt, it has been lessened because of considerable variations in usage and precise meaning of the terms used. The frequency with which the above four categories of leaf and bud behaviour occur provides a basis for improving the physiognomic classification of tropical forests. A useful distinction has been made,for instance, between facultatively deciduous trees, such as *Ochroma pyramidalis* and *Tectona grandis*; and obligately deciduous species,such as *Ceiba pentandra*, *Cordia alliodora* and *Enterolobium cyclocarpum*.

Experimental studies on the relationships between leaf abscission and the growth and development of other organs may also be stimulated if these classes are used, for example the effects of defoliation on bud-break in type B and C shoots. Although not discussed in detail in this chapter, one of the most important consequences of all the different strategies of leaf development and maintenance will be the overall influence they have upon dry weight accumulation in the tree and forest crops. However, some assimilates will be used up in producing the organs discussed in the ramaining sections.

Cambial Activity

A further dimension to growth and development is the increase in thickness of organs through the activity of the vascular cambium. Indeed,the secondary tissues produced by this meristem constitute much the greater part of the plant body in large perennials, except for palms and bamboos. In time, individual tropical trees can become extremely large, with extensive crowns and massive root systems, and diameters at the base of the trunk (exclusive of any buttresses) exceeding 2 m. Details of some big specimens are given in Richards (1952), including an *Entandrophragma cylindricum* in Nigeria a record of a 5.4 m diameter at the base. There is also a record of a 4.1m diameter *Bertholletia* spp. in Brazil, and a 3.8 m *Balanocarpus heimii* in Malaysia. Such trees may have total dry weights of the order of 100 t, and (although not as large as some of the giant redwoods, eucalypts and *Agathis*), they rank amongst the largest living organisms in the world.

Yearly increases in main stem diameter vary greatly in different species, and from one individual to the other. As the trunk grows in size, the cross-sectional area of new tissue required for a given diameter increment gets larger. Thus, for example, what is produced by a 0.1 m diameter tree when it grows 10mm would only increase a 1 m diameter tree by about 1 mm. In middle age, trees in natural forest that have their leaves mostly in the euphotic layer may add around 5-20 mm, but in very old emergents the increment is generally lower, Amazonian rain forest, it was predicted from the stand structure that the average tree in the diameter class 0.25-0.35 m would be growing at about 8 mm a year; for larger trees the increment would decline below 4 mm. In a wet tropical forest in Costa Rica, analysis of 13-year increments of 45 tree and one liane species by growth simulation showed a range of median annual diameter increases between 0.3 mm and 13.4 mm.

Young saplings and pole-stage trees often make slow radial growth, suppressed by shade cast by the canopy trees and competition from their root systems. In gaps and along forest margins, however, the same species may be capable of rapid growth, while colonisers may be even faster growing. In plantation conditions, too, substantial annual diameter increments can be sustained; for example approximately 45 year old *Triplochiton scleroxylon* in line enrichment plantings at Mabak, Cameroon, were still producing 12 mm. Exceptionally in young, dominant individuals, values of 20-60 mm can occur, and 90 mm has even been recorded in *Ochroma lagopus*.

Although there is little problem in relating approximate average diameter increments with age in fast-growing plantations, accurate measurement of the rate and duration of radial growth is usually difficult in tropical woody plants. Increase in thickness is generally the result of activity by two different lateral meristems, the vascular cambium and the phellogen (cork cambium), both of which may be cutting off cells towards the outside of the trunk and towards the inside. Moreover, successive girth or diameter readings do not necessarily show the combined acitivity of the two cambia, even if a standard measuring position is marked on the trunks (generally 1.3 m, on the upper side of the tree if there is an appreciable slope).

In the first place, this is because there may be irregularities present such as ridges or spines,branch and leaf bases or scars, or major convolutions in outline. Secondly, bark may be abscised in various amounts and fashions,sometimes in whole sheets, as in *Eucalyptus* spp, and *Tristania* spp. of the Myrtaceae. Thirdly, tree trunks regularly shrink appreciably during a hot, sunny day, because the xylem sap is under considerable negative pressure. Accurate measurement therefore requires continuously recording dendrometeres, unless all the manual recording can be completed by 0600 h ! The existence of tension in the conducting elements of the xylem in the middle of the day can be simply demonstrated in *Dillenia* spp. and some lianes if the crown has been in sunshine for some time. If a clean cut is made into the stem with a machete, a distinct sound can often be heard as air is drawn in.

In many cases, interest centres on the major component, the porduction of xylem by the vascular cambium. Increment cores or anatomical examination are needed to provide accurate information, and a further problem arises in natural forest from the frequent lack of clear and reliable annual growth rings in many tree species.

The problem of knowing how old tropical trees are introduces inaccuracies, not just for calculating cambial growth rates, but also for every other aspect of the analysis of growth and stand structure in tropical forests.

There is some prospect of newer methods being developed,such as modelling and detailed examination of the transverse surface for small changes in chemical or physical structure. For instance, the $^{14}C/^{12}C$ ratio in carbon dioxide changed during the last 30 years, because of the atmospheric testing of nuclear weapons in the late 1950s and early 1960s, before this was banned by international agreement. This left a characteristic 'thumbprint' in the permanent structural cell components of all woody plants, which varies only slightly according to latitude. For many purposes, simpler methods may be more appropriate, such as the marking of the xylem at a particular points by dyes or damage with pins or nails, or by opening a small annual 'window'. The living cells in the wood often ramain free of starch for a considerble time after they have been formed, which could be used to determine the increment in numbers.

In order to understand more about the physiology of cambial activity, it is probably best to concentrate on species with clearcut, annual rings. There are more of these than is generally realised, including *Homalium tomentosum*, *Pterocarpus indicus* and *Tectona grandis* in South-East Asia; *Bursera simarouba*, *Goethalsia meiantha*, *Swistenia mahogani* and *Taxodium distichum* in Central America; *Ceiba pentandra*, *Chlorophora excelsa*, *Entandrophragma* spp., and Terminalia spp. in West Africa.

Most of these are deciduous or leaf-exchanging species and the former group were shown many decades ago to exhibit dormancy of the cambium. By a combination of careful measurement, experiments and anatomical studies, Coster (1927-28) demonstrated the cessation of cambial activity at the beginning of the leafless period, with divisions occurring again when flushing took place. He also showed that the stimulus for renewed cambial acitivity in leafless plants must arise from the expanding buds, since the cambium remained dormant if they were removed, or if they were isolated from the portion of cambium under study by complete ringing. The influence of the expanding buds was not exerted through photosynthesis, for the cambium was still activated if the plants were kept in the dark, so Coster proposed a hormonal explanation at a time when plant growth substances had only recently been discovered.

Thus it is likely that auxin,moving basipetally in the phloem from expanding buds, is primarily responsible for initiating cambial activity. Other hormones are probably involved as well, since for instance the cambium in dormant, leafless teak shoots did not respond to IAA unless bud dormancy was artificially broken with ethylene chlorhydrin. After leaf expansion has been completed, it is generally considered that there may be sufficient auxin produced by the mature leaves for cambial activity to be maintained, but when they all senesce the supply dwindles and the cambium becomes dormant. Indeed, Coster was able to cause the cessation of cambial activity by ringing the stem between the mature leaves and the part of the cambium being studied.

Conversely, there is recent evidence that the living cells of the wood may influence primary shoot growth. The vertical parenchyma cells, especially those associated with the vessels show high levels of acid phosphatase enzymes, particularly in the pits linking the two types of cells. It seems likely that sugars are being secreted into the vessles, as occurs in the temperate zone sugar maple tree, but for much longer periods, and in more species in the tropics.

The timing of the periods of cambial activity and rest is thus likely to be controlled indirectly by the same external and internal factors which determine bud-break and leaf-fall. One exception is probably provided by trees that flower during the leafless period, where opening floral buds and developing fruits could provide the stimulus for an earlier resumption of cambial activity. This is one of the many possible reason for 'flase' rings in the xylem, which mean that not every deciduous tree shows clear-cut annual growth patterns in its wood.

In the leaf-exchanging trees of type B, which are leafless for at most a few days, the cambium may not become fully dormant. Indistinct rings were formed in *Khaya grandifoliola*, but clear annual rings in *Dillenia indica* and *Ficus variegata*. In evergreen trees of type C, continuous cambial activity is probably the rule, although a few samples of cambial dormancy and even annual rings have been reported. More generally, they are indistinct, irregular in occurrence or absent, espcially when seasonality of climate is slight. The clear-cut growth rings in *Avicennia germinans* consist of alternating bands of wood and softer phloem-containing tissue, and do not indicate age. The most vigorous shoots can contain up to 5 rings, all of uniform width. In evergreen trees of type D, including young saplings that are continuously expanding new leaves, the cambium is almost certainly active throughout the year.

Whether cambial activity is continuous or periodic, there are often large fluctuations in the rate at which cells are formed and differentiated. As generally recognized, there is a clear positive correlation between the rate of diameter increment and the amount of height growth being made by a tree, and there is often an even closer link with its crown dimensions. Cambial activity may in some cases by most rapid which shoots are extending, but in cocoa it was slow while flushing proceeded, and reached a peak during the rainy rseason,. Similar fluctuation has been recorded for other tropical tree species.

The factors which control the rate of cambial activity have been little studied experimentally, in spite of their obvious importance of forestry. Some external influences may be expected to act directly on the cambial initials or the differentiating cells, as for example temperature and moisture stress, or the effect of gravity in branches and leaning main stems. Others, for example photoperiod and mineral nutrition, probably affect the cambial zone indirectly, perhaps by modifying growth or internal control systems, such as the partitioning of assimilates between various active meristems or into storage sites. High temperatures between 35°C and 40°C, which reduced the rate of shoot extension in *Ceiba pentandra* compared with somewhat cooler conditions,also significantly depressed diameter increments. Water stress was implicated in a study of cocoa, in which plants supplied with the equivalent of the annual rainfall (1,830 mm) spread evenly over the whole year made about 30 percent more diameter growth than those receiving natural precipitation.

Considerable genetic differences within species in their diameter increment have frequently been reported between provenances of tropical forest trees. Pronounced clonal variation also occurs : for example, both the radial increment of xylem and the number of cells produced radiallly was more than twice as great in one clone of *Terminalia superba* than in another. In terms of cross-sectional are (basal area) or volume production of wood,such differences become even larger, since the calculations involve squaring the data.

Significant clonal differences also occurred in *T. superba* in the radial dimensions of vessels and of fibres + parenchyma cells, which may be expected to alter the wood properties. These will clearly be affected as well by the proportion of the different types of cell that are produced. However, although the patterns of different tissues have been extensively described, and are often of value in identification of timbers, there is as yet little understanding of what determines the type and size of cell produced. How, for instance,do the narrow,tangential bands of parenchyma and tracheids which mark cessation of cambial acitivity at a growth ring differ from those which are often formed during continuing wood production? Scanning electron microscopy in association with detailed studies on clonal material growing in controlled environments might clarify this. Moreover, there is still much to be done in relating the structure of wood to its properties, and here the work of botanists and those in the timber trade may be assisted and integrated by a useful handbook by Chudnoff (1984), which covers the extremely confusing area of local and botanical names for 374 different timbers, as well as many aspects of their utilization.

Root Formation and Development

The first roots to be formed in a young seedling are usually a group of first-order laterals that arise from the tap-root just below ground level. Like most roots, these do not grow vertically downwards but elongate more or less horizontally. If the tip of the primary root of young cocoa seedlings is decapitated, some of the laterals nearest to the cut curve downwards and become positively geotropic. Thus, even at this very early stage, it is clear that there are dominance relationships within the root system, and these presumably lie behind the various characteristic organizational patterns found in older trees. Because of the difficulties involved in following individual subterranean roots, the majority of work on branching has been done using aerial roots. Unless damaged, these typically show little or no branching while in the air, but lateral roots start to appear if the tip makes contact with the ground. In *Cissus*, branching is induced if the root is placed in water, but in *Rhizophora* darkness appears to be the chief stimulus.

The initiation of aerial roots takes place within the shoot system, providing an opportunity to examine the formation of new root apices above-ground. Trees with stillroots, stranglers and epiphytes may all provide potential material, while climbers such as *Cissus aralioides* and *Urera* spp., actually spread vegetatively through the formation of adventitious roots. *Anthonotha macrophylla*, *Scaphopetalum amoenum* and *Sleotiopsis* usambarensis form new roots during the natural layering of their drooping branches; while conversely the rarer case of the formation of shoot apices on root tissue could be studied in plants forming root suckers, such as *Bumelia reclinata*, *Cedrela serrata*, *Chlorophora excelsa*, *Cordia alliodora*, *Drypetes diversifolia*, *Melia azedarach*, *Millettia thonningii* and *Trema* spp.

Despite these and other instances, vegetative spread is thought to be relatively uncommon amongst the woody plants of the tropical forest. However, the capacity to produce roots in stem cuttings appears to be widespread; particularly if the material originates from young trees and so perhaps natural vegetative reproduction may turn out to be commoner than expected. Since many commercially important plantation species can be rooted from stem cuttings, this has a number of important implications for research, tree improvement and clonal forestry.

It also provides favourable circumstances for studying the formation and out growth of roots, since large numbers of cuttings can be set and removed at intervals from relatively uniform propagation beds. Few anatomical investigations have been done on tropical trees, but there is a growing literature on environmental, harmonal and other effects on root initiation. For instance, rooting of cuttings of *Triplochiton scleroxylon* under automatic mist was promoted when the temperature of the propagation medium was raised to 25°-30°C, compared with 20°C and these conditions appear to be suitable for many other genera.

There are often substantial differences in rooting ability between clones; and also amongst cuttings that are of different sizes,originate from different parts of a tree,or from tree pretreated in various ways. The addition of auxins may increase the rate of rooting, and the number of root initials formed,but when supra-optimal can inhibit root elongation. Rooting with and without auxin in the medium has even been achieved with teak in aseptitic culture,using small shoots produced by meristem proliferation.

The daily course of root elongation was followed in aerial roots of a large liane, *Cissus adnata*, whole foliage was high in the forest canopy. The roots showed little or no growth when the sun was shining or when a drying wind blew during the night, but immediately after rain the rate of elongation rose temporarily to the very high value of 30 mm/h or more. Coster showed that the chief factor involved was the water status of the plant, and found that the roots could actually shrink by as much as 7 percent in length when subjected to drying. In contrast, the aerial roots of *C. sicyoides* grew at a constant 4 mm/h throughout the night and a sunny day.

Seasonal changes in the rate of root growth no doubt occur widely, but have been little studied. Nor is it clear whether there is any general cessation of root acitivity at certain times of year, as there usually is in the shoot system. Only in extreme cases does temperature appear to stop elongation : *Citrus* roots for instance grow only above 10°C. However, it may well be that root and shoot growth are linked by internal control systems as well as being influenced by their separate external environments. For instance, in controlled environments the root and shoot growth of cocoa are both rhythmic, with peaks and troughs of acitivity alternating. The rapidly expanding leaves of the flush apparently inhibit the roots, because if the foliage was all removed immediately after unfolding, the roots showed a high constant growth rate during this period.

Coster (1932) grew about 70 tropical (mainly woody)species at a site in Java which had a deep, permeable soil and regular rainfall. After 6 months, he dug out the seedling root systems carefully and found that in general the main root was longer than the main stem, and the spread of the horizontal surface roots was greater than that of the crown. Average root elongation rates were over 20mm per day for the most rapidly growing trees, which is faster than that found in most temperate trees. The total length of the main root plus the primary or first-order lateral roots had actually exceeded 30 m in *Melia azedarach* and *Sesbania sesban*, though others had grown much more slowly.

Just how extensive root systems can become is vividly illustrated by the fact that coffee plants can produce roots totalling 25 km in length by the time they are three years old. At La Selva, Costa Rica, it was estimated that more than 250 gm^{-2} dry weight of fine roots was produced in the first year by woody regrowth. This was roughly comparable to that present in the 5-year-old secondary forest that had been cleared, indicating a very rapid restoration of the nutrient uptake capacity. Fine roots often have a short life-span: it took much longer for the new structural root systems to develop. A major problem that hampers studies of the dynamics and turnover of root systems is the long, hard and tedious work involved indigging out, separating and assessing the different types of root.

It is seldom possible to attempt sufficient replicates to assign variation reliably to any factor,but one exception is provided by a study of the maximum rooting depth in tea. This indicates a significant difference between 8 clones, the deepest growing of which had reached 2½ times the depth of the shallowest in the same time. Cambial activity in roots is another area which has been little studied. When aerial roots of *Rhizophora mangle* reach the ground, their extraordinary rapid extension rates slow, and they cease to possess an elongating zone averaging 12 mm in length. Soon afterwards cambial acitivity becomes very marked and the roots thicken.

An unusual feature is the joining together through anastamoses and secondary thickening of two roots of the same, tree termed self-grafting by Ng (1975), and noted in several species in addition to *Ficus* spp. including *Shorea leprosula* and *Swietenia macrophylla.* Grafting between different trees was found in *Gmelina arborea* and *Shorea curtisli*, and Leroy Deval (1974) considers that dominant trees in stands of the gregarious species *Aucoumea klaineana* acutally 'annex' the root-systems of dominated trees to which they are fused.

Physiological Changes with Age

One of the distinctive features of long-lived perennial plants is that they usually show variation in structure and function as they grow older. Some of the changes which occur are gradual, while others appear suddely, but in either case shoots which are collected from the top of an emergent tree seldom resemble closely those produced by the same species as seedlings in the undergrowth. The leaf of *Celtis mildbraedii* from the canopy is a quarter of the length and only 9 percent of the area of that taken from the under growth, while in the liane *Hugonia platysepala* the canopy leaf is 40 percent and 30 percent of the shaded one. Sometimes leaf pairs of this kind are so unlike each other that most botanists would assume that they belonged to two different species or genera. It is one thing to recognize that such within-tree variation occurs, but quite another to determine its cause. The leaves could be dissimilar for a large number of reasons that can be considered in four broad groupings:

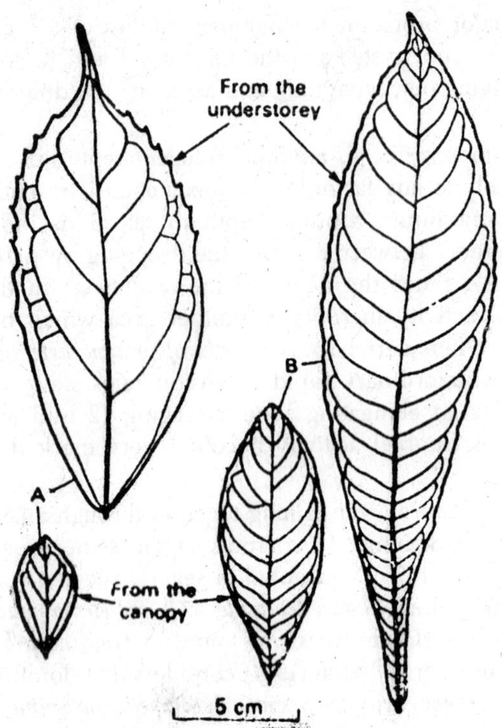

Fig. 6.7. Leaves from the canopy and the understorey of Ghanaian forest: A–Celtis mildbraedii, a large tree; and B–Hugonia platysepala, a liane.

Firstly, they have grown in very differing environments, so that factors such as light intensity and quality, temperature and moisture stress may have modified them, directly or indirectly. Although it is difficult for logistic reasons to carry out experiments using emergents, it is possible in plots of saplings or batches of potted seedlings to investigate how much their leaf size and shape very in response to treatment differences.

Secondly, they are normally bound to be of a different genotype. The extent of genetic variation is often great, especially in outbreeding species of forest trees that are 'undomesticated'. It is well known that this can influence many aspects of growth, and thus affect leaf development directly and indirectly. With the increasing use of vegetative propagation in forest tree research and for clonal foresty, the strongly inherent characteristics of clones are likely to become as easily recognized as in *Hibiscus* or

Bougainvillea, citrus or mango. For instance, potted plants of *Triplochiton scleroxylon* clone 8036 could be readily distinguished from other clones by the pronounced secondary lobing of the leaves.

Thirdly, the leaves are borne on trees of a very different size, with much greater internal competition between potential sites for growth and storage in the big tree (*i.e.,* many more 'sinks'), and longer distances for translocation. This question has been approached by experimental disbudding, using the short-lived woody plant cassava, with its relatively simple crown. This follows Leeuwenberg's model with regular and repeated increase in the number of actively growing shoot species, and associated decline in their vigour of growth. Since removal of all but one of the growing points more than doubled the rate of growth of a 'singled' apex by the end of the first week, and the leaves produced later were as much as 35 percent longer than in unpruned controls, internal competition can indeed influence leaf size. Differences in this catetgory,which are due to ageing as a tree gets older., can typically be reversed by appropriate treatment.

Fourthly, the variation between the leaves may arise because of phase-chage. These changes also occur with age, so that shoots originating from an older crown often exhibit more permanent differences from those taken from young seedlings. The change from the 'juvenile' to the 'mature' phase is termed maturation, and the features are still exhibited when the mature tissue has been vegetatively propagated and the two phases are growing is similar-sized plants in the same environment. Ideally, investigation of phase change should be done with material that is also of the same genotype, which is in fact possible if young coppice shoots can be obtained, as these are generally regarded as still being juvenile. When such detailed investigations are not possible, a reasonable compromise might be to compare the crown foliage of several large trees with leaves taken from regeneration in a large gap or clearing nearby, where (away from the margins) the microclimate is relatively similar to that at the top of the canopy.

Differences that were clearly due to phase-change were found in similar-sized mature grafts and juivenile seedlings of *Cedrela odorata* that were grown in controlled environments under different photoperiods. The rate of shoot elongation was much more strongly influenced by day-length in the mature plants, which also ceased extension for a time in both treatments. In contrast, the terminal buds of the juvenile trees only became dormant under short-days. On the other hand, differences which are definitely not due to calender age are shown in, since all the trees in the photograph are 5 years old. Because of the favourable external environment, one of the trees has far outgrown the others and has entered the category A phase with periodic growth, a rapid flush of height growth and a clear-cut deciduous period. The small trees remained ever-green, and grew continuously or for long periods, though at a very slow rate.

In the uncontrolled setting of the tropical forest, it is generally difficult or impossible to separate environmental and genetic factors on the one hand from physiological changes with age (ageing and maturation) on the other. Differences of various kinds between the foliage of large and small trees are often reported, the leaves of the former generally being smaller with drip-tips generally absent. Compound leaves often have fewer leaflets, as in the Meliaceae, for example. In some instances, the leaf shape, and sometimes the phyllotaxy, may alter strikingly with age; for example young seedlings of *Acacia mangium* have the normal pinnate leaves typical of the Leguminosae, but these quickly give way to phyllodes, with a few intervening nodes bearing transitional structures. *Araucaria cunninghamii*, *Triphyophyllum peltatum* and some *Eucalyptus* spp. have at least three contrasting types of foliage.

As trees become larger, they often cease producing leaves continuously, and the leaf-exchanging or deciduous habits may appear. In genera which show periodic growth from the start, such as *Agathis*, *Barringtonia* and *Mangifera*, longer periods of terminal bud dormancy may be found. In addition, there may be a tendency for individuals of a species to become more synchronized in their development. For instance, the typically out-of-phase condition of a 2-3-year-old plantation of *Terminalia ivorensis* contrasts with the comparative regularity of a 20-year-old stand.

As a seedling grows into a large tree, the characteristic branching patterns also emerge. Sometimes branches start to grow almost at once, while in other species there can be a period of a few months or years without any. In young saplings of some species in the Meliaceae, the rachis of large compound leaves borne on the main stem acts instead of a branch in supporting the photosynthetic surfaces. Palms that follow Corner's model, such as *Elaeis guineensis* and *Cocos nucifera*, produce very large fronds from their trunks, the only branches being reproductive and determinate. Conversely, the terminal growing point of the main stem loses its meristematic nature and disappears in cocoa, signalling the production of a 'jorquette' of plagiotropically growing branches. Neither branches nor main axis are produced during the 'grass stage' of *Pinus merkusii*, most of the growth of the young tree going into the development of the root system. Only later do the main stem and branches develop as in other pines. The grass stage is thought to be an adaptation by which the terminal bud is protected from 'light' fires by the cluster of long needles; it occurs in this species in Vietnam, but not in Sumatra.

Some interesting theoretical modelling of branching is reviewed by Fisher (1984), but the amount of experimental work on branch development has not been extensive. The formation of the first primary branches was delayed by more than 3 months, and no second-order branches were formed, when *Rauvolfia vomitoria* seedlings were grown under 8 h rather than 12 h day-lengths. Applied gibberellin promoted the elongation of both orthotropic and plagiotropic branches in cocoa, but did not change their phyllotaxy or geotropic orientation. *Triplochiton scleroxylon* clones differed strongly in the number of branches per metre of main stem and in their thickness and persistence after they had become shaded by later growth. Some clones showed a tendency to produce 'multistems' (several strong vertical shoots from near the ground). Decapitation studies with small potted plants have shown that clones vary in the number of buds which are released from apical dominance, and that mineral nutrition and gibberellins, for example, can affect the re-imposition of dominance by the uppermost lateral. Experiments with *Terminalia ivorensis* seedlings involving removal of terminal or lateral shoots showed that branches can affect the growth of the leading shoot, as well as *vice versa*.

Besides environmental and genetic effects, there appear to be various endogenous and ontogenetic factors at work : for instance, a clone of rubber trees usually formed the first branches after the terminal bud had flushed nine times. In some tropical trees there are two buds in each axil, and typically these behave differently. For example, the upper one in *Nauclea diderrichii* produces plagiotropic branches, while the accessory bud may produce an orthotropic shoot after damage or decapitation; and these differences persist incuttings taken from the two types of shoot. There can be several accessory or serial by treatment with various growth regulators, including the cytokinin benxyladenine, the anti-auxin tri-ioidobenzoic acid and compounds releasing the gaseous hormone ethylene in the tissues. The natural or induced sprouts from these accessory bud produces plagiogropic, flowering shoots. However, the upper three buds can form flowers sequentially, but the fourth is often vegetative.

In many ways, the most fascinating change in the life of a tree is the onset of reproducutive activity. Some species, notably pioneers, start flowering profusely after about 2-5 years, although the great majority remain purely vegetative for much longer than this. Nevertheless, specific precocious individuals are often noted in the literature as producing flowers before they are two years old, for example. *Anthocephalus chinensis*, *Dipterocarpus oblongifolius*, *Citrus* spp., *Funtumia africana*, *Hildergardia barteri*, *Monodora tenuifolia*, *Pinus caribaea* var, *hondurensis*, *Trema guineensis*, and many mangrove species. If such precocity is primarily genetic in origin, clones made from early-flowering individuals might provide an elegant research tool for miniaturising the study of flowering. for commercial plantations, although early flowering may be an advantage in fruit trees it is probably undesirable in species that are grown for vegetative yields. There is some evidence that there has been unconscious negative selection in forestry, perhaps because seed is more conveniently collected from precocious and prolific individuals or stands. In teak, in West Africa, for example, entire young plantations may be seen flowering or fruiting copiously and occasional individuals may even do so in the nursery when only 3 months old. Since the inflorescences are large, and terminate the main stem, this may be expected to cause poor form and reduced vegetative growth.

In most forest trees, however, there is a more extended juvenile period through which they pass before flowering commences. It is common for this interval to average between 10 and 30 years, although there is considerble variation between sites and amongst individuals. In general, it tends to be shorter in plantations than in natural forest, probably for a combination of genetic and environmental reasons; and it is also typically a briefer stage for pioneer species than for climax species. Thus Ashton (1969), for example, considers that it is often about 60 years before majority of the dipterocarps of the upper canopy of Malaysian forests start to flower. In one group of plants the onset of flowering is of particular significance–monocarpic species which flower once and then die.

Agave is a well-known example, in which the life-cylce is about 10 years, while in *Tachigalia versicolor* it ends after many years. Thus monocarpy is not confined to species such as *Corypha* palms, which exhibit Corner's model, but can occur in emergent trees with many shoot apices. Another striking feature of this group is that they often flower gregariously and since clones that have been dispersed to different regions still tend to flower and die at the same time, some long-term endogenous 'calendar' is implied. Some but not all bamboos are monocarpic, with the life-cycles often rather clearly known, ranging from a few years in *Chusquea tonduzii* to 80 years in *Rhipidocladum pittieri.* Such very long juvenile periods are thought to have evolved as extreme instances of deprivation followed by saturation of seed predators.

More experimental studies are needed on the various physiological factors which may be involved in controlling the length of time a tree produces only vegetative shoots, particularly because it is often desirable to be able to shorten this in crop species. However, it is known that coffee is a short-day plant in its floral initiation, whereas *Stylosanthes guianensis* is a long/short day plant. In *Bougainvilleo* spp, and *Euphorbia pulcherrima,* the first flower production in these vegetatively propagated ornamentals is affected by an interaction between photoperiod and temperature.

Cuttings originating from young seedlings of *Triplochiton scleroxylon* have flowered on several occassions under warm, long-day conditions in glasshouses in Scotland, at an age of 2-6 years from germination. Observations on the nut-bearing *Shorea penanga* in a nutritional trial in Sarawak suggested that profuse flowering and fruiting at 6 years occurred particularly in plots receiving all the fertilizers. Immature mango shoots have also been induced to flower and fruit by treating them with potassium nitrate.

The most generally applicable treatment for stimulating forest trees to flower may prove to be growing them as quickly as possible, which appears to hasten the processes involved in maturation. However, various additional techniques may be needed once the trees have reached 'ripeness-to-flower', to provide them with the 'opportunity-to-flower'. Methods which have had some effect in seedlings include removal of the terminal bud in mango, decapitation in coffee just after the first branches are formed, and defoliation in guava. Bark-ringing has also been successfully used in a number of tropical as well as temperate woody plants.

Some indication of possible hormonal effects is suggested by the induction of flowering in some 10 year old cocoa trees, that had never been seen to flower previously, by transplanting into their bark small clusters of flower buds from very floriferous trees. Treatment with gibberellins stimulates cone initiation in young trees in the Cupressaceae, and the techniques may be applicable to other conifers. Other plant growth regulators may also be invloved; for example compounds which release ethylene in the tissues have stimulated the onset of flowering in mango. The herb *Geophila renaris* remains vegetative as long as the soil moisture content is near its maximum value. If it falls below the permanent wilting percentage flowering is stimulated, and it continues afterwards even if the soil again becomes saturated with water.

Flowering

Flowers often attract the attention, of people as well as of potential pollinators. However, many open only for short periods, and may be hard to see high up in the euphotic zone. There is an extraordinary range of shapes, sizes and distribution of flowers in the tropical forest with diverse colours, markings and often scents. *Raffesia arnoldi* forms a single, foulsmelling flower about a metre in diameter; cocoa bears undreds of small, cauliflorous blossoms; *Bombax buonopozense* unfolds red, waxy 'ponds' high up in the canopy, in which aquatic invertebrates and vertebrates can flourish; while a large dipterocarp tree might remain purely vegetative for many years, and then suddenly produce a million flowers that all open within a week or two. With diversity on this scale, it is obviously difficult to generalise, although fortunately there is now a growing literature on floral biology at the community and species levels.

The critical first stage in reproduction is the conversion of an apex from producing only vegetative primordia to forming an inflorescence, cone or single flower instead of or in addition to foliage leaves. The first detectable signs are often an alteration in the shape of the apical dome, and in the growth rates of leaf and bud primordia, as well as inhibition of the sub-apical meristems which produce and extend the internodes. Floral development may continue without a break from initiation until the opening of the flowers, as for example in *Hibiscus* and herbaceous species. However, in many trees there is an interval during which the initials remain within a dormant flower bud, which can vary from a few weeks to many months, or even 2½ years in some *Eucalyptus* spp. It is therefore necessary to distinguish carefully between the timing of floral initiation, and that of flower opening, since the factors which influence these two processes may not be the same.

More or less *continuous* flowering is found in some evergrowing secondary tree species such as *Dillenia suffruticosa*, *Harungana madagascariensis*, and *Trema orientalis*. The common mangrove *Rhizophora mangle* also flowers continuously, although in Southern Florida it shows a peak in June/July. In the great majority of cases, however, the flowers emerge *periodically*, either regularly or irregularly, though there is usually sufficient overlap for the forest as a whole never to be without flowers. Individual trees may also show variation within the crown, with some parts in flower, other branches perhaps in fruit, and others again producing only leaves.

On the other hand, synchronized flowering can occur over many hectares of forest, as in *Pterocarpus indicus* in Malaysia, *Tabebuia serratifolia* in Surinam and theWest African liane *Calycobolus heudelotii*. A few species that flower annually are so reliable in their timing that they have been used as a signal for the planting of a food crop–for example., *Erythrina orientalis* for yams in the New Hebrides, *Sandoricum koetjape* for rice in West Malaysia and *Trichilia heudelotii* for the second planting of maize in Ghana. Perhaps they have proved more useful than astronomically based calendars, since they co-ordinate the agriculture cycle with current environmental conditions rather than with fixed dates, which do not allow for year-to-year variation in temperature and precipitation.

The frequency with which flowering periods recur varies from about 3-4 months in *Ficus sumatrana*, for example, to 10-15 years in *Homalium grandiflorum* Guava can flower twice a year, while fairly regular annual flowering occurs in the fruit trees *Baccaurea parviflora* and *Xerospermum intermedium* and in a number of common forestry plantation species, including *Cedrela odorta*, *Gmelina arborea*, *Tectona grandis*, *Terminalia ivorensis* and *Pinus kesiya*. Biennial tendencies appear to be characteristic of most species reaching the canopy in Surinam forests and may also become marked in older mango trees. Although dipterocarps may occasionally flower again after 1-2 years, the interval is more commonly about 3-8 years. In these irregularly spaced 'mast' flowering and fruiting seasons, flower profusely and gregariously over a period of a few months, often accompanied by heavier than usual reproductive acitivity by trees in other families.

In a detailed study of six related *Shorea* spp, which involved frequent assessments high up in the canopy, Chan (1977) found that each species occupied a specific 'time-niche', with the flowers open for about 15 days in the first to flower and for about 25 days in the last. Peak flowering times of the 6 species did not overlap at all. This remarkable characteristic, which has also been found in the New World tropics for 4 *Guarea* spp. and in the Bignoniaceae is thought to have evolved through competition for a limited number of pollinators. There is considerable speculation about the nature of the physiological trigger for irregualrly spaced, but community-wide flower initiation. Perhaps the most likely hypothesis is that an increase in the diurnal fluctuation of temperature, associated with drier spells of weather with clearer skies, stimulates floral initiation at the same time in all the six *Shorea* spp. which then vary in their rates of subsequent flower development.

Because some tree species show reproductive activity that is clearly or probably related with seasonal changes does not mean that all them do so. The proportion of 'net clearly seasonal' species was 14 percent in south Florida, 26 percent in a detailed study at Barro Colorado Island, Panama, that also included herbaceous plants, and about 33 percent in Ghana. Some examples of a seasonal tree species are *Anisophyllea corneri*, *Anthonotha macrophylla*, *Blighia sapida*, *Genipa caruto*, *Santiria laevigata* and *Trophis racemosa*.

Considering the forest as a whole, however, there are usually peaks of flowering at certain times of year, even in rather uniform climates. Comparisons between data from different regions can be made using a *seasonal flowering index*, which is the ratio between the number of species flowering in the 6 month period with the most flowering, and that with the least. Values between 1.2 and 4.9 have been calculated for consecutive 6 month periods, the degree of seasonality depending both on the magnitude of the difference and the amount by which the number of flowering species exceeds zero. A seasonality at the community level has been claimed for a forest in West Malaysia, but it is interesting that if the data presented are recalculated, there is 25 percent more flowering in the months March-May and September-November than in June- August and

December-February. This difference is statistically significant, and gives a seasonal flowering index (on a 3 month + 3 month basis) of 1.25.

Not surprisingly, there tend to be closer links between reproductive phenology and particular times of year in forests experiencing definite climatic seasons. Very often the flowering peaks occur during the dry season and the early part of the rains. Deciduous species tend to flower in the dry season, expanding flower buds that do not also contain foliage leaves (e.g. *Ceiba*, *Hildegardia*), or mixed floral and leaf buds at the usual time of vegetative bud-break. Peak flowering in evergreen species occurred during the transition to the wet season, the inflorescences or single flower often emerging with the new leaves, or perhaps being produced cauliflorously.

There is more variation found between different forests when climate is relatively uniform. Most flowering occurred from March to July in a six year study of 56 individuals (42 species) in West Malaysia, whereas June and July were not good flowering months in a four-year study of 131 trees (62 species) only about 50km away. At Pasoh Forest Reserve, only about 150 km away, a 3½ year study of over 400 trees (19 fruit tree species) showed a clear predominance of flowering in the earlier part of the year. A comprehensive study of the forests of Panama, invloving three years fieldwork and the examination of more than 50000 herbarium specimens, suggests that the different life-forms do not all flower at the same time. For instance, trees, lianes and epiphytes showed peak flowering in the dry season, whereas shrubs and herbs growing in the undergrowth or in small gaps flowered especially in the rainy season.

The *timing of floral initiation* has been little investigated, although it is clearly an important topic that is relatively easy to study by dissection if suitable material can be collected regularly. This would be easier in species such as *Ateramnus lucidus*, *Bosqueia anagolensis*, *Monodora tenuifolia* and members of the Cupressaceae, in which reproductive and vegetative shoots can be distinguished several months before flower opening occurs, because the floral buds are larger or the developing cones appear different. Besides clarifying the questions of seasonality of flowering discussed above, knowledge of when and where flower formation occurs in a tree can greatly assist physiological studies into the factors which promote floral initiation. For instance, if it is known that terminal or lateral reproductive organs start to form just as shoot elongation ceases, experimental treatment can be applied with much greater temporal and spatial precision.

Coffee, tea, bougainvillea and poinsettia are all facultative or obligate short-day plants, and the suppression of flowering in *Hildegardia barteri* growing close to street lights has been interpreted in the same way. However, it should not be concluded that all tropical trees will be short-day plants or day-neutral, for a substantial minority of tropical plants have been reported to be long-day species, while temperature can modify effects of photoperiod. Drought stress has long been claimed to be a factor stimulating flower initiation, although the evidenceis often lacking for tropical trees.

The most effective hormonal treatment for stimulating reproductive activity is undoubtedly the application of gibberellin (GA_3) to members of the Cupressaceae. Large numbers of male and female cones can be stimulated, particularly if doses of about 1-100 mg are injected in alcohol into small holes gibberellins ($GA_{4/7}$) appear to be promising tools for promoting cone initiation in *Pinus* spp. but in broad leaved trees it appears that gibberellins often inhibit floral initiation. Three different growth retardants promoted the formation of flowers in mango during an 'off' year, especially in combination with bark-ringing. Pronounced stimulation of flowering following a soil drench application of another inhibitor was found in the vine *Clerodendrum thomsonae.*

Other treatments which have been found to modify flower initiation include bark-ringing,which is stimulatory in *Faurea speciosa, Lonchocarpus utilis, Pinus elliottii* var. *elliottii, Terminalia ivorensis,* mango and rubber. It may also be possible to modify the sex of inflorescences of *Elaeis guineensis* towards male by removing the older leaves, although there is a lapse of two years before this is seen because of the long period of development after initiation. In *Carica papaya,* which is dioecious, there is some evidence that male trees can produce functional female flowers if the stem has been damaged.

Progress in the physiological study of flower initiation is likely to come from a sustained effort to develop miniaturised systems in which flowering can be studied in depth. In particular, the use of selected clones that can initiate flowers under controlled environments could transform our knowledge of the effects and interactions of external factors, and also elucidate the internal links and balances between the formation and growth of flowers, leaves, stem units, roots, etc. There is even the prospect nowadays of studying reproduction in aseptic cultures, in which even the hormonal balance of an organ can be carefully controlled.

The *opening of flowers* has been studied extensively in coffee, and also in trees and shrubs in a Costa Rican forest with a pronounced dry season. *Bernardia nicaraguensis* and *Croton reflexifolius* initiate and develop floral buds, and then become leafless during the dry season. Just 2-5 days after the first rains, they flower gregariously, whether this happens to be in March, April or May. Other woody species also appeared to be triggered by rainfall to open their flowers, taking longer to respond, but tending to occur each year in the same sequence. The shrub *Hybanthus prunifolius* similarly flowers synchronously is Barro Colorado Island, Panama, a few days after the first rainfall of 12 mm or more,provided that the preceding dry period has been sufficiently intense and prolonged. Watering a plant after dry weather will stimulate it to flower, but watering it during the dry season prevents it from flowering.

A rapid drop in temperature normally accompanies a sudden tropical thunderstorm, and in coffee either a temperature-drop of at least 3°C in 45 minutes, or the relieving of water stress by the rain, can trigger flower opening after a dry spell. In the epiphytic orchid *Dendrobium crumenatum*, the change in temperature appears to be the critical factor; whereas with drybulbs of the 'rain-flowers' *Pancratium* and *Zephyranthes*, which flower 3 days after heavy rain, a direct effect or moisture is indicated by their reponse to either warm or cool water. The environmental factors and changes in endogenous promoters and inhibitors that are associated with coffee flower opening are discussed by Browning (1977).

A suprisingly large proportion of flower buds may abort and be abscised before opening, particularly when there are large numbers present, as in *Shorea* spp, and *Triplochiton scleroxylon.* Sometimes the flowers on a plant all open within a week or so, as in the gregraious examples already quoted. In other species, such as *Jacaratia dolichaula* and *Pentagonia macrophylla*, a few flower buds may continue to open for a long for a long as 3-4 months. Individual flowers may last for days, as in mangroves or only for hours, as in many other tree species, and they frequently have very precise opening times that are clearly related to the behaviour of the appropriate pollinator. Thus, for example, anthesis starts around 1600 h and is completed by 2000h in *Durio zibethinus*, while pollination occurs mainly between 200 and 0100 h. The flowers of *Cieba pentandra* open as dusk falls, and are pollinated by bats that take the nectar and distribute pollen on their fur.

Many other physiological and structural aspects of flowers and pollination are covered by Tomlinson (1980), Bawa (1982), Tanner (1982) and Baker *et.al* (1983), amongst others. An important point is that around 30-40 percent of tropical trees have been found to produce separate male and female flowers, the majority of these being dioecious, with separate male and female trees. Various intermediate categories also occur quite frequently, serving as a reminder of the great variability in reproductive biology which exists in the tropical forest.

Fruit Development

Except in parthenocarpic plants such as banana, plantain and seedless varieties of *Citrus*, pollination must take place if fruit-set is to occur. Unpollinated flowers are usually abscised, and it is common for this to occur to many pollinated flowers as well. High proportions of flower abscission have been found for instance in *Aucomea klaineana* and *Shorea* spp. Indeed, there can be several thousand flowers on a single inflorescence of *Parkia clappertoniana* or mango, and yet only 1-5 of them normally set fruit.

One reason for this may be that flowers which are fertilized first often inhibit fruit-set in those that are pollinated later. On the other hand, in species such as epiphytic orchids that produce relatively few flowers, it may be noticeable that the individual flowers appear to set fruit independently of each other. External factors can also affect the setting of fruits : for example, water stress and insect attack are both likely to increase flower abscission. Either high or low temperatures partially or completely inhibited fruit-set in coffee, the optima being 23°C day/17°C night, or 26°C day/20°C night.

The subsequent enlargement and development of fruits takes place at rates that are characteristic of the species, modified both by the prevailing environmental conditions and by mutual competition between fruits and with other 'sinks'. By no means every fruit that is set reaches maturity, partly because fruitlet abscission continues to 'thin out' the crop remaining on the tree, even if there is no abnormal stress. Insect predators or mammalian herbivores may damage seeds or eat fruits before thay are ripe, while diseases such as cherelle wilt in cocoa may cause a high proportion of the young fruits to shrivel on the tree.

The final stages in fruit enlargement are often relatively slow, but then ripening can occur quite rapidly. During the latter process, a sudden rise in respiratory rates can frequently be observed in fleshy fruits, associated with the metabolic be observed in fleshy fruits, associated with the metabolic changes leading to softening and the development of the characteristic colour and flavour. The onset of ripening may be stimulated by an increase in ethylene levels within the fruit, associated with changes in auxins. In avocado for instance it has been shown that low levels of applied indoleacetic acid suppressed ethylene production and delayed ripening, whereas 100-1000 micromolar applications stimulated ethylene production, respiration rates and ripening. Ripening can also be stimulated by picking the green, fully grown fruits.

Shortly afterwards,most fleshy fruits are abscised, unless they have already been taken by primates, birds, bats, etc. On the ground, they are subject to consumption, dispersal and decay through the activities of many other forest organisms. Once again, only a relatively small proportion of the total survive, and sometimes the entire fruit crop may fail to produce a single viable seed. The same can occur if seed insects or fungi are widespread, as frequently occurs in *Triplochiton scleroxylon*, where the fruit is non-fleshy and single-seeded. Where such fruits contain many seeds, the ripening process involves drying, and abscission layers in the fruit wall generally allow the seeds be shed while the fruit is still attached to the tree. In conifers, the ripe female cones open their scales as they dry.

The period from anthesis to fruit ripening has been studied in West Malaysia from the literature and observations over a 2 year period. The average interval was 4.3 months for 86 indigenous and 7 exotic species, with *Pterocymbium javanicum* showing the shortest period. *Firmiana malayana* and *Dillenia suffruticosa* also ripened their fruits rapidly. Those taking 9 months or more to ripen included *Diospyros maingayi*, *Palaquium hispidum* and *Vatica ridleyana*. The New World species *Bertholletia excelsa*, *Enterolobium cyclocarpum* and *Pithecellobium saman* have also been reported to ripen slowly. In West Africa, *Lovoa klaineana* has a short fruiting season, and *Mimusops heckelii* a long one. *Bosqueia angolensis* and *Celtis* spp, may fruit twice a year, while *Trema guineensis* fruits continuously. Although there may be a good deal of within-species variation, data of this kind can help in planning fruit collections, based on the time since flowering, which is often relatively easy to detect. However, not every flowering event is followed by a fruit crop.

There are fruits growing and ripening all the year round in tropical forests, which is important for the survival of frugivorous bats, in particular, since they have to maintain their body temperature by regualr feedling. However, there are usually *fruiting seasons* when a higher proportion of the species and individuals is in fruit. In Ghana, for instance,this is clearly during the dry season, with a fairly steady build-up to March. Most seeds will therefore be dispersed before the start of the rains. In a forest at 850 m altitude in eastern Zaire, the 200 species showed least fruiting (8 percent) in May, at the end of the 9 month wet season, and most fruiting (46 percent) in February.

Two dispersal peaks in the year were found by Foster for the canopy trees at Barro Colorado Island, Panama, one in March/ June and one in September/October. Lianes ripened fruit once a year, primarily between March an May, while understory trees and shrubs produced most ripe fruit in November/December. In discussing the significance of these seasonal peaks. Foster recognizes shorter and longer fruit development periods associated with species that flower at different times, such that they may disperse their seeds together. Also noted is the effect of whether variation, greatly affecting the animals normallly depending on this food supply.

In West Malaysia, ripe fruits were most frequently seen in September and October, and were least common from January to May. Of the 6 related *Shorea* spp that flowered successively the slowest rate of fruit development was found in the first to flower, and *vice versa*, with the result that all species bore ripening fruit at about the same time. Besides being a remarkable example of precise physiological control over the timing of reproduction, this result suggests that there may well be selective advantage in dispersing seeds at a particular time of year, even where the climate is realtively uniform.

However, by no means every genus distributes its fruiting periods in the same fashion- as with flowering there are instances of long-sustained fruit ripening, and of regular or irregular dispersal at intervals from a few months to many years. In West Africa, for example, fruits of *Terminalia ivorensis* have been collected from different sites in most months of the year, but on the other hand the fruiting of *Claoxylon hexandrum* is so regular that it is used to mark the date of a festival.

Seed Germination and Dormancy

A number of different types of seed germination can be recognized, base both on structure and function. In a study of more than 200 West Malaysian tree species, representing about 8 percent of the indigenous tree flora, Ng(1978) found that the hypocotyl elongated in 72 percent, carrying the cotyledons well above the ground surface. Most of these were the normal *epigeal* type, with cotyledons exposed and green, as in *Shorea leprosula*, *Koompassia malaccensis* and *Podocarpus nerifolius*, and many species that colonzie clearings and margins. Two modifications were noted: in *Diospyros* spp. the cotyledons are abscised early, before turning green; and a *durian* category is recongized, for example, in *Durio zibethinus*, *Dipterocarpus oblongifolius* and *Strombosia javanica*, in which the cotyledons remain within the endosperm and are abscised later. They also stay inside the seed coat in the normal *hypogeal* type, in which the hypocotyl does not elongate significantly, as for instance in *Anisophyllea*, *Garcinia* and *Lithocarpusl.* Here again, a modification is reported, with trees such as *Eugenia grandis*, *Lansium domesticum* and *Pithecellobium* spp. showing *semi-hypogeal* germination, with the cotyledons exposed at ground level.

Germination can also be classified in other ways. In a few species, the embryo has already developed into a viviparous seedling before becoming detached from the parent plant : *Dryobalanops aromatica* and several mangroves are examples of such *prior* germination. A much larger number exhibit *prompt* germination, in which the radicle emerges within a few days or weeks of dispersal. This is characteristic of dipterocarps, for example, and many other tree species,particularly those having large seeds that contain considerable food reserves.

Another class of prompt germinators are those species whole seeds germinate more rapidly and to a higher final percentage after thay have passed through the gut of a herbivore. Such stimulation was found for instance in *Azadirchta indica*, *Nauclea latifolia*, and especially strongly in *Securinega virosa* seeds germinated from baboon dung in a dry forest/savanna woodland area in Ghana, compared with seeds taken from fruits that had not been ingested.

The simplest reason for *delayed* germination is that the seeds were formed in fruits that dried on repening, and since being shed have not yet come in contact with moist soil or rain.

All the other categories involve some type of dormancy, such that most or all of the seeds fail to germinate in moist conditions at normal ambient temperatures. This relatively simple idea, of seeds being 'dormant', is frequently confused in the literature with their being 'viable', or 'not germinating', concepts which are related but not synonymous. The testa, and sometimes the indehiscent pericarp, may be impermeable to water until weakened by the combined effects of repeated cycles of heating and cooling, mechanical abrasion by mouth-parts or soil particles, chemical attack by micro-organisms. Once water can penetrate, such seeds become imbibed and usually germinate at once, but the key ecological significance is often that the germination of a cohort of seeds is spread out over a period of months, years, or even decades, persumably increasing the chances of some of the progeny surviving. 'Hard' seeds are common in the Leguminosae, although they are not formed for instance by *Koompassia malaccensis.*

Scarification is often used to pre-treat such seeds before sowing, and this increased germination of *Ochroma lagopus* from 3.5 percent to 84 percent. Surprisingly, effective techniques also included placing the balsa in boiling water for up to a quarter of an hour,or subjecting them to dry heat between about 60°C and 120°C. It appears that this type of breaking of dormancy could be one of the reasons why balsa seedlings frequently spring up after the passage of 'light' fires. Other seeds are freely able to take up water, but need to remain for a time in a dry condition for after-ripening to take place. Oil-palm, and freshly harvested cereals, including rice, generally do not germinate readily, but will do so after being kept indoors for a period.

In the common annual grass *Dactyloctenium aegyptium*, dry storage at 20°C for 10 weeks partially relieved this type of dormancy, while 40°C was much more effective. Vazquez-Yanes and his colleagues have elegantly demonstrated and investigated light-sensitivity in colonizing tree species in Mexico. The seeds of *Cecropia obtusifolia* and *Piper auritum* can remian dormant for more than a year if the imbided seeds are kept in the dark. They respond to light by germinating, but unlike some of the classical responses of herbaceous species a substantial period of illumination is required.

Nevertheless, the usual red/far red system involving the pigment wavelength and its stimulatory influence can readily be counteracted by following it with far- red irradiation. Even if red light was given for several hours, the germination percentage of *C. obtusifolia* only reached about 25 percent, but values of around 90 percent were achieved when seeds of either species were given 2 h of red light on each of 4 successive days.

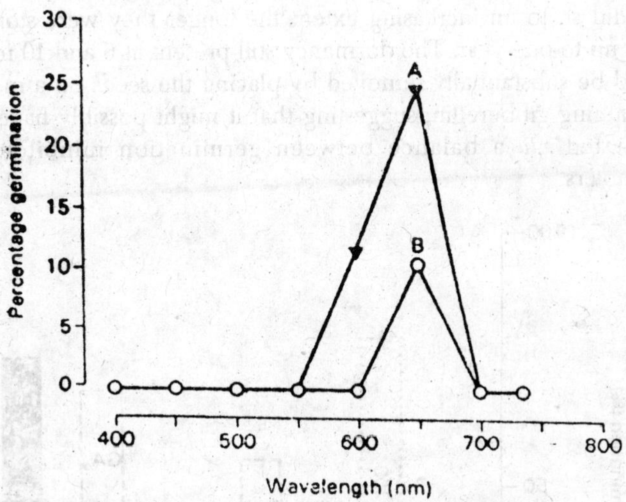

Fig. 11.8. Wavelength dependence of the breaking of seed dormancy by light, in (A) Cecropia obtusifolia; and (B) Piper auritum. Seeds were kept in the dark, except for 6 h irradiation at different wavelengths, and the final germination percentages were recorded after 1 month.

As discussed low red/far ratios probably act to keep such seeds dormant within the forest or in small gaps, but the higher ratios found away from the margins of larger clearings permit germination. The same group have demonstrated that a requirement for diurnal temperature fluctuation may be another mechanism involved in the breaking of seed dormancy of colonizer species in large gaps. For instance, imbided seeds of *Heliocarpus donnellsmithii* mostly remained dormant when kept under constant temperatures in the laboratory or buried in soil within the forest at about 25°C. High germination percentages were obtained under fluctuating laboratory temperatures, especially with a 6 h period at 32-39°C, and 18 h at levels 10°C cooler. Field germination was also enhanced near the middle of a sizeable gap, where the surface soil temperatures showed substantial diurnal fluctuations. Futher research can be expected to show whether other types of seed dormancy operate in tropical forest trees.

Combinations of more than one type of class tend to occur in the same seeds, and it is not unusual for dormancies to change during storage. Nor is it always easy to correlate the results obtained in the laboratory with conditions in the forest. For example, *Trema guineensis* seeds stored at 22°C would not germinate initially, but did so to an increasing extent the longer they were stored, at least up to one year. The dormancy still present at 6 and 10 months could be substantially removed by placing the seeds on a medium containing gibberellin, suggesting that it might possibly have been mediated *via* a balance between germination inhibitors and promoters.

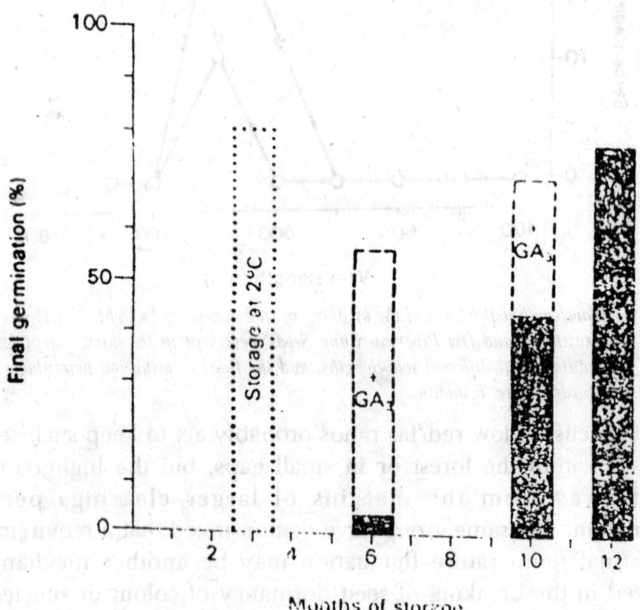

Fig. 11.9. Effects of storage time and temperature on the germination of Trema guineensis seeds, with and without gibberellin. Black column–storage at 22°C; dotted column–storage at 2°C; dashed column–promotion when seeds were germinated on agar plates containing 500 ppm gibberellic acid (GA_3).

Surprisingly, storage at 2°C markedly promoted germination, rather as winter chilling breaks the dormancy of tree seeds in north temperature zone forests. Additionally, there was an obligate requirement for light, with no germination whatsoever being obtained in the dark series. The germination temperature itself is also important, since all plants are thought to have minimum, optimum and maximum temepratures for this process, and germination rates are also strongly influenced temperature. For instance, germination of coffee took three months at 17°C, but only three weeks at 30°C.

Germination tests are generally carried out in convenient, standard conditions, for example at 25°C or 30°C, but as already mentioned these may not be appropriate for all species. The same applies to the question of suitable storage conditions: sometimes the dormant state may remain unchanged for considerable periods, as for instance in seeds requiring after ripening which may retain their dormancy if stored in a deepfreeze. More commonly, there is a continual modification of the degree of dormancy with the passage of time. Storage at 50 percent and 70 percent relative humidity at 22°C led to the development of a secondary type of dormancy in *Hildegardia barteri*, which was not the case when the fruits were stored at 90 percent relative humidity. This effect could be largely removed by placing the drier seeds into a moister atmosphere, and it apparently invloves changes in the permeability of the structures surrounding the seed itself. If may be that this mechanism operates under natural conditions on the stoop, rocky slopes colonised by this species.

During the dry season, the majority of seeds may perhaps remain unresponsive to occassional rain or dew, but in the moister conditions of the rainy season they may lose their dormancy, and germinate with a greater chance of survival. Storage temperatures and humidities are therefore the main considerations, together with avoidance of pests and diseases, when seeds are to be kept for use at a later date, particularly if long-term storage is planned for conserving gene resources.

An unusually extended longevity of nearly half a century has been reported for *Ochroma lagopus* seeds from herbarium sheets. However, in 'recalcitrant' species, which include many of the prompt germinator class, viability can be quickly lost if the seeds are dried, and they either germinate or go mouldy if kept at high humidity. Nevertheless, it now seems likely that more thorough research may indicate ways in which some of these species may be stored. For example, viability can be preserved for a month or more if dipterocarp fruits are disinfected and stored at about 95 percent relative humidity in closed containers. Temperatures must be above 15°C for *Shorea ovalis* and above 4°C for several other species, of which *S. talura* was the most tolerant. For oil palm seeds, formerly considered to be recalcitrant, the excised embryos can be thoroughly dried and kept at–196°C, where it appears that viability may be preserved indefinitely. *Agathis hunsteinii* seeds kept better at a moisture content above 32 percent, and die if dried to about 14 percent, but *A. cunninghamii* appears to be relatively easy to store. In *Triplochiton scleroxylon*, the speed of drying is an improtant consideration : when fruits were dried to 8 percent moisture content at a rate not exceeding 1 percent of the original weight lost per hours, they could be stored at–18°C for at least 18 months.

12

Hormones in Growth

Among the substances which markedly influence the reactions and metabolism of plants are those internally synthesized compounds called *hormones.* In general, the term "hormone" is used to designate certain organic compounds which exert important regulatory effects upon the metabolism of an organism when present in only minute quantities. In the animal body it is generally considered that a further characteristic of a hormone is that it exerts its effects at a site remote from the locus of its synthesis. Adrenalin, for example, secreted in higher animals by the adrenal gland, has pronounced effects upon the heart and vascular system. Many, but not all, plant hormones similarly exert their effects in cells at some distance from those in which they are synthesized, after translocation from the latter to the former.

Included among the plant hormones are certain substances usually classified as vitamins in considerations of animal physiology. In the higher plants those vitamins which are essential exhibit essentially the same general type of behaviour as hormones and may be regarded as falling into the same general category of substances. In addition to the naturally occurring hormones in plants, a number of organic compounds are known which, when introduced into plants in relatively small quantities, induce effects which are similar to, and often apparently indentical with, those induced by naturally occurring hormones.

By an extension of the original concept, such substances are also often called plant hormones, although the more rigorous definition of the term would strict it to naturally occurring compounds. Other terms commonly used to designate plant hormones are *phyto-hormones*, *growth hormones*, *growth substances*, and *growth regulators* (a term which includes both *growth activators* and *growth inhibitors*).

Emergence of the Hormone Concept

Soding (1923), after Paal's deduction that specific substances produced in the coleoptile tip were responsible for phototropism, established that these very substances were capable of stimulating straight growth as well. Among the workers attracted by the demonstration of a correlation carrier in oat tips were Cholodny (1927) in Russia and Went (1928) at Utrecht, who independently extended the correlation carrier theory to both phototropism and geotropism. They concluded that all tropisms were mediated by a growth hormone system which was essential to all plant growth.

Went, while working on the role of auxin in plant growth discovered that hormone could be collected in an agar block by diffusion, and also presented a technique for obtaining the hormone from a great variety of plant materials. Using oat coleoptile he established a test so accurate and reproducible that it still stands as the best auxin assay technique, the *Avena* test. The sound basis which Went gave to auxin physiology resulted into a great deal of very productive work in subsequent years, and in a short eight-year period, from 1928 to 1936, three auxins were isolated characterized and identified, the quantitative relationships of auxin to tropisms of roots and shoots were established, and at the end of this period half of the significant and major functions of auxin in growth and development had also been discovered.

Kogl (1933) and his co-workers found two materials, strongly active in the *Avena* test, which they named auxin *a* and auxin *b*. Auxin *a* has a molecular weight of 328, which is very close to Went's figure for diffused auxin from *Avena*, i.e. 372. These compounds, today are only of historical interest since they have never been positively isolated from growing plant tissue. Kogl group (1934), as a result of further research, discovered another auxin compound which was identical to indole 3-acetic acid. The presence of this auxin was demonstrated in *Rhizopus* cultures by Thimann (1935). It is now generally accepted that indole acetic acid is the common growth hormone in higher plants.

Since the time of Went several physiological roles of auxins have been clarified. Went (1934) himself discovered that auxins stimulated the formation of adventitious roots (throwing light on another correlative effect). The developmental importance of auxin in morphological differentiation came to a full realization when Skoog and Tsui (1948) found that relative auxin levels in plant tissues play a crucial role in determining growth pattern.

La Rue (1936) found that auxins applied to leaves could retard leaf abscission. Later researches have proved that abscission of all plant organs (leaves, flowers, fruits, etc.) is correlated with low auxin content. The development of knowledge of the auxins has had a remarkable effect on the agricultural sciences as well. A historical outline of the early and recent contributions to our knowledge of plant growth hormones has been given by Boysen-Jensen (1936).

Definitions and Consideration of Nomenclature

It will be convenient at this point to turn to a consideration of nomenclature, since a wide range of terms have been used to characterize plant growth regulating substances. It is generally agreed that an ideal system of nomenclature must be based on their biochemical and physiological attributes, irrespective of their mode and site of production and processes involved in transporting them to the places where they act. Such an ideal nomenclature must therefore embrace not only naturally occurring compounds, but also all artificially made substances having the same action in the cell. A recent treatise by van Overbeek (1950) has given a very broad definition of plant hormones as "organic compounds which regulate plant-physiological processes regardless of whether these compounds are naturally occurring and/or synthetic, stimulating and/or inhibitory, local activators or substances which act at a distance from the place where they are formed.

Similarly, the term auxin is made to include all synthetic compounds having the specific auxin action. Before entering into an extensive discussion of these compounds, it will be well to define these terms. Out of the recent definitions of a plant hormone is given by Thimann (Pincus and Thimann, 1948) as "An organic substance produced naturally in higher plants, controlling growth or other physiological functions at a site remote from its place of production and active in minute amounts "Thimann favours the term "Phytohormone" (Greek : Phyton = a plant) which literally means a plant hormone. *Growth hormone* is the phytohormone involved in growth. The term auxin was originally suggested to refer to substances which were capable of promoting growth in the manner of the growth hormone. The term *growth regulator* refers to organic compounds other than nutrients, small amounts of which are capable of modifying growth. Thimann Pincus and Thimann, (1948) has defined auxin as "an organic substance which promotes growth (*i.e.*, irreversible increase in volume) along the longitudinal axis when applied in low concentrations to shoots of plants freed as far as practicable from their own inherent growth-promoting substances. Auxins may, and generally do , have other properties, but this one is critical." This definition is widely accepted among plant physiologists today.

The Extraction, Detection, and Estimation by Bioassay of Auxins

Although a detailed account of extraction and estimation techniques of auxin is beyond the scope of this text, a brief account may however be given. All procedures for the determination of auxin content of plant materials do not measure the same constituents. There auxin concepts are, however, not entirely simple. There are many examples were bound auxin is released in free form during extraction. Consequently it appears that the free and bound forms are in a dynamic stage and their strictly separate measurement is often difficult. There are two well defined methods for the extraction of auxins from material.

Diffusion method

Diffusion method, the simplest of all, consists of obtaining the growth hormone from plant material by diffusion into agar. The procedure consists of severing the growing tip or other organ to be tested under conditions which discourage transpiration and placing the cut surface for a period of an hour or so on a block of agar, usually of 1.5 per cent concentration. Three major difficulties may arise in the operation of diffusion techniques :

(a) The excessive loss of water, or a negative tension in the vascular system can prevent the accumulation of the diffusate in the agar block.

(b) The destruction of the auxin at the cut surface frequently interferes with the quantitative yield. Such destruction is enzymatic and is attributed in some cases, to polyphenol oxidase and in other cases to peroxidase.

(c) Growth inhibitors, whose existence is common in many green plants may prevent the effective use of diffusion techniques.

Solvent extraction method

Several solvents have been employed for the extraction of auxins from plant material. Among the solvent used are, chloroform, diethylether, ethyl alcohol and even water. A serious drawback to the use of chloroform, however, is the slow accumulation of a toxic substance, perhaps an auxin in activator, thought to be chlorine Thimann and Skoog (1940). Boysen-Jensen(1936) demonstrated that diethylether may serve as the most satisfactory solvent. But before use the ether should be redistilled over ferrous sulphate and calcium oxide in a small amount of water. This is done in order to avoid an oxidation of easily oxidizable auxin by the presence of spontaneously formed peroxide. It has been amply demonstrated that the formation of new auxin during extraction may be a source of serious error. This can be successfully avoided by using Gustafson's (1941) technique which consists essentially of boiling the plant material a short time prior to solvent extraction. The main features of his technique are as under:

1. Freezing the tissue rapidly on carbon dioxide ice.
2. Grinding the tissue with mortar and pestle.
3. Dropping the tissue into boiling water and allowing one minute of active boiling.
4. Collecting the plant material on a filter paper and extracting with ether for 16 hours, using 3 changes of ether.

Gustafson's technique has some limitations, since as pointed out by Thimann, Skoog and Bayer (1942), certain amount of the free auxin must be destroyed during heating, and by Van Overbeek, *et al.* (1945) that heating may cause the release of inhibitors which interfere with auxin assay. The formation of new auxin during extraction is also avoided by the use of freezing and lyophilization technique first described by Wildman and Muir in 1949. Their technique may be summarized as follows:

1. Plunging the plant material into liquid air or dry ice and acetone for rapid freezing.
2. Drying by lyophilization.
3. Grinding to 40 mesh in Wiley mill.
4. Storing in vacuo over phosphorus penta-oxide, in darkness, until use.
5. Extracting with peroxide-free ether (95 percent) at 0°C for four half-hour intervals.

6. Combining ether extracts and reducing in volume to a few ml.
7. Transferring qualitatively to agar for *Avena* assay. Some plant materials may not require either boiling or lyophilization for solvent extraction of free auxin.

Van Overbbeek *et .al* (1945) have developed a technique for obtaining free auxin using short term extraction. It may be epitomized as below:

1. Freeze plant material on CO_2 ice.
2. Slice bulky tissues into 2-5 mm slices.
3. Extract with peroxide-free ether at 0°C for two half-hour intervals.
4. Combine the ether extracts and reduce volume by evaporation to a few ml.
5. Transfer quantitatively to agar for *Avena* assay.

All the above techniques have been designed for the extraction of free auxin. Bound auxins can be extracted by using either of the latter two methods by carrying on extraction over a period of time.

Relation of Auxins to Growth of the Oat Coleoptile

The *auxins* have been the most comprehensively investigated group of plant hormones. Their action was first clearly demonstrated in the leaf sheath or *coleoptile* of the oat plant (*Avena sativa*). This is a tubular, leaf-like structure, closed at the top, which is the first part of the plant to emerge from the soil. Similar coleoptiles develop during early seedling growth of other members of the grass family. The coleoptile encloses the initial leaf and is eventually

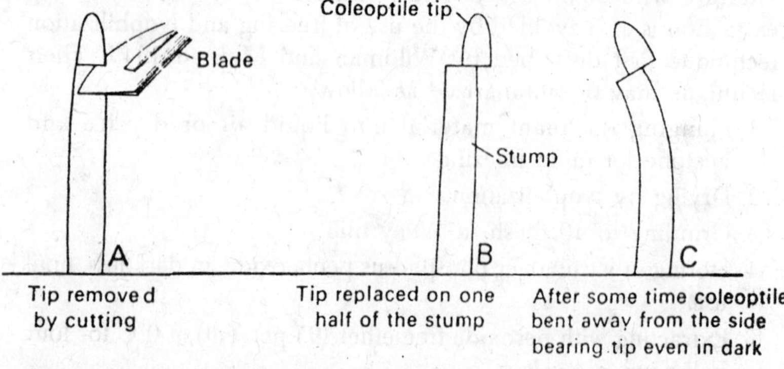

Fig. 12.1. Discovery of auxins. Experiment of Paal-eccentric placement of tip caused curvature.

pierced at the tip as a result of the growth of this leaf, soon after which all growth in length of the coleoptile ceases. Oat coleoptiles are approximately 1.5 mm. in diameter, and when illuminated seldom attain lengths of more than 2 cm. In the dark they may attain height ranging up to 6 cm. Cell divisions cease relatively early in the life history of an oat coleoptile, and during approximately the last three- fourths of its growth period all increase in its length results from cell elongation. If the tip of a coleoptile is removed by a clean cut made several millimeteres below the apex, the rate of elongation of the stump is immediately retarded.

If, however, the cut-off tip of the coleoptile or a similar tip from another coleoptile is affixed on the stump, its elongation will be resumed, and may nearly regain the original rate. Retipping the coleoptile with a short segment cut out of another coleoptile somewhat below the apex results in little or no increase in elongation rate. Such experiments indicate that the elongation of a coleoptile, which occurs in the more basal regions, is maintained only under the influence of some sort of a "stimulus": orginating in the tip, whence it is transmitted basipetally (apex to base) through the coleoptile. Went (1928, 1935) placed the cut-off tips of oat coleoptiles on a thin layer of 3 percent agar and after 1 hr, removed them and sliced the agar into a number of equal-sized small blocks. If

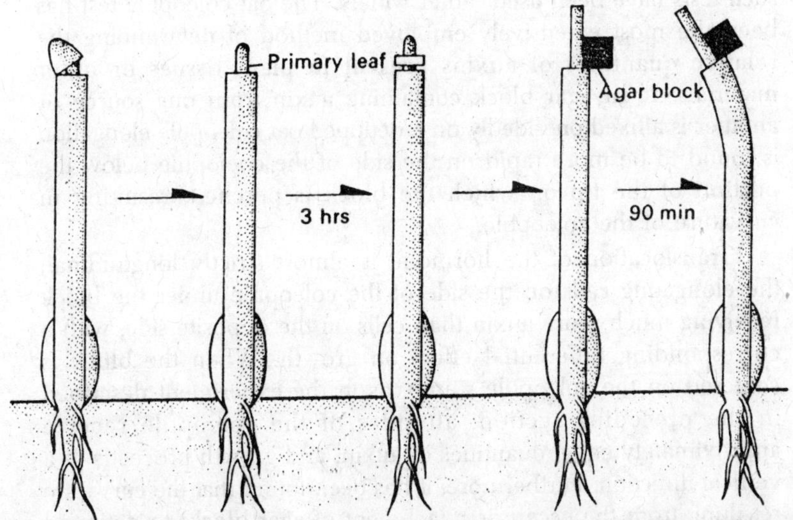

Fig. 12.2. Discovery of aux'ns. Experiments of F.W. Went–Avena Curvature test.

one of these blocks was placed upon the stump of a detipped coleoptile, the rate of elongation was accelerated just as if the stump had been capped with a fresh coleoptile tip. On the other hand, retipping a coleoptile with a block of pure agar had no appreciable accelerating effect on elongation.

It seems evident from these results that some substance or substances were transported out of the cut-off tip into the agar block, and subsequently out of the block into the coleoptile stump, whence they were translocated downward to the elongating region of the coleoptile. The substances which induce such a reaction now are classed as auxins. Many other plant organs such as stems, petioles, flower stalks, and coleoptiles of other species behave similarly upon removal of the apical region. Elongation is stopped or retarded by such a treatment but will be resumed if the excised apex of the organ is carefully relocated on the cut surface of the sump.

Biological Tests for Auxins

Auxins are known to be of widespread distribution in plants. They occur in such small quantities, however, that detection of their presence in an organic material by chemical methods is usually impossible. Recourse is had, therefore, to sensitive biological tests in order to demonstrate the presence of these substances. Several such tests have been used rather widely. The oat coleoptile test has been the most extensively employed method of determining the relative quantities of auxins present in plant tissues or other materials. If an agar block containing auxin from one source or another is affixed onesidedly on a detipped oat coleoptile elongation is found to be more rapid on the side of the coeloptile below the portion of the tip on which the block is perched, resulting in curvature of the coleoptile.

Translocation of the hormone is almost strictly longitudinal, the elongating cells on the side of the coleoptile under the block receiving much more auxin than cells on the opposite side, with a corresponding differential effect on growth. When the block is centered on the coleoptile stump, as in the experiment described in the preceding section, all side's of the coleoptile receive approximately equal quantities of auxin, and growth proceeds in a vertical direction. Furthermore, it has been found that the curvature resulting from the eccentric attachment of agar blocks to detipped oat coleoptiles is proportional, within the range of about 0 to 20

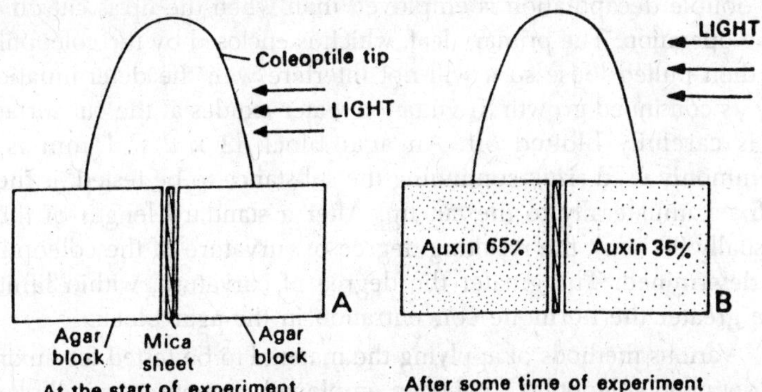

Fig. 12.3. Differential distribution of auxin in coleoptile tip under the influence of unilateral light.

degrees, to the concentration of the auxin in the agar block. It is this proportionality between hormone concentration and curvature that makes possible the use of oat coleoptiles as living test objects in the quantitative estimation of the auxin content of plant tissues or other materials.

Quantitative measurements of auxins by the oat coleoptile technique must be carried out under carefully standardized conditions. The oat seedlings (usually from a genetically uniform variety) are grown in a dark room at a temperature of 25°C, and a relative humidity of 90 percent. All manipulations are performed under phototropically inactive orange or red light. The Coleoptiles are used when about 2.3 to 4cm in length. The extreme tip of the coleoptile is first cut off, and after 3 hr, the topmost 4mm of the stump is removed. For reasons which cannot be considered in a brief discussion, the coleoptiles are more sensitive when this method

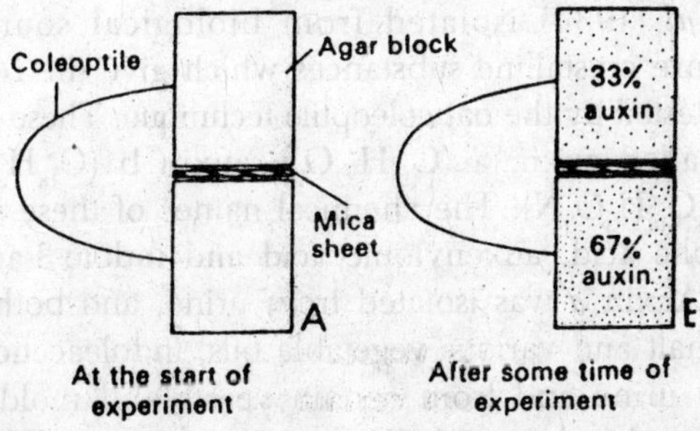

Fig. 12.4. Differential distribution of auxin from coleoptile under the influence of gravitational force.

of double decaptitation is employed than when the tip is cut off in one operation. The primary leaf, which is enclosed by the coleoptile, is then pulled loose so it will not interfere with the determination by its continued growth. If guttation water exudes at the cut surface it is carefully blotted off. An agar block (2 x 2 x 1 mm is a commonly used size), continuing the substance to be tested, is then affixed unilaterally to the cut tip. After a standard length of time (usually 90 min), the resulting degree of curvature of the coleoptile is determined. The greater the degree of curvature, within limits, the greater the hormone concentration in the agar block.

Various methods of applying the material to be tested for auxins to detipped coleoptiles have been employed. Sometimes small plant organs or pieces of plant organs are affixed directly to the cut surface of the coleoptile. More commonly plant tissues are placed in contact with 3 percent agar into which the hormone will move. The tissue is generally left in contact with the agar for about 2 hr. The agar is then cut into block of standard size each of which is then affixed unilaterally to a detipped coleoptile. The effects of pure chemicals or extracts can be tested by first dispersing them in agar, and them, after solidification, determining the influence of standard sized blocks of this agar on curvature of the coleoptiles. Or agar blocks can be soaked in a solution of the substances and then used in the coleoptile test for auxins.

Another commonly used biological test for auxins is to immerse sections of young oat coleoptiles in the solution to be tested and measure their increase in length over a certain period of time (commonly 6=25 hr.) as compared with increase in length of similar section immersed in water or an auxin free solution.

Chemical Constitution of the Naturally-occurring Auxins

Kogl *et al* (1934) isolated from biological sources three chemically pure crystalline substances which give the reactions of auxins when tested by the oat coleoptile technique. These substances have been called auxin a ($C_{18}H_{32}O_5$), auxin b ($C_{18}H_{30}O_4$), and heteroauxin ($C_{10}H_3O_2N$). The chemical names of these substances are auxentriolic acid, auxenylonic acid and indole-3-acetic acid, respectively. Auxin *a* was isolated from urine, and both auxins *a* and *b* from malt and various vegetable oils, indoleacetic acid was isolated from urine and from certain yeasts and molds. At one time it was considered possible that indoleacetic acid did not occur in the tissues of higher plants, but more recently it has been isolated from corn grains and there are numerous indications that it occurs in many other tissues of higher plants including oat coleoptiles. It seems probable that this will prove to be the principal naturally-occurring auxin.

The Occurrence and Synthesis of Naturally-occurring Auxins in Plants

Auxins appear to be universally present in plants, and their occurrence has actually been demonstrated in a wide variety of species. Furthermore, the auxins are nonspecific in their action, i.e., the same auxin, chemically speaking, which influences growth phenomena in one species usually also influences the same or similar phenomena in other species.

Auxins are present in plant cells in several different forms: free auxins, auxin precursors, and bound auxins. Various methods of extracting the "total" auxins (all forms) from plant tissues have been devised and more or less successfully employed. Other investigators have devised methods which they believe extract only the free auxins from plant tissues.

Only in the free form is an auxin readily diffusable, and only in this form is it effective in the oat coleoptile tests. Since, however, free auxins may be continuously formed from bound or precursor auxins, considerably more free auxin than originally present may diffuse into agar blocks from plant tissues. In most tissues thus far investigated the auxins present in inactive forms are many times greater than the free auxins present; the later seldom constituting more than 10 percent of the potentially available auxins.

The amino acid tryptophane has been shown to be a precursor of indoleacetic acid in plants. Conversion of the former compound into the latter apparently takes place through the intermediate stop of indoleacetaldehyde, which is therefore an even more immediate precursor of indoleacetic acid. The transformation of tryptophane to indoleacetic acid is catalyzed by a specific enzyme system which has been found in a number of plant tissues. The presence of zinc is necessary for tryptophane synthesis, and one of the indirect effects of zinc deficiency is a reduction in the quantity of auxin present. Following are the structural formulas for tryptophane and indoleacetic acids.

Many, but not all, of the bound auxins are auxin-protein complexes. Since, by suitable treatments, active auxins can be released from bound auxins, the latter are sometimes referred to as "auxin precursors," although they would not be so regarded in the usual sense of the term.

The auxins naturally present in plants are synthetic products of plant metabolism. The principal centers of auxin synthesis are apical meristematic tissues of aerial organs such as opening buds, young leaves, and flowers or inflorescences on growing flower stalks. Small quantities of auxins are also synthesized in apural root meristems although much of the auxin present in roots probably comes from aerial organs of the plants. The auxin synthesized in one tissue is frequently translocated to other organs of the plants. The concentration of auxin may vary greatly from one tissue to another; in general auxin is found in greatest concentration in the tissues in which it is synthesized or stored. Temperature is a factor in auxin synthesis, but the optimum for this process probably is not the same in all plants and tissues. In coleoptile tips the auxin is apparently synthesized from a precursor which is translocated atropetally (base to apex) through the coleoptile from the grain.

Auxins are not only synthesized in plant cells but are also inactivated in them. Inactivation may be brought about in various ways. Among other inactivation agents is a specific enzyme which breaks down indoleacetic acid; this enzyme has been isolated from the epicotyls of etiolated pea seedlings.

The Role of Auxins in Cell Elongation

Auxins play a role in the elongation phase of growth in many other plant organs similar to that described for the oat coleoptile. It is generally considered that cell elongation occurs only in the presence of auxins, and that with increase in auxin concentration there is an increase in the rate of elongation if no other factors are limiting. The optimum range of concentration for cell elongation varies greatly with different tissues, and relatively high concentrations usually exert an inhibiting, effect upon this phase of growth.

If the extreme tip of a maize or lupine root is cut off, its rate of elongation increases, although not greatly. Replacement of the root tip in maize plants results in a retandation in elongation rate at compared with detipped roots. Furthermore, attachment of coleoptile tips of maize to detipped root tips of the same plant results in a retardation in the elongation rate of the root tip. These results suggest that the same concentrations of auxin which accelerate elongation in coleoptiles and other aerial organs retard elongation in roots.

This supposition has been confirmed by experiments in which the roots of oat seedlings were immersed in pure solutions of auxins. The growth of the roots was found to be retarded in proportion to the concentration of auxin used. However, when roots which contain either no auxin at all or virtually none, are treated with auxin solutions of very low concentration, acceleration of growth as compared with similar but untreated roots often results.

The apparently contrasting effects of auxins upon elongation of roots and aerial organs may be explained by assuming that roots, buds, and stems all react in a comparable way to auxins their growth being inhibited by relatively high, and promoted by relatively low, auxin concentrations. Elongation of roots is favoured only at very low concentrations; at all higher concentrations their growth is checked. Stems and coleoptiles show a similar behaviour except that the optimum range of concentrations for elongation is much higher than for roots. The same concentrations of auxins which favour stem elongation result in retardation of root elongation. Briefly, therefore, whether an auxin will exert an accelerating or an inhibiting effect upon growth seems to depend in part upon its concentration and in part upon the specific tissue involved.

Shortly after the discovery of the naturally-occurring auxins various investigators showed that certain compounds, not known to occur naturally in plants, induce reactions in plants similar to those evoked by the naturally-occurring auxins. The list of such "synthetic" auxins has become quite extensive. The best known of these substances are α-naphthalene acetic acid, indolebutyric acid, 2,4-dichlorophenoxyacetic acid, and α-naphthoxyacetic acid. Most of these compounds induce curvature of oat coleoptiles when tested by the standard technique, although many of them are not as effective in causing this reaction as the naturally occurring auxins.

The term "auxin" has been used in different senses by different authorities. Most commonly, however, an auxin is considered to be any organic compound that promotes growth along a longitudinal axis when applied in low concentrations to organs which are initially low in their content of such growth-promoting substances but are under conditions which are otherwise favourable for elongation growth. The phrase "low concentrations" is a somewhat vague one, but in general can be taken to refer to concentrations of less than 10^{-3} molar. In spite of the apparent diversity of the compounds which act as auxins, certain similarities in molecular structure are common to all of them. These are : a ring system with a side chain containing at least one carbon atom between a terminal carboxyl or potential carboxyl group and the ring, a double bond in the ring adjacent to the side chain, and a definite space relationship between the carboxyl group and the ring system. *Cf.* the structural formula for indoleacetic acid given earlier.

Translocation of Auxins

If a block of agar containing auxin is affixed to the morphologically upper end of a segment of oat coleoptile, and a block of pure agar to the lower end, auxin will move into and accumulate in the lower block. The final concentration of auxin in the basally attached block may greatly exceed that in the one affixed to the apex. If the agar block containing auxin is affixed to the morphologically basal end, no translocation of auxin will occur. Translocation of auxin in oat coleoptiles apparently takes place through the paraenchyma tissues.

The results of such experiments show that transport of auxin in the oat coleoptile is *polar* i.e., occurs only basipetally, and that it can occur against a concentration gradient since the auxin

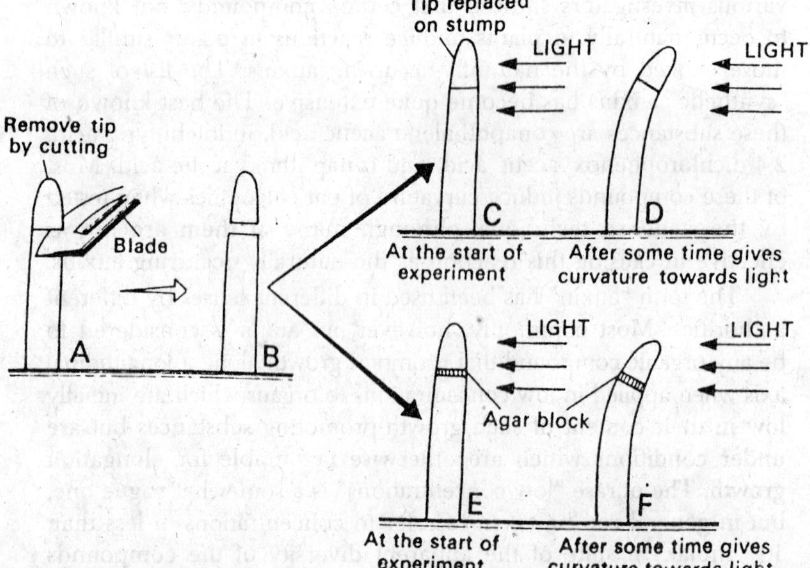

Fig. 12.5. Discovery of auxins. Experiments of Boysen-Jensen.

accumulates in the lower block. There is evidence that a similar basipetal translocation of auxin, either naturally-occurring or introduced, occurs in many other plant tissues or organs. Among these are coleoptiles of other species than oat, the veins and petioles of leaves, hypocotyls, herbaceous stems, and woody stems. In such organs the polarized downward movement occurs in parenchyma or phloem tissues. In roots on the other hand, movement of auxins appears to be nonpolar and there are undoubtedly other tissues in which this is also true. The mechanism of the polar movement of auxins is unknown. Etherization stops the transport of auxin, except insofar as it can be accounted for by diffusion, and destroys its polarity. This indicates that living cells are involved in the process, but tells nothing of the manner in which they operate.

It has been suggested that the polarity in the movement of auxins may result from differences of electrical potential in the tissues, but experiments designed to test this hypothesis have not yielded evidence in its support. Upward translocation of auxins can occur in plants, at least under certain conditions. If an auxin is applied to the soil or to the basal parts of entire plants or to cuttings in adequate concentrations, their absorption and upward movement through the plant can be demonstrated. Such upward transport apparently takes place only when the auxin molecules pass into the transpiration stream. As soon as the auxin molecules move back into the living tissues of the stem or leaf their polar basipetal movement is resumed.

Role of Auxins in Root Formation

It has been known for many years that the presence of buds on a cutting favours development of roots when the basal portion of the cutting is introduced into a suitable rooting medium. Developing buds are more effective in promoting root formation than quiescent buds. Leaves, especially if young, also often favour the production of roots on cuttings. These observations suggest that root initiation on cuttings is favoured by hormones which are synthesized in the buds and young leaves and are subsequently translocated to the basal part of the cutting. Soon after the identification of them as naturally occurring auxins it was found that auxin *b* and indolesacetic acid are active in inducing root formation.

There is good evidence that other hormones besides auxins are also necessary for root formation or atleast for their continued development, whether the roots are initiated on other roots, on stems, or on leaves. The effect of auxins on root *formation* should be clearly distinguished from their effect on root *elongation.* In general, the concentrations required for the former process are much greater than for the latter. A number of the "auxins" not known to occur naturally in plants have also been found to be effective in promoting root formation in many species.

Extensive experiments have been carried out on the suitability of treatments with various auxins as a practical method of aiding in the rooting of cuttings. Such treatments are not effective with all kinds of plants, but with cuttings of many species they lead to a speeding up of the process of root formation and to the development of a greater number of roots per cutting. Hormones do not induce root formation, however, on cuttings of species on which at least some roots do not develop without their application. Various techniques are used to introduce hormones into cuttings. Formerly the most favoured procedure was to immense the basal end of the cuttings in a dilute (10-200 p.p.m) solution of the hormone for periods ranging up to 24 hr, before setting them in the rooting medium. Laterally this method has been largely superseded by two less time-consuming procedures. In one of these the hormone is applied in a day form, being first mixed with an inert powder such as tale, most commonly in proportion of 500-2000 parts of the hormones to 1,000,000 parts of tale.

The basal end of the cutting is first dipped in water, and then in the powder before insertion into the rooting medium. In the "quick dip" method the basal ends of the cutting are dipped momentarily (for about 5 sec) into a relatively concentrated solution of the hormone (4,000-10,000 p.p.m, in water or 50 percent ethyl alcohol) before being a set in the cutting bench. Auxins many also be applied to stems or cuttings in lanolin or an vapours. The compounds most commonly employed in all of these methods, either singly or in mixtures, are α-napthaleneacetic acid, napthalene acetamide, and indolebutyric acid. The most effective treatment varies according to species; extensive tabulations of the effects of various compounds and methods of treatment on numerous species are given by Avery and Johnson (1947) and Mitchell and Marth (1947).

Other Effects of Auxins

One of the most remarkable facts about the auxins is the multiplicity of the growth reactions in which they participate. Only

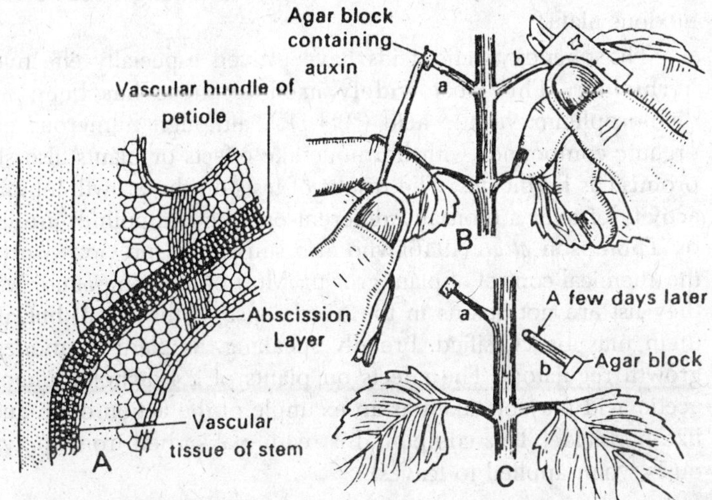

Fig. 12.6. Effect of auxins on abscission. A–A normal abscission layer at the base of petiole. B–(a) agar block contains auxin is fixed on debladed petiole (b) agar block without auxin is being fixed on debladed petiole. C–(a) There is retardation of abscission of petiole (b) Petiole abscised.

a few of their roles are discussed in this chapter; other important growth phenomena in which they play a part include the development of fruits, abcission of leaves and fruits, activation of cambial cells, apical dominance, flower initiation, geotropism and phototropism. These various roles of the auxins will be discussed in subsequent chapters.

Toxic Effects of Auxins

Reference has already been made to the inhibitory effect of relatively high concentrations of auxins on the process of cell elongation. Applications of auxins in relatively high concentrations also result in various kinds of growth malformations in plants such as distortions of leaves, stems and roots, discolourations of leaves, inhibitions of stem or root elongation or flower opening, and the formation of tumors. It should be emphasized that the term "relatively high concentration" as used in this discussion refers to concentrations which, in absolute terms, are very low, of the general order of magnitude of 1000 p.p.m. In somewhat higher concentrations, but in absolute terms still very low, these compounds often result in death of the plant. Realization that auxins, when applied in relatively high but actually very low concentrations, exert toxic or lethal effects on plants led to the suggestion that they could be employed for the purpose of killing weeds or other noxious plants.

The phenoxyacetic acids have proved especially effective as herbicides. The most widely used of these has been 2,4 - dichlorophenoxyacetic acid ("2,4-D"), although numerous other organic compounds with hormone-like effects or plants also show promise as herbicides. The results of tests on the growth-inhibiting activity of over a thousand different organic compounds are listed by Thompson *et al.* (1946),who also summarize previous work on the chemical control of plant growth. Most of the compounds which they list are not auxins in the usual sense of the word, but all of them may be classified, broadly speaking, as plant hormones or growth regulators. The effects on plants of 2,4-dicholorophenoxy acetic acid may be taken as an example of the action of an auxin-like herbicide. This compound is readily absorbed from sprays or dusts when applied to leaves.

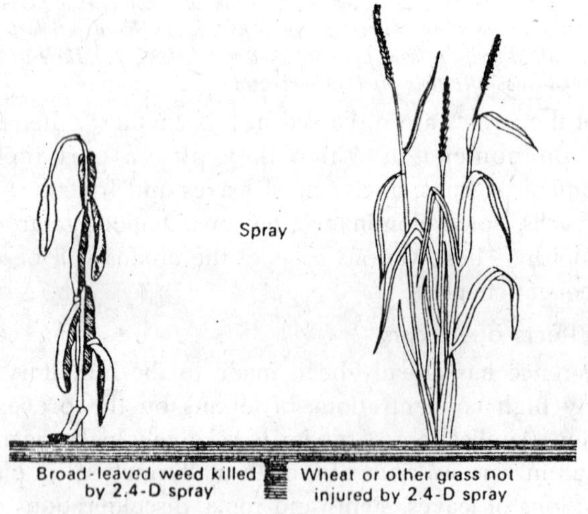

Fig. 12.7. Destruction of weeds by 2, 4-D.

It is quickly translocated to other parts of the plant and affects especially the meristems. The rapid distribution of this compound throughout the plant contributes greatly to its effectiveness as a toxicant. Death results from demagements of metabolism, especially in the meristems. Different kinds of plants differ markedly in their reactions to applications of 2,4 -dichlophenoxyacetic acid. Cereal

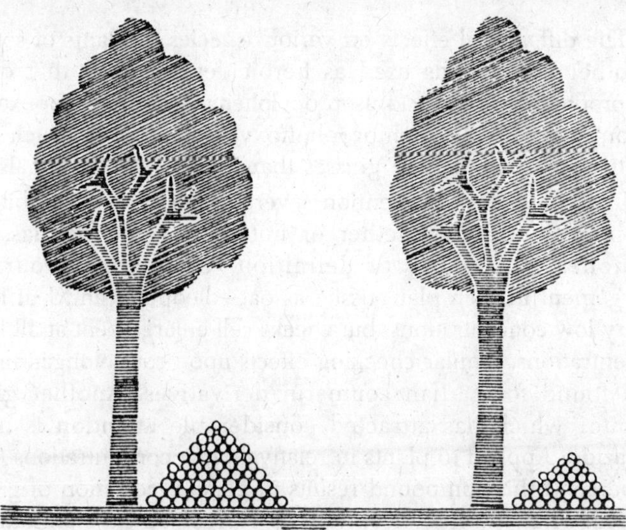

Fig. 12.8. Auxin spray on a plant prevents premature fruit abscission and increases the yield. A–Unsprayed, B–Sprayed.

grains and most other grasses are less susceptible, for exmple, than most broad-leaved annuals, and most woody plants are less susceptible than most herbaceous species. This very selectivity of effect is one of the advantages of this compound, and of many similar ones, as herbicides. Broad-leaved weeds can be elinimated from sugar cane fields or from lawns, for example, by application of a spray in the proper concentration and at the proper volume per square foot. Besides its selectivity of effect, 2,4-dichlorophenoxyacetic acid has certain other advantages when used as herbicide. The absolute concentrations required are very low.

Many plants can be killed, for example, by spraying with an 0.1 percent solution at the rate of 5 gal.per 100 ft^2. In the concentrations generally employed this compound is harmless to men and animals. Under most conditions, 2,4-dichlorophenoxyacetic acid disintegrates rapidly in soils. The possibility of detrimental effects on plants which subsequently grow on soils in which this compound was once present thus disappears rapidly. Great care must be exercised, however, in the use of this herbicide, that the spray does not drift to plants other than those being treated. Accidental damage or destruction of valuable plants as a result of careless application has been the chief disadvantage in the use of 2,4-dichlorophenoxyacetic acid as an herbicide.

The differential effects on various species of plants of some of the other compounds used as herbicides is unlike that of 2,4-dichlorophenoxyacetic acid. Isopropylpheny learbamate, for example, in contrast with 2,4-dichlorophenoxyacetic acid is much more effective in killing certain grasses than broad-leaved annuals

It is convenient to mention several other growth inhibitors at this point, although whether or not they would be classed as hormones is a matter of definition. Coumarin favours cell enlargement in such plant tissue as oat coleoptiles and leaf blades at very low concentrations, but checks cell enlargement at all higher concentrations. Similar checking effects upon root elongation have been found for certian coumarin derivatives. Another growth inhibitor which has attracted considerable attention is maleic hydrazide. Applied to plants in relatively low concentrations (about 0.4 percent) this compound results in general cessation of growth. At somewhat lower concentrations loss of apical dominance is evident, and axillary buds start to grow soon after treatment. At certain concentrations flowering appears to be largely suppressed without much suppression of vegetative development. Plants of various species seem to be affected in essentially the same way by this compound.

Mechanism of Auxin Action

The earlier attempts to elucidate the mechanism of auxin section mostly centered around proposed explanations of the part played by such compounds in cell elongation. Auxins appear to have two main effects in this process: they cause an increase in the plasticity of the wall and they participate, directly and indirectly, in the reactions whereby additional cellulose molecules are deposited within the wall. With the growing relization, however, of the large number of growth reactions conditioned by auxins, it has become obvious that these compounds must play a role in some pivotal metabolic process. The apparent necessity of specific chemical structure for a compound to act as an auxin, mentioned earlier, also suggests a particular metabolic role. The effect of auxins on cell wall development is, therefore, now generally considered to be, not a direct one, but only one possible end expression of fundamental auxin-conditioned or regulated metabolic processes.

There is considerable evidence that the auxins act primarily in a catalytic or regulatory capacity in some phase of the carbohydrate metabolism of plants. A suggestive finding in this connection is that introduction of auxins into leaves or cuttings induces marked hydrolysis of starch. Auxin also appears to participate in some part of the respiratory process but the exact nature of this relationship has not yet been traced. A significant fact in this connection is that auxins exert their effects only under aerobic conditions. The discovery by Wildman and Bonner (1947), that certain auxin protein complexes from spinach leaves possess phosphatase activity, may prove to be a significant one in allocating a specific role to the auxins. Although the question of the metabolic role of the naturally-occurring auxins as yet cludes a definite answer, it seems practically certain that they operate as carriers, coenzymes, or prosthetic groups in come fundamental enzyme system which plays a part in carbohydrate or organic acid metabolism.

The inhibiting as well as the accelerating effects of the auxins must be accounted for in any hypothesis of the mechanism of their action. If we accept the likely assumption that an auxin is active when it operates as the prosthetic group of an enzyme, it is reasonable to postulate that two fundamental properties of its molecules would be: (1) a specific group which reacts with the substrate molecules, and (2) the necessary structural configuration to combine in some manner with the protein portion of the enzyme.

Viewed in the light to this hypothesis (which incidentally is also applicable to other enzyme systems), relatively high concentrations of an active auxin may result in inhibitory effects as a result of some of the auxin molecules preempting all free positions on the protein component of the enzyme, while others become chemically attached to substrate molecules. This effectively blocks the reaction which can occur only when the protein and auxin operate as a catalytic team, so that substrate molecules become linked temporarily to the protein through auxin bridges

Inhibition of auxin-controlled reactions may also result from the presence in the metabolic system of growth-inhibitors with an auxin-like structure which possess strongly only one of the two properties listed above and the other one only weakly or not at all. If such a compound possesses only the capacity of combining with the protein, it may act as an inhibitor by occupying positions in the protein complex which would otherwise be taken by more active auxins. On the other hand, if the auxin like compound possesses only the property of reacting with the substrate, it may block the overall reaction which can only take place if the compound can act as a chemical bridge between the protein and substrate molecules.

Certain compounds, often called *antiauxins*, are known which offset or counteract the usual growth-enhancing effect of auxin in plants. Among these are coumarin and maleic hydrazide both mentioned earlier in this chapter as growth inhibitors. Another antiauxin is *trans* -cinnamic acid. The effect to this compound is of especial interest in view of the fact that its isomer *cis* -cinnamic acid has the properties of an auxin.

Other Plant Hormones

The auxins are only one kind of a number of hormones occurring in plants. Some of the other compounds in this category have been isolated from plant tissues; the existence of others is inferred from the occurrence of various physiological reactions; other such substances doubtless are as yet undiscovered or unsuspected.

Traumatic acid

Haberlandt (1921) showed that if freshly cut plant tissue is immediately rinsed with water, very few cell divisions occur in the cells adjacent to the would. However, if the wounded area is smeared with finely ground tissue of the same species cell division takes place. This result led him to postulate the existence of substances which he called "wound hormones" in injured tissues, which are required if cell division is to be engendered in the cells bordering a wound.

More recently English et al. (1939) have succeeded in extracting from bean pods and purifying a compound called traumatic acid which acts like a "wound hormone", This compound has the following formula:

$$HOOC.CH = CH.(CH_2)_8.COOH$$

Traumatic acid induces extensive wound periderm formation in washed disks of potato tuber. It is probable that there are also other "wound hormones" present in plants.

Calines

Went (1938) has postulated the existence in plants of a group of hormones which he calls the *calines* : (1) rhizocaline, made in the leaves and necessary for root formation; (2) canlocaline, synthesized in roots, but necessary for elongation of stems; and (3) phyllocaline, made or at least stored in cotyledons, and necessary for leaf growth. It is also considered that rhizocaline and caulocaline may be stored in seeds. Most of the evidence for the existence of such hormones in plants is indirect and the effects attributed to these hypothetical hormones may actually turn out to result from the activity of substances already well known to have important influences on the growth or metabolism of plants. Glaston and Hand (1949), for example, have obtained evidence that adenine has effects on plants analogous to those of the postulated calines. The calines are discussed further in the following chapter.

Hormones of reproduction

Several hormones are known or believed to play a role in the reproductive process of plants. Among these are the ubiquitous auxins, the "embryo growth factor" and the postulated "florigen".

Vitamins

From the standpoint of human physiology the vitamins constitute a group of specific organic substances which must be supplied in diet, but which are required in relatively small, often only minute; quantities compared with carbohydrates, fats, and proteins. From the standpoint of living organisms in general, for reasons for discussed later, vitamins must be regarded simply as one rather arbitrarily delimited group of growth and metabolism regulators which cannot be sharply distinguished from the hormones. The principal substances recognized as vitamins in human and animal nutrition constitute a very heterogeneous group of compounds.

All of the vitamins, or at least their immediate precursors, are sythesized in green plants. The human body, on the other hand is dependent on sources outside its own metabolism for most essential vitamins and obtains a large proportion of them in green plants used as food. One partial exception is vitamin D which can be synthesized in the human body under irradiation from appropriate precursor which come from foods. Absence of deficiency of any one of the necessary vitamins in the human body results in physiological malfunctioning often evidenced as a specific disease. Some of the vitamin-deficiency diseases of man, such as scurvy, resulting from ascorbic acid deficiency, and beri-beri, resulting from thiamine deficiency, have been known for centuries, although only a comparatively recent years have their causes been recognized. Other higher animals appear to have vitamin requirements very similar to those of man.

The latest addition to the group of B vitamins, is vitamin B_{12}. This vitamin is of especial interest because cobalt is a constituent of the molecule. Whether of not this vitamin plays any role in plant metabolism is not known. Vitamin requirements differ from one kind of living organism to another, but no plant or animal is known for which at least some of them are not essential. The first fundamental question to be answered regarding the vitamin physiology of any organism is: Does the organism require this specific vitamin in its metabolism? Certain of the vitamins appear to be necessary in the metabolism of all plants and animals. This is true of many, if not all, of the vitamins of the so-called B complex. On the other hand, certain vitamins seem to be essential in the metabolism of only certain kinds of organism.

Table 12.1. The Vitamins.

"Alphabetical" Designation	*Name*	*Molecular Formula*
Vitamin A	Hydrolytic product of carotene	$C_{20}H_{29}OH$
Vitamin B (Complex)	Thiamine chloride hydrochloride (B_1)	$C_{12}H_{18}N_4SOCl_2$
	Riboflavin (B_{12} Vitamin G)	$C_{17}H_{20}N_4O_6$
	Nicotinic acid (niacin)	C_5H_4N-COOH
	Pyridoxine(B_6)	$C_8H_{11}NO_3$
	Pantothenic acid	$C_9H_{17}NO_5$
	Inositol	$C_6H_{12}NO_6$
	Biotin(vitamin H)	$C_{10}H_{16}O_2N_2S$
	Folic acid	$C_{19}H_{19}N_7O_6$
	p-amino benzoic acid	$NH_2C_6H_4COOH$
Vitamin C	Ascorbic acid	$C_6H_8O_6$
Vitamin D	Derivatives of ergosterol, cholesterol and other sterols	$C_{28}H_{43}OH$ (activated dehydrocholesterol)
Vitamin E	α-tocopherol (and others)	$C_{29}H_{50}O_2$
Vitamin K	K_1 : 2-methyl, 3 phytyl, 1,4-naphthoquinone	$C_{31}H_{46}O_2$
	K_2 : 2-methyl, 3-difarnesyl, 1,4-naphthoquinone.	

A second fundamental question to be answered regarding any essential vitamin for a given organism is: Does the organism synthesize this vitamin in quantities adequate to meet its own metabolic requirements? The answer to this question for the human organism is essentially "no" for all of the corresponding question for green plants is "yes". For bacteria and fungi no categorical answer is possible since the vitamin- synthesizing capacities of such organisms vary greatly from one species to another, and even from one variety of a given species to another.

For many years it has been known that, in the culturing of bacteria or fungi on artificial media, it is necessary to introduce some organic material such as potatoes, peptone, oat or corn meal, yeast extract, dung, or wood into the medium in addition to sugar and other pure chemicals if growth of the organism is to occur. Only in comparatively recent years has it been recognized that these organic materials are sources of essential growth substances, some of which, at least, are identical with the known vitamins.

Direct experimentation has demonstrated that many bacteria and fungi cannot synthesize certain of the vitamins which they require, or at least cannot synthesize them in sufficient quantities to permit optimum growth even when all other growth conditions are favourable. If the substrate on which they are growing is deficient or lacking in one or more of the necessary vitamins, development of the organism will be retarded and may cease entirely. By systematic addition of the various vitamins to a vitamin-free medium on which an organism is cultured it is possible to ascertain which of the essential vitamins that organism must obtain from its substrate.

It is probable that the vitamin requirements of all fungi are very similar; but different species differ greatly in their vitamin-synthesizing capacity. Many species, for example, cannot synthesize thiamine in adequate quantities. Hence, unless this compound is present in the substrate, the fungus will be retarded in development as a result of the thiamine deficiency. In capacity to synthesize adequate quantities of biotin, pyridoxine, pantothenic acid, have also been demonstrated for many species of the fungi.

It has already been mentioned that all of the known vitamins, or at least their immediate precursors, are synthesized in higher green plants. Whether or not all such compounds are essential in the metabolism of green plants is an open question. The evidence regarding this point must of necessity be somewhat indirect. There are no good reasons, however, for believing this to be true of vitamins A, D, E and K. On the other hand, thiamine, nicotinic acid, riboflavin and pyridoxine are known to be constituents of the prosthetic groups or coenzymes of important enzymes found in vascular green plants and are probably necessary in the metabolism of all living cells in such organisms. It is highly probable that some, and perhaps all, of the other B vitamins are required by green plants and ascorbic acid may also be an essential green plant vitamin.

Although the higher green plants appear to be self-sufficing with regard to necessary vitamins, and such compounds are widely distributed through the organs of such plants, this does not necessarily mean that all cells of all tissues synthesize these compounds in adequate quantities. Some parts of a plant are dependent upon other parts for certain necessary vitamins. Experiments on excised roots of certain plants growing in sterile culture show, for example, that thiamine must be added to the culture medium if normal growth is to continue. Obviously roots, at least of some species, do not synthesize enough thiamine to meet their own needs, and irract plants are presumably dependent upon down-ward translocation of this substance from leaves in which it is synthesized. Such downward translocation of thiamine, and also of pantothenic acid and pyridoxine, has been demonstrated in the tomato. This is probably also true of nicotinic acid. Upward translocation of thiamine from older to younger leaves also occurs. There is also evidence for a similar downward translocation of thiamine from older to younger leaves also occurs. There is also evidence for a similar downward translocation of ascorbic acid in plants. On the other had, roots of at least some species appear to synthesize certain other vitamins in such quantities as not to be dependent upon the tops for a supply of these compounds.

The growing embryo and endosperm in the developing seed of at least some kinds of plants appear, like roots, to be dependent upon translocation from other parts of the plant for necessary thiamine. In wheat, for example, as the content of this hormone in the developing grains increases, there is a corresponding diminution in the thiamine content of vegetative parts. It is possible that young developing embryos of some species may similarly be dependent upon vegetative tissues for other essential vitamins.

From the foregoing discussion it should be evident that no sharp distinction can be drawn between "vitamins" and "hormones" in considerations of the metabolism of higher plants. Some of the so-called hormones, such as traumatic acid, operate in or close to the cells in which they are synthesized;' others, such as the auxins, are frequently translocated to other, often distant cells, in which they influence metabolic processes. Similarly, some of the so-called vitamins, such as riboflavin, appear to operate principally in cells in which they are synthesized, whereas others, such as thiamine, are often present in some tissues largely or entirely as a result of translocation from other organs of the plant. From the standpoint of plan metabolism, all such compounds fall into the one comprehensive category of "growth-regulating substances" or "hormones".

The Effect of External Factors on Auxin Content

It is difficult to make generalizations of aid to plant physiologists owing to sparse literature on this aspect of auxin physiology. Nevertheless several such external factors may be listed, which influence the auxin content of plants:

Temperature

Few, if any, effects of temperature are interpretable in terms of auxins. Several workers in the past have attempted to study the relation between auxin content of the plant and temperature prevailing at the time of experimentation, but only a few of the recent ones are mentioned. Burton (1956) found no consistent relationship between storage temperature and ether - extractable auxin content in potato tubers. Blommaert (1955) and Hendershott and Bailey (1955) reported a decrease or total absence of auxin in winter-dormant buds of peach and marked resurgence in auxin content immediately prior to and during the onset of growth in the spring.

Mineral Nutrition

The relation between auxin content of intact plants and mineral nutrition has been studied in detail, only for nitrogen and zinc; deficiencies of either of the element lead to a lowered auxin content. The effect of nitrogen is considered to be an indirect one, since its effects appear after the appearance of visible deficiency symptoms or a decline in growth. Zinc probably affects auxin content possibly by reducing tryptophan synthesis.

Like nitrogen, effects of copper and manganese on auxin content of the plant appears to be indirect, because in copper and manganese deficient plants decreased auxin content is found only after the appearance of visible deficiency of these elements. Eaton (1940) suggested that boron deficiency reduced auxin level in cotton plants but MacVicar and Tottingham (1947) failed to confirm these results with cotton, tomato, sunflower, soybean and tobacco plants. The effect of other mineral elements on the endogenous auxin level is little studied and the facts thus, are incoherent.

Hydrogen-ion Concentration

Rufelt and Fransson (1956) studied the effect of pH on the auxin content of wheat roots, with solutions at pH 5.0-7.5. They noticed a short-term increase in auxin content, with a maximum at about six hours after transfer, and a subsequent restoration of the original level after 24 hours in the new solution. The mechanism of this effect is, however, obscure.

Ionizing Radiations and Ultraviolet

The publications on this aspect clearly indicate that X-irradiation can markedly and rapidly lower auxin content probably by interfering with auxin synthesis.

Skoog (1935) reported that moderate X-ray doses immediately reduced the amount of diffusible auxin obtainable from apices of *Pisum*, *Vicia* and *Avena* plants. Gordon's work (1956, 1957) has confirmed Skoog's observations of immediately decreased auxin content following X-irradiation but stresses the significance of reduced auxin synthesis. Spencer and Cabanillas (1956) reported that exposure of *Indigofera endecaphylla* seeds to either X-rays (80,000r) or thermal neutrons, both significantly reduce the auxin content of plants emerging from the treated seeds. The effect of ultraviolet irradiation has also been found to decrease endogenous auxin level.

Visible Light

A considerable work on the effects of visible radiations on the auxin content of plants had been undertaken. High light intensity, length of photoperiod and effects on etiolated plants, all have been investigated in detail. The photoinactivation of auxins has already been discussed in this chapter. It may be concluded that visible light has the effect of inactivating the auxin present in the plants.

Chemicals

Certain externally applied chemicals, such as ethylene, maleic hydrazide, and 2, 4-dichlorophenoxyacetic acid have been found to exet influence on the auxin content of intact plants. Lockhart and Weintraub (1957) showed that 2,4-D decreases the free auxin content of terminal shoot of *Phaseolus* seedlings. It is unanimously but unusually concluded that growth inhibitory properties of maleic hydrazided are not due to any effect on auxin content. Exposure of *Avena* coleoptile tips to ethylene has been found to decrease the diffusible auxin content by van der Laan (1934).

Gibberellic Acid

Nitsch (1957) reported that application of gibberellic acid to certain woody plants results in an increase in extractable auxin, Galston's observation that feeding of gibberellic acid (GA) to pea plants raises the level of an auxin inhibitor of the enzyme IAA oxidase leads us to believe that the overall effect of GA is that it has an auxin sparing action. Similar view was expressed by Stutz and Watenabe (1957) with *Lupinus albus.* Both gibberellic acid and long photoperiods were found to depress overall initial IAA oxidase activity, thus probably leading to a "sparing" of auxin by gibberellic acid.

Miscellaneous

Several other factors, in addition to those discussed above may be responsible for altering both the level and metabolism of the endogenous auxin in the plant. Among these may be mentioned: (i) abnormal growth and pathology (ii) root-nodule bacteria, (iii) crown gall, (iv) virus diseases, and (v) other pathogens. All these, in some way or the other, are known to alter auxin level of intact plants and plant parts.

Theories of the Mechanism of Auxin Action

It would appear to be over-simplication if we ascribe the full responsibility for growth to any one function in the plant. This is so because the action of auxins in the control of growth is a complex of many functions. Although the mechanism of these functions it may be concluded that none of them known till today can entirely account for the effects of auxin on growth.

The numerous extremely interesting physiological and biological responses to plant hormones, observed in the last 25 years account for the intricate bearing on their mechanism. Gordon (1952) enumerated some of the many responses which may be briefly summarized.

1. increase in plasticity of the shoot and elasticity of root cell walls,
2. increased permeability of the cell to water,

3. enhanced capacity to retain water taken up,
4. "active"; uptake of water and solutes,
5. decreased photoplasmic viscosity,
6. increased respiratory rate,
7. more rapid synthesis of proteins with lowered levels of free amino acids in the free amino acid pool,
8. increased accumulation of monosaccharides at the expense of polysaccharides, and
9. a stimulation of the activity of many enzymes, such as alcohol and malic dehydrogenase, catalase, phosphatase and ascorbic acid oxidase, and also several other responses.

Several theories which have been put forward to explain the mechanism of auxin action consider one of the above effects as the basic response on which other phenomena rest. For the sake of convenience, discussions of the hypotheses concerning the mechanism of auxin action will be grouped into five heads.

Molecular Reaction Theories

A molecular reaction into which auxin might enter in causing growth was first suggested by Skoog et.al (1942) who postulated that auxins may act as coenzymes, serving as a point of attachment for some substrate onto an enzyme controlling growth. A different suggestion for the role of auxins was forwarded by Veldstra (1955). He opined that the degree of fat solubility as influenced by the ring structure and water solubility as influenced by the side chain structure could be correlated with auxin activity. His conception was that the action of auxin was something of a physical binding of some fatty material to some more aqueous phase.

Muir *et al.* (1949) gave a third suggestion that hormones of the phenoxy acid type may combine with some material (presumably proteinaceous) at the ortho position of the ring. They conceived that hormones react with some material in the cell at two positions. viz (i) at some position in the ring (ortho position in the phenoxy acids), and (ii) at the acid group of the side chain. On the basis of these Foster *et al* (1952) advanced a theory of hormone activity by two-point attachment and brought forward variety of kinetic evidences to support their hypothesis.

Theories of Enzyme Effects

It has been observed that growing tissues when treated with auxin show an increased activity, either directly or indirectly, of a number of enzymes. This fact has given rise to a theory of auxin action through enzyme mechanism. Northen (1942) found that auxins cause decrease in cytoplasmic viscosity and suggested that they bring about dissociation of the proteins constituents of the cytoplasm. This dissociation effect would increase water permeability, increase the osmotic value of the cytoplasm, and possibly also bring about increased enzymatic dissociation can sometimes activate enzymes and such dissociation effect would increase water permeability, increase the osmotic value of the cytoplasm, and possibly also bring about increased enzymatic activity. It has been shown by Hand (1939) that gentle protein dissociation can some times activate enzymes and such dissociation might increase the availability of substrates for the enzymes as well. As a consequence of this an increased respiratory activity and growth might follow.

Thimann (1951) postulated that auxins may act, not as enzyme activating agents, but as agents protecting growth enzymes from inactivation. He points out that organic acid metabolism is primarily responsible for growth. One of the enzymes (succinic dehydrogenase) in this system, which has a sulfhydril structure, is a critical link and may be very intimately involved in growth. Poisons which attack sulfhydril enzymes also antagonize auxin action and growth.

Bonner (1949) and Bonner and Bandurski (1952) have suggested that auxins may serve in some way to couple or mesh together the respiratory process with growth processes, and thus they act in a manner which would make the energy formed in respiration available to growth processes.

Theories of Osmotic Effects

Water uptake, serves as the basis of growth through increase in cell volume. The uptake of water could be due to changes in the cytoplasm itself, specially changes in osmotic value; or it could be due to changes in the properties of the wall and the cell membranes, particularly with respect to extensibility and permeability. Each of these two factors in water uptake has been defended as a possible mechanism through which auxins may bring about growth.

Czaza (1935) suggested that auxins may increase the osmotic value of the cell sap which would result directly in water uptake and growth. A basic limitation to Czaza's concept lies in the fact that growth is not necessarily associated with an increase in osmotic value. Actually van Overbeek (1944) and Hackett (1951) have been able to obtain growth with an associated decrease in osmotic value.

Commoner et al. (1942, 1943) have proposed that auxin may influence the respiratory uptake of salt and the resultant increase in the osmotic value of the cell sap would cause water uptake and hence growth. Reinders (1942), however, has shown that auxin induced growth can occur in pure water and in the absence of salt uptake.

Reinders (1942) and Mackett and Thimann (1952) have collected evidence that water uptake in response to auxin treatment is dependent upon oxidative metabolism. Bonner et al, (1953) suggest that phosphorylation reactions provide energy for water uptake phenomenon.

Theories of Cell Wall Effects

Direct measurements of the effect of auxins upon the stretching properties of the cell walls have been made by Heyn (1932, 1933). He found that auxin application causes an increase in the flexibility and extensibility of cell walls. According to this, increased extensibility would result in a drop of wall pressure around the cell and would allow water uptake due to this simple drop in turgor pressure. This concept attributes growth primarily to the dynamic function of the cell wall.

Burstrom (1953) after a critical study of water uptake concluded that cell wall attains a plastic quality during growth and that water uptake only follows the changes in the cell wall. Thus water uptake is a consequence and not the cause of growth.

Toxic Metabolism

It is known that auxin, besides being growth promoters, can also cause inhibition of elongation. In shoots this inhibition requires high concentrations of auxin and is not prominent. In roots the regular response to exogenous auxin is an inhibition of elongation even at relatively low auxin concentrations. Very little is known concerning the mechanisms of this specific inhibition. According to one view it is conceived that the inhibition is due to excess auxin molecules inactivating the sites of auxin action and thus preventing maximum growth response. This mechanism although plausible, is supported by limited decisive evidences.

It may be concluded that the mechanism of auxin action remains yet unsolved, but a variety of promising lines of evidence seem to be emerging. The different angles from which the problem is being tackled may provide valuable lines of approach.

THE SYNTHETIC GROWTH HORMONE

A natural consequence of the discovery of auxin activity was the isolation and characterization of the auxin molecule. As soon as this was accomplished, an intensive search began for compounds chemically similar to IAA and with similar activity. Before long, the results of this search brought forth other indole derivatives, such as indole-3-propionic acid, indole-3-butyric acid and indolepyruvic acid all of which demonstrate physiological activity similar to that of IAA. Other compounds, similar in activity but not in chemical structure of IAA were also discovered. Of these, the more important ones are α- and β-naphthylacetic acids, phenylacetic acid naphthoxyacetic acid and phenoxyacetic acid. The structures of the above-mentioned compounds are given.

Molecular Structure and Auxin Activity

The chemical characterization of physiologically active compounds never fails to generate interest in the relationship between the structure of the compound and its physiological activity. It was interest such as this that led to the listing of certain minimal requirements needed by a compound for auxin activity. These requirements are:

1. An unsaturated ring system.
2. An acid side chain.
3. Separation of the carboxyl group (–COOH) from the ring (several exceptions).
4. A particular spatial arrangement between the ring system and the acid side chain.

The above requirements are minimal for auxin activity. However, the degree of substitution in the ring and side chain, nature of the ring (indole, phenyl, anthracene, etc.), and length of the side chain all are factors that will influence auxin activity.

Nature of the Ring System

After IAA had been isolated and characterized, it was soon found that the nitrogen of the indole ring was not essential for auxin activity. When either a carbon or an oxygen atom was substituted for nitrogen activity, although considerably diminished, was still observed. One might have anticipated this due to the fact that ring sizes ranging from the small phenyl ring to the relatively lare anthracene ring have been found in compounds having auxin activity. Nitrogen is not found in the phenyl or anthracene ring. There seems to be considerable evidence supporting the requirements that the ring be unsaturated. Activity decrease with hydrogenation of the double bonds of the ring, with complete cessation of activity occurring in the saturated ring.

It was not until the auxin activity of the phenoxyacetic acid series was discovered that the profound effect of the substitution of various groups onto the ring or side chain was truly appreciated. The nature of the group substituted and the location of the substitution was found to influence the activity of the compound. A striking example of this may be seen in the substitution of the chlorine atom at various positions on the phenyl ring of phenoxyacetic acid.

This was clearly demonstrated by Muir *et al.* (1949) when they experimented with the substitution of halogens and methyl groups in the 2, 4, and 6 positions of the phenyl ring. They found that substitution of both the 2 and the 6 positions with chlorine atoms resulted in complete loss of activity. However, substitution of the 3 or 4 position alone increased activity. Chlorine atoms attached to the 2 and 4 positions of the phenyl ring of the phenoxyacetic acid molecule gave the greatest auxin activity. Instead, 2, 4 - dichlorophenoxyacetic acid (2,4-D) is one of the most widely used synthetic auxins at the present time. The fact that there is a complete loss of activity when both the 2 and the 6 positions are chlorinated, led these investigators to hypothesize that the position on the phenyl ring adjacent to the point of attachment of the side chain is involved in the growth reaction.

Nature of the Acid Side Chain

According to the work of Koepfli et al. (1938) and others, the position and length of the acid side chain has a striking influence on auxin acitivity. Side chains that have the carboxyl group separated from the ring by a carbon or a carbon and oxygen given optimal activity. For example, the acid side chains of IAA and 2,4-D, two highly active auxins, fulfill these requirements.

As the length of the side chain in the phenoxyacetic acid series in increased, there is a falling off of activity. This drop is erratic, activity falling much lower when the side chain contains an odd number of carbons. That is, 4-dichlorophenoxybutyric acid (4 carbons) is more active than 2, 4-dichlorophenoxy propionic acid

(3 carbons). A possible explanation for the above phenomenon may be found in the work of Synerholm and Zimmerman (1947) and Fawcett et al. (1952). Apparently, side chains containing an odd number of carbons are metabolized in living tissues to the inactive phenol, where as the even-numbered chams are broken down to the active phenoxyacetic acid.

Substitution of different groups on to the side chain also affects activity. Thus, substitution of a methyl group on the carbon of the side chain of phenylacetic acid does not take away auxin activity. However, substitution of two methyl groups on the carbon eliminates auxin activity entirely. Later, when we discuss the mechanism of auxin action, we will learn why the substitution of "bulky methyl" groups on the acid side chain inhibits auxin activity.

Hydorxyl substitution in the side chain also can eliminate auxin activity. For example, substitution of a hydroxyl group or an alcohol group on the carbon of phenyl acetic acid produces two inactive derivatives.

Earlier, it was thought that separation of the carboxyl group from the side chain was essential for auxin activity. However, many exceptions to this rule have been found. For example 2,3, 6-trichlorobenzoic acid exhibits a strong auxin activity.

Spatial Arrangement

The spatial relationship between the ring and side chain is an important factor in the activity of an auxin molecule. For example, cis-cinnamic acid demonstrates good auxin activity in the *Avena* straight growth test, and its *trans* isomer does not. Veldstra (1944) suggested, after a study of the relation between structure and activity, that in order for a molecule to have auxin activity the –COOH group and the ring should lie in different planes. This theory was supported by observation of the activity of *cis* and *trans* forms of tetrahydronaphtylideneacetic acid and the above *cis* and *trans* forms of cinamic acid.

Antiauxins

If one assumes that the chemical configuration of a compound is responsible for its effect on physiological processes, then we must also recognize that there can be an interference with such action by similar, but not identical, compounds . Many antiauxins have been discovered and, generally, these combination with auxin, inhibit its acitivity.

Actually, the term antiauxin should define only those compounds that will complete with auxin for a reactive site in the growing cell. What do we mean by a reactive site? Because of the relationship of molecular structure and configuration with degree of physiological activity, most theories on auxin action are based on the attachment of the auxin molecule to some substance in the cell (e.g. protein). This complex (bound auxin)can then induce auxin activity. True antiauxins, then, are compounds which, because of the molecular similarity to auxin, would become attached to reactive site, thus neutralizing them for action in growth . However, if true competition for reactive sites is involved here, increasing applications of auxin should eventually swamp the reactive sites with auxin molecules and overcome the antiauxin effect.

There is general agreement that in order for an auxin molecule to be active, it must make a *two -point* attachment at a reactive site. In addition, it is thought that the two points of attachment are made through a position on the unsaturated ring and by the carboxyl group of the side chain. In the phenoxyacetic acids, the position on the ring through which attachment is made is the *ortho* position.

Based on the two-point attachment theory, McRae and Bonner (1953) have drawn up a rigorous classification for true, antiauxins. Using 2, 4-D, a synthetic auxin, as representative molecule, they have shown where "analogs" of the 2, 4-D moleucle containing some, but not all, of the structural properties of the auxin are capable of competing for reactive sites. The antiauxin makes a cone-point instead of the two-point attachment necessary for growth action. McRae and Bonner have listed three ways by which a modification of the 2,4-D molecule can lead to antiauxin activity.

1. Elimination of the essential carboxyl group.
2. Elimination of the essential reactive *ortho* group.
3. Elimination of proper spatial relationships between the ring and the carboxyl group as by introduction of bulky groups in the side chain.

Although not as popular as the two-point attachment theory, a three-point attachment theory has been proposed as necessary for auxin activity. According to this theory, an auxin molecule in order to be active must contain the following necessary structural, properties; an unsaturated ring, a carboxyl group, and at least one α-hydrogen. A further stipulation is that all three must be correctly oriented in space with each other. Of the isomers of 2, 4-dichlorophenoxy-α-propionic acid, only the "+" form is active. The "–" form is not properly oriented, thus failing to accomplish the three -point attachment necessary for auxin activity.

Contact is made by the auxin with the reactive site simultaneously at three positions on the auxin molecules. If only one site or even two positions are occupied, no activity ensues. In fact, molecules that occupy only one or two of the positions on the reactive site can be considered antiauxins.

A type of antiauxin which has not, as yet, been considered is the compound having weak auxin activity. A weak auxin can make the necessary two-point (or three-point) attachment and initiate a stimulation of growth. However, this stimulation is small, and at the same time active sites involved with the weak auxin cannot be occupied by a strong auxin. An example of a weak auxin having antiauxin characteristics is phenylbutyric acid.

Kinetics of Auxin Activity

A valuable contribution to the study of auxin-induced growth reactions has been the application of the Lineweaver-Burk kinetic analysis of competitive inhibition. McRae and Bonner (1953) demonstrated that auxin-induced growth reactions could be treated by the methods of classical enzyme kinetics.

It is generally accepted that in enzyme reactions, an intermediate complex is formed between the substrate and enzyme and that this complex is formed at an active site on the enzyme. The complex is converted to the enzyme and reaction products.

$$E + S = ES \rightarrow E + P$$

Using the above scheme, McRae and Bonner demonstrated that stimulation of straight growth of *Avena* coleoptile sections by auxin could be mathematically analyzed. In their scheme, enzyme (E) is the auxin receptor, substrate (S) is the auxin applied, complex (ES) is the receptor-auxin attachment, and the product in this case is growth. Let us rewrite the above equation using A for auxin, R for receptor, RA for the intermediate complex, and G for growth.

$$R + A = RA \rightarrow R + G$$

Now, in classical enzyme study, a competitive inhibitor is thought of as a compound that will compete with the normal substrate of an enzyme for active sites on that enzyme. The formation of an enzyme-inhibitor complex may be illustrated in the following manner.

$$E + I = EI$$

The fact that the formation of EI is reversible is important since this allows for competition by the substrate with the inhibitor for active sites. Therefore, by increasing the substrate, inhibition by a competitive inhibitor may be overcome.

The effect of a competitive inhibitor may be observed in a decrease in the rate of an enzyme reaction. However, if the substrate concentration is increased until all active sites on the enzyme are "swamped," then maximum velocity (V_{max}) or rate will be obtained. This maximum velocity will be identical to that of the same reaction without a competitive inhibitor. In other words, increasing the concentration of a substrate decreases the amount of inhibition, and conversely, decreasing the substrate concentration increases inhibition.

We have mentioned that antiauxins compete with auxins for active sites on an auxin receptor or growth center. This situation,of course, is analogous to the concept of competitive inhibition in enzyme study.

Now, using the Lineweaver-Burk plotting method, one can measure the velocity of an auxin reaction (in this case, rate of growth) and, at the same time, calculate the effect of an antiauxin on this velocity. By plotting the reciprocal of the velocity (I/V) of the auxin reaction against the reciprocal of the concentration of auxin (I/[A]) applied, one obtains a straight line relationship. Maximum velocity (V_{max}) may be obtained by extending this line to the ordinate, the intercept being I/V_{max}.

Note that the concentration of inhibitor will affect the slope of the line, but not the intercept. A mathematical analysis of the interaction of 2,4-D and the antiauxins 4-chlorophenoxyisobutyric acid, 2,6-dichlorophenoxyacetic acid, and 2, 4- dichloroanisole has been under taken. The compound 4-chlorophenoxyisobutyric acid is an antiauxin because of bulky methyl groups in the side chain interfering with attachment of the carboxyl group to the auxin receptor. 2,6-Dichlorophenoxyacetic acid owes its antiauxin properties to block age of the reactive *ortho* positions on the ring by chlorineatomes. 2, 4-Dichloroanisole has to carboxyl group and therefore cannot make the necessary two-point attachment.

Inactivation of Auxin

Just as the production and subsequent physiological effects of auxin have a profound influence on plant development, the inactivation of auxin appears also to be of significance in this respect. For example, the inactivation of auxin is important in phototropisms, control of cell elongation, and in the aging of plant tissues. We will discuss the mechanisms involved in the destruction of auxin and the influence of its destruction on cell elongation and aging.

Mechanisms of Auxin Inactivation

A preponderance of literature has developed on the mechanisms of auxin inactivation since the isolation in 1947 by Tang and Bonner (1947) of an enzyme capable of oxidizing IAA. This enzyme is now called *IAA oxidase*. Recently an IAA oxidase was extracted from tobacco roots and partially purified. Although IAA oxidase represents one way for a plant to destroy auxin, other natural means of auxin inactivation have been discovered. However, two systems for the destruction of auxin in the plant appear to dominate, and these are (a) enzymatic oxidation and (b) photo-oxidation.

Enzymatic Oxidation

Enzyme systems that oxidize IAA have been found in several plant tissues. However, as is generally true, one system is usually studied in much more detail than the others, and in this case it is the enzyme system present in extracts of etiolated pea epicotyls. It appears that in this system a flavoprotein must be present, which gives rise to hydrogen peroxide. The oxidation of IAA by hydrogen peroxide is catalyzed by a peroxidase to yield some inactive product, probably indolealdehyde. In the inactivation of IAA by this system, 1 mole of O_2 is consumed for every mole of IAA inactivated and CO_2 released.

In addition to indolealdehyde, other breakdown products have been suggested. However, it is generally accepted that the logical end product of IAA oxidation is indolealdehyde. As mentioned before, IAA oxidase systems have been found in many plants, and in several cases these have differed from the original IAA oxidase system found in the pea plant. Perhaps in these, different end products of IAA oxidation are obtained.

An inverse relationship between IAA oxidase activity and IAA content in the plant has been found. That is, where IAA content is high, IAA oxidase activity is low and vice versa. Meristematic regions that have a high auxin content have been found to below in IAA oxidase activity. The root, which is generally thought to below in auxin content, has been found to be generally thought to be low in auxin content, has been found to be high in IAA oxidase activity. In fact, Galston (1956) has found (at least in the pea plant) that as cells age, their IAA oxidase activity increases, and their auxin content drops.

Increased capacity to inactivate IAA has been demonstrated by young plant tissues treated with synthetic IAA or analogs of the IAA molecule. It appears, then, that IAA is capable of inducing the formation of the enzyme that destroys it. This is particularly interesting since IAA, which initiates growth, also puts into action the mechanisms leading to the termination of growth. Galston (1956), a leading figure in research on plant growth, has this to say: It seems possible that the decreased sensitivity to auxin of older cells is a consequence of their higher IAA oxidase activity, which in turn is a consequence of prior induction by IAA. According to this scheme,. the administration of IAA to a young cell not only initiates growth, but also sets in motion a chain of events leading to the diminution and eventual culmination of growth.

Photooxidation

It has long been known that IAA can be inactivated by ionizing radiation. Skoog (1934; 1935) demonstrated that rapid inactivation of pure IAA takes place when it is subjected to x-and gamma-radiation. He also noted that little, if any, inactivation takes place in a nitrogen atmosphere, suggesting that inactivation is due to oxidation by peroxides formed during irradiation. There is some evidence that only a small amount of IAA is inactivated or oxidized in this manner, most of the detrimental effect of this type of irradiation to IAA being of an indirect nature. For example, Gordon (1956) has claimed that the major effect of ionizing radiation on auxin metabolism may be found in the destructive effect of the radiation on the enzyme system converting tryptophan to IAA.

Ultraviolet light also inactivates IAA. This might have been predicted because of the ring structure of the IAA molecule, which absorbs to some extent in ultraviolet (maximum absorption at about 280 mμ). Here, there is a direct effect on the IAA molecule due to the absorption of ultraviolet light. Determinations of auxin content before and after irradiation with ultraviolet light have shown that this type of irradiation reduces auxin levels in plants.

GIBBERELLINS

The gibberellins are tetracarbocyclic compounds whose small amounts profoundly stimulate the growth of many plants.

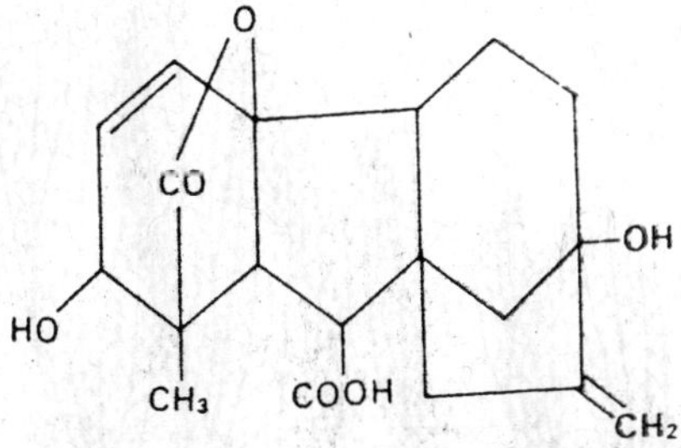

Fig. 12.9. Gibberellic acid.

Gibberellins were originally regarded as unusual and interesting fungal metabolites, but their importance in the regulation of plant growth has recently been emphasized by the widespread occurrence of gibberellin-like substances in the plant kingdom. The chemical identification of some of these has been established. The discovery of gibberellins is very interesting and rather accidental. These growth regulating substances were discovered as a direct consequence of interest in the "bakanae" disease of *Oryza sativa*, a disease which had serious economic effects on the rice industry in Japan. The disease has two important symptoms:

1. seedlings assumed long, thin, and pale green appearance than those of healthy plants, and
2. mature plants, sometimes taller than their healthy counterparts often failed to set fruit.

The "bakanae"symptoms were first described by the plant pathologist S.Hori in 1898, although its technical account appeared in the literature much later. Early in the twentieth century, extensive studies by Japanese plant pathologists conclusively showed the "bakanae"disease to be resulting due to certain strains of a species of fungus now known as *Fusarium monilliforme* Sheld (*Gibberella fujikuroi*). Sawada (1912) suggested that the seedling symptoms of the disease could be due to a "substance" secreted by the fungus. This was experimentally proved by Kurosawa in 1926. He demonstrated the appearance of bakanae symptoms in *Oryza sativa* fungus. This observation attracted little attention outside Japan, but the work on isolation and purification of the growth-promoter was initiated. Progress was slow but eventually, in 1939, Yabuta and Hayashi isolated a crystalline 'A'.

More recently Japanese chemists have suggested that this substance is mixture of several different growth promoters collectively known

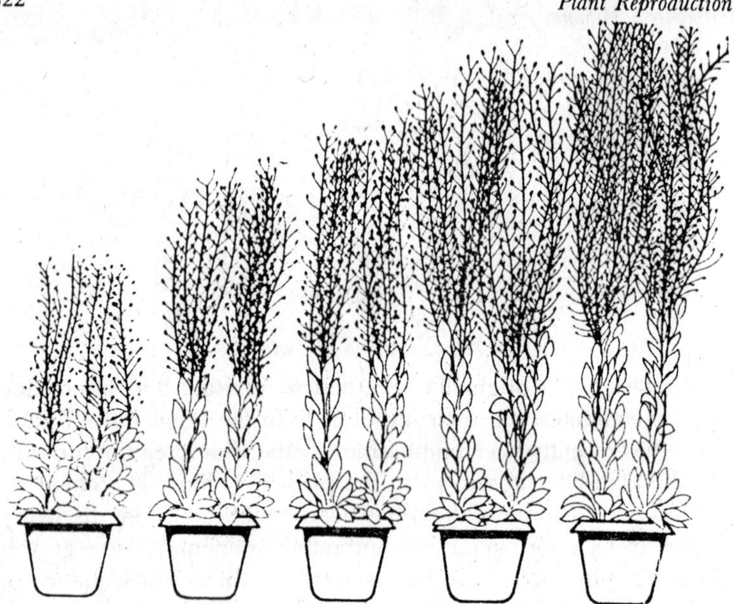

Fig. 12.10. Primrose require long days to flower. Plants were treated with increasing amount of Gibberellic acid, all plant flowered without long day conditions.

as gibberellins. Pure gibberellin was first obtained in England. Many years after the description of gibberellin 'A' in Japan, interest developed in other countries. All the Japanese work, from Kurosawa's observations of 1926 to the isolation of an active material by Yabuta and Hayashi in 1939 was promptly abstracted in "Chemical Abstracts" and the "Review of Applied Mycology".

In attempting to repeat Japanese work, Curtis and Cross isolated, in 1954, a substance obviously chemically and physiologically similar to the Japanese gibberellin 'A', but nevertheless distinct, they called this substance gibberellic acid. Shortly afterwards Stodola, in United States, described a new gibberllin which he called gibberellin 'X', this was found to be identical with gibberellic acid. Several pure gibberellins are now known. They are probably fairly closely related chemically and all have similar effects on plant growth.

Terminology

Following the suggestions of Phinney and West (1960), two terms, gibberellin and gibberellin-like are maintained. The former term is restricted to substances defined both by biological and chemical properties whereas the latter term defined by biological properties only. The term, gibberellins will therefore be reserved for substances biologically active in stimulating an increase in the size of a plant organ or parts thereof and known to consist chemically of a carbon skeleton identical with,or closely related to, that of gibberellin 'A_3' (gibberellic acid). The five gibberellins identified are : gibberellin 'A_1' (GA_1), gibberellin 'A_2' (GA_2) gibberellin 'A_3' (GA_3), gibberellin 'A_4' (GA_4), and gibberellin 'A_5' (GA_5).

Assays

Depending on the physical or chemical properties of the gibberellins rather than on their physiological activities, a number of qualitative and quantitative assays have been developed.

Physico-chemical assays

With the use of filter paper chromatography a variety of detection methods have been employed to locate gibberellins.

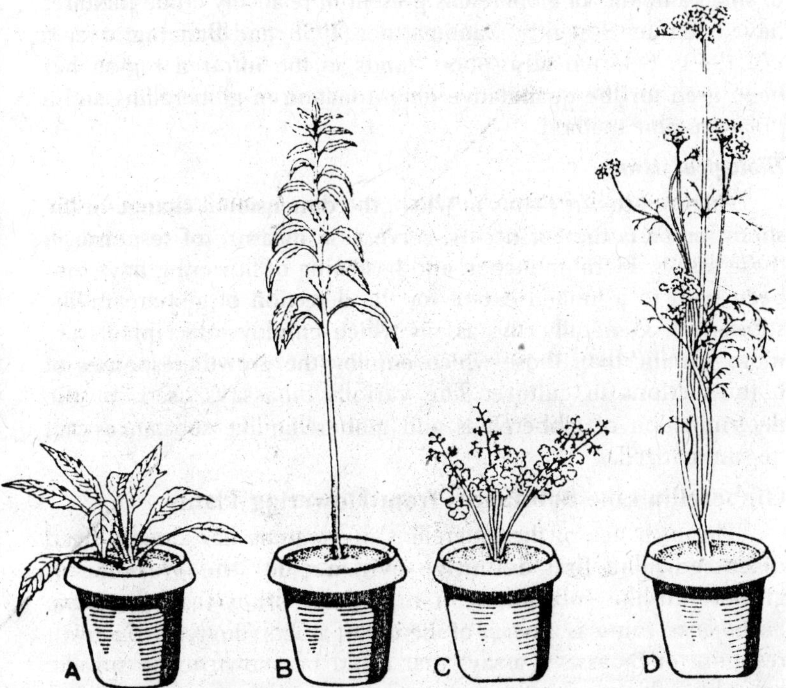

Fig. 12.11. A-B–Henbane plants which usually require a period of cold treatment and long days for flowering-flowered when they were treated with Gibberellic acid as in B (No low temp. or long day were required), C-D–Carrot plants require low temperature for flowering, C–Plant without low temperature but with gibberellin spray flowered, D–Plant provided with low temperature.

Gibberellin A_3 is defected by a characteristic fluorescence developed in sulfuric acid. The substance, by this method, could be detected in amounts as low as one microgram. The fluorescence of the fluorogen formed from gibberellin 'A_3' in sulfuric acid solution has been used as the basis for a quantitative analysis. A linear calibration curve is obtained for gibberellin 'A_3' concentrations in the range of 0.00625 to 3.2 micrograms per ml. The amount of fluorescene formed from a sample is subject to a number of variables which must be controlled to get reproducible gibberellins, GA_1, GA_2, GA_4 and GA_5. Other spray reagents used include 0.5 percent potassium permanganate reagent and an acidic periodate-permanganate reagent.

A number of useful isotopic labelling methods, for the assay of small amounts of gibberellins present in relatively crude mixtures have been developed by Baumgartner (1958) and Banumgratner *et al.* (1959). Selected absorption bands in the infrared region has been used for the quantitative determination of gibberellins and a polarographic method.

Biological assays

Angiospermous plants in which the dimensional change of the shoot, or parts thereof occurs, serve as a measure of response in these assays. Floral induction and hastening of flowering have also been used to a limited extent for the detection of gibberellin-like substances. Generally, the assays which employ intact plants are more specific than those which employ the growth responses of plant sections in culture. The various bioassays used for the determination of gibberellins and gibberellin-like substances can be summarized.

Gibberellin-Like Substances from Flowering Plants

The presence of the giberellins in *Fusarium* has already been discussed. The first definitive evidence for the presence of gibberellin-like substances in organisms other than *Fusarium moniliforme*, came as a result of the use of gibberellin-type of growth response in bioassays, assays that could be shown to be specific for gibberellins only. Several worker groups have reported the widespread occurrence of gibberellins in flowering plants.

Gibberellin-like substances have been reported from 64 genera reporting 15 different families of angiosperms, from one member of the mosses, from two genera of fungi and also from two members of Actinomyces. These gibberellin-like materials have been reported

Table 12.2. Species of plants used as assays for gibberellins and gibberellins-like substances.

Plant species	*Plant part*	*Specificity*
1. *Avena sativa* L.	leaf section	non-specific
2. *Perilla ocymoides* L.	intact seedling	specific
3. *Pharbitis Nil chois*	intact seedling	specific
4. *Phaseolus vulgaris* L.	intact seedling	non-specific
5. *Pisum sativum* L.	intact seedling	specific
6. *Oryza sativa* L.	intact seedling	specific
	leaf section	non-specific
7. *Rudbeckla bicolo* Nutt	rosetted plants	specific
8. *Triticum vulgare* Vill	leaf sections	non-specific
	excised coleoptile	non-specific
9. *Zea mays* L.	intact seedling	specific

from the root and all organs of the shoot, and, from both young and old tissues of flowering plants. Macmillan and Suter (1958) have isolated gibberellin A_1 in small quantities from immature seed of *Phaseolus vulgaris.* West and Phinney (1959) extracted a substance, bean factor I, from immature seed of *Phaseolous vulgaris* and which subsequently was shown to be identical with gibberellin A_1, Kawarada and Sumiki (1959) have isolated gibberellin A_1 from "water sprouts" of *Citrus unshuii.* West and Phinney (1959) obtained a second substance, bean factor II, form extracts of immature seed of *Phaseolus vulgaris,* which had gibberellin like biological properties but differed in chemical and physical properties from the reported fungal gibberellin. Macmillan, Seaton and Suter (1959) also obtained a second gibberellin, which they name gibberellin A_5, from their extracts of *Phaseolus multiflorus* seed. A comparison of the infrared spectra of gibberellin A_5 and bean factor II, and their methyulesters, has shown that these substances are identical. The name gibberellin A_5 has been accepted by both groups. gibberellin A_3, the gibberellin produced in largest amounts in most fermentations has not been isolated from flowering plants.

CHEMISTRY OF GIBBERELLINS

Structural Studies on the Fungal Gibberellins

The two fungal gibberellins, gibberellin A_3 and gibberellin A_1 are structurally known as reported by the British group at Akers Research :Laboratories of ICI. The results of more recent investigations of the Tokyo group on gibberellin A_1 and gibberellin A_2 fractionated from the gibberellin A mixture, have led to proposals IB and IIB for gibberellin A_3 and gibberellin A_1 respectively. These differ from the above proposal only in the positions of attachment of the lactone ring in A.

The relationship between gibberellin A_1 and gibberellin A_3 has been established. Grove *et al.* (1958) could partially hydrogenate methyl gibberellate ($C_{20}H_{24}O_6$) and separate from the mixture of products two isomeric methyl ester ($C_{20}H_{26}O_6$). One of these methyl-α-dihydrogibberellate, was identical with the methyl ester of gibberellin A_1. Since gibberellin A_1 still possesses an exocyclic methylene group as indicated by ozonization studies the two hydrogens must have added to saturate the A ring. A few of the products which have been derived by chemical treatment of gibberellin A_2 and gibberellin A_1 can be summarized.

Stodola et.al (1957) and Cross (1954) indicated the molecular formula $C_{19}H_{22}O_6$ for gibberellin A_3. The functional groups in this were identified as a carboxyl, two ethylenic double bonds, two alcoholic hydroxyl groups (one secondary and one presumably tertiary), a saturated lactone, and probably one methyl group. This indicated a compound with four carbocyclic rings. Acid hydrolysis yielded a number of products, such as allogibberic acid (identical with gibberellin B of Japanese) and gibberic acid.

Cross et al. (1956 ,1958) assigned the structure 'V' to gibberic acid. The Japanese group obtained the hydrocarbon, gibberene, as a product of the selenium catalyzed hydrogenation of gibberellin A or gibberellin B or gibberic acid. Yabuta et al. (1941) identified this hydrocarbon as a substituted fluorene. Mulholand and Ward (1954) also isolated gibberene as a hydrogenation product of giberric acid and showed it to be 1,7 dimethyl-fluorene. These probable presence of a substituted, perhydrofuorene nucleus is gibberic acid. Gibberic acid was also converted by Cross et al (1958) to methyl -1, 7-dimethyl fluorene-9-carboxylate by appropriate series of reactions. This has established the position of the carboxyl group. By a step-wise degradation procedure gibberic acid was converted to the methyl ester of a tetracarboxylic acid by Cross et. al.(1958).

Cross et al. (1956), and Mulholand (1958) assigned the structure IV to allogibberic acid. Like gibberellin A_2, it gave formaldehyde of ozonolysis indicating the presence of a terminal methylene group. Allogibberic acid was converted to 7-hydroxyl-1-methylfluorene by a series of reactions including a selenium dehydrogenation. This located the hydroxyl group substitution on the nucleus.

Gerzon, Bird and Woolf (1957) assigned structure IIIa or IIIb to gibberellenic acid–another product obtain from gibberellin A_3. Gibberellenic acid is readily converted in acidic solution to allogibberic acid, which in turn is readily converted to gibberic acid, an end product of the acid hydrolysis.

The establishment of structure of allogibberic acid and gibberic acid also fixed the structure in the B, C and D rings of gibberellin A_2. It still remained to be established how the lactone, secondary hydroxyl and ethylenic double bonds were accommodated in the ring A. The position of the secondary hydroxyl was indicated by the conversion of the methyl ester of gibberellin A_1 through a series of intermediates to 2- hydroxyl 1, 7-dimethyl fluorene or 2-hydroxyl-1- methylfluorene. There seems at present no satisfactory way, however, of reconciling all the reported degradation products with one structure of gibberellin A_2 and gibberellin A_1. Structure has been proposed by Kitamura et al. (1958, 1959) for gibberellin A_2 and structure for gibberellin A_4. There structures are similar to that proposed for gibberellin A_1 (ring A is shown in the form proposed by the Japanese) except for the position of substitution of the tertiary hydroxyl and the absence of unsaturation in A_2 and the absence of the tertiary hydroxyl in A_4.

Structural Studies on Gibberellins from Angiosperms

The chemical nature of the active principles in angiosperms has been determined in only a few instances. Some of the most important sources from which the gibberellins have been isolated in angiosperms has already been mentioned. Structure has been assigned to gibberellin A_5 by MacMillan *et al.* (1959). Thus, gibberellin A_5 is dehydrogibberellin A_1 in which the secondary hydroxyl grop is eliminated from the ring A. A key compound in the structural proof was the keto-acid formed from gibberellin A_5 on acid treatment. Thus at least two gibberellins, one identical to a fungus-produced gibberellin and the other very closely related to the fungal gibberellins chemically, have been found to occur naturally in flowering plants.

The Relation of Chemical Structure to Biological Activity of Gibberellins

In an assessment of the relation of chemical structure to biological activity of gibberellins many factors influence the interpretation of results. These include (i) the types of assay methods used; (ii) the method of application of the test material, and (iii) accompanying permeability problems and the range of concentration of the test material. Studies with the naturally occurring gibberellins have shown gibberellin A_3 to be the most active followed by gibberellin A_1, gibberellin A_4 and gibberellin A_2 in decreasing order of activity in most assay used. But this is not true always because the activity of gibberellin A_4 stimulating growth of some cucurbits at levels at which the other gibberellins are inactive, has been found. Also depending on the particular dwarf mutant used for assay, gibberellin A_5 is as effective as gibberellin A_2 or less than 10 percent as active as gibberellin A_1.

Gerzon et al. (1957) showed that gibberellenic acid formed from gibberellin A_3 in acid solution, is devoid of biological activity. Allogibberic acid and gibberic acid which are formed from gibberellin A_3 on more vigorous acid treatment are likewise devoid of biological activity. Besides, gibberellin C, formed on acid catalyzed rearrangement of gibberellin A_1, retains some activity. The isomeric products formed from gibberellin A_1 and gibberellin A_2 by mild alkaline treatment are biologically inactive. Under some condition which do not open the fact one ring, gibberellin A_1 is inactivated; this is believed to be due to the formation of an primer (pseudogibberellin A_1) in which the secondary hydroxyl group in the ring A is inverted from an axial to a more stable equatorial configuration. On mild alkaline hydrolysis of gibberellin A_3, the diacid results by an allylic rearrangement of the hydroxyl group formed on opening the lactone ring.

The free acid and salts both, of gibberellin A_3 are biologically active. It has been found that conversion of the carboxyl group of gibberellin A_3 to the methyl, ethyl, butyl, or acetyl esters results in a complete loss of biological activity. Generally, those products derived from the gibberellins which involve modification of the A or B ring structures are inactive. Sumike and J Kawarada (1941) found the derivative of gibberellin A_2 in which the alcohol is oxidized to a Keto Group, to be inactive. Many of the products derived from the gibberellins which involve structural modification only in the C or D rings, such as gibberellin C, are active.

Biosynthesis of Gibberellins

The broad outlines of the gibberellin biosynthesis have been indicated but very little detailed knowledge has been gained. Cross et al. (1958) advanced "biogenetic grounds" as one of their reasons for favouring the carbon skeleton they proposed for gibberellin A_3, since it could be considered as an "isoprenoid" (isopentane) derivative. Birch et al. (1958), and Birch and Smith (1959) have proposed a general route from acetate and mevalonate (β-δ-dihydroxy-β- methyl-valerate) to gibberellin A_2 *via* diterpenoid intermediates. Some key features of this proposal may be summarized. Mevalonate (B) which can be biosynthesized from acetate, has been shown to be a precursor of sterols and terpenes.

Condensation of four molecules of a mevalonate derivative in a "head- to-tail" fashion leads to an acyclic diterpene with the carbon skeleton on C. The cyclization of the acyclic precursor leads to a tricyclic diterpene of carbon skeleton D.Subsequent modification of the skeleton to that of gibberellin A_3 (E) would occur by (1) elimination of the C_{17} methyl group, (2) a ring contraction of the B ring, and (3) a rearrangement involving the C ring and its attached vinyl group to produce the phyllocladene type C-D rings of gibberellin A_3.

This general proposal has been supported by the labelling studies. The previous figure shows one of the ways in which gibberellin A_3 would become labelled from acetate-1-C^{14} or mevalonate -2-C^{14}. Zweig and Cosens (1959) have also presented C^{14} labelling evidence for the incorporation of both the methyl and carboxyl carbons of acetate into gibberellin A_3 by the fungus. Little, direct information bearing on the biosysnthesis of gibberellins in flowering plants is available at the present time. It is presumed that the route in the fungus and in flowering plants will be very similar.

Anatomical Effects of Gibberellins

It is now well established that changes in size and form of plants take place due to gibberellin treatment. Hence it raises the problem of whether these effects are the result of changes in the net cell number, net cell size (including form of the cell) or combinations of the two. In general, it was showed by Japanese that increased cell length rice seedlings. Nevertheless since some of the early studies showed increased cell elongation to be insufficient to account for all of the elongation observed at the organ level, increased cell number was also implicated but only by inference and only to a minor degree. The discovery by yabuta and Hayashi (1939) that gibberellin increased the "reproductive rate"of *Lemna paucicstata* gave a signal that the effect must have been associated with increases in cell number.

It is now realized that gibberellins may markedly increase both number and length of cells, the effect, however, depending on the nature of the plant material and the conditions under which the material is grown. Lang (1956) pointed out that stem elongation in gibberellin treated *Hyoscyamus niger* must be due to an appreciable increase in cell number since the internodes are essentially absent in the non-treated iosetted forms. In 1956 Lona reported that cell length increases would account only for about 13 percent of the elongation of *Perilla ocymoides* stems following gibberellin application. He thus concluded that gibberellin treatement must result in an appreciable increase in cell number in the stem of this plant. Sachs and Lang (1957) and Sachs, Breitz and Lang (1959) have shown a very marked increase in the frequency of dividing cells in gibberellin induced elongation, of *Hyoscyamus niger* and *Samolus parviflorus.* Bradley and Crane (1957) showed an increase in cell number in the meristems of spur shoots of gibberellin treated *Prunus armeniaca.* Gundersen (1958) has observed increase in both cell number and cell lengths in certain of the internodes of *Begonia*, and both cell number and cell lengths are factors in the gibberellin induced shoot growth of *Phaseolus vulgaris.* This is also true for petiole elongation of *Fragaria* sp and seedling elongation of *Zea mays* and *Pharbitis nil.*

Increased cell number is indeed a reflection of an increase in mitotic activity. In angiosperms this increase due to gibberellin treatment is most marked in meristems. Differential growth is also a predominant feature of the gibberellin response at cellular level. The planes of mistoses are such that cell progenies are almost exclusively formed in one direction, followed by cell elongation in the same direction. The response in cell number and cell length would suggest that the effect of gibberellins may be to control a physiological mechanism common to both. While it appears that mitotic activity can precede cell elongation there are cases of gibberellin-induced elongation of pollen tubes which occur without any preceding flush of mitotic activity. Haber and Luippold (1960) have shown that the gibberellin-induced growth of *Triticum vulgare* seedlings, which come from irradiated grains, must be due to increases in cell lengths only, since the irradiation completely inhibits mitosis. Thus it appears that the two types of responses at the cellular level can occur independently of one another. It is possible that the response of cell to the application of gibberellin will depend on the physiological age of the cell, i.e., enhanced cell division (mitotic activity) in meristems, primarily cell elongation in regions of elongation, and no response or very limited response in regions which consist of differentiated cells.

Gibberellin-Induced Growth

A number of types of growth responses now known to be associated with the gibberellins have been reported. Some of the "bakanae" symptoms have already been described. Here an attempt will be made describe some of the general growth responses induced by gibberellins.

Among the various growth responses induced by gibberellins shoot growth is the most apparent response of flowering plants. At low dosage level the plant size is generally increased, while little or no changes in form take place. At higher dosage levels the response may become one of growth, primarily, in the dimension of normal elongation of the plant. excessive treatment with gibberellins may result in marked differential growth to produce plants with leaves and long and thin stems. The ability to respond is dependent on the genotype of the plant and the environment in which the plant is grown. Field grown plants are often more responsive to gibberellin treatment when grown under environmental stresses. Gibberellin treatment may result in inhibition of lateral buds and thus lateral branching. Abscission has also been reported to be affected by gibberellin treatment. The increased stem thickness often seen is indeed a reflection of the stimulation of the cambium and its immediate cell progeny.

The type of leaf response is also associated with the developmental pattern of the leaf. Thus leaves of grasses with intercalary meristems respond mainly by elongation. Leaves of dicotyledons may respond by increases in area of the blade. Both acceleration and retardation of abscission have been reported.

Generally roots of intact plants show inhibition or no response as a result of gibberellin treatment, although exceptions are found in seedlings where elongation of the radicle has been reported, for example, among the gymnosperms, *Pseudotsuga menziesii* and *Pinus lambertiana.* Rooting of cuttings has been observed to be inhibited and certain cases of modulation of roots are inhibited of a consequence of gibberellin treatment.

Physiological Role of Gibberellins

The specificity of response of many flowering plants to exogenous gibberellin, and the widespread occurrence in flowering plants of gibberellin-like substances of which some have been isolated and chemically identified as true gibberellins, establishes the fact that gibberellins may be regarded as native plant growth regulators. Critical evidence for a causal relationship between endogenous gibberellin and growth has come from the positive correlation of certain types of growth with the level of endogenous gibberellin-like substances. Although synergism between gibberellins and auxins have been demonstrated for a number of cases, nevertheless the mechanistic interpretations of these data are questionable beyond the idea that numerous growth regulators must be present in optimum amounts for maximum growth of an organism. At present, however, the more convincing examples of growth which may be dependent on a gibberellin mechanism are those which exhibit a specified response, i.e. they respond to gibberellins but not to auxin, other known growth regulators.

Shoot Growth

Many cases of appreciable shoot growth resulting specifically from gibberellin treatment are now known. These include; the bolting and flowering responses of numerous photoperiodic and cold-requiring plants, normal growth of certain single gene dwarf mutants, elongation of red-light-inhibited plants, and possibly the breaking of certain types of dormancy and shortening of the after-ripening time required by certain seeds. Generally the effects seem to be not of differentiation of tissues but more of enhanced growth of preexisting cell types, with changes in form being consequences of differential growth. Whenever examples of primordial initiation are recorded, they are apparently indirect consequences of a gibberellin-induced general growth stimulation.

Bolting and flowering

Most rosetted long-day plants and rosetted cold-requiring winter biennials bolt and flower in response to exogenous gibberellin,.and these effects can be brought about under an environment which would normally maintain the rosetted habit of growth. With appropriate gibberellin dosages the bolted and flowering plants appear very similar to those induced by photo induction or cold treatment. At limiting dosages of gibberellin, however, it is possible to separate shoot elongation (bolting) from floral differentiation so that bolting will occur without flowering. Such experiments lead us to the view that the flowering response is only an indirect effect of the gibberellin treatment, as a consequence of increased growth of the shoot; the endogenous factors than gibberellins are elaborated and ultimately result in the differentiation of floral primordia. That the action of gibberellins in flowering is an indirect one is also evidenced by the fact that short-day plants do not flower even though the gibberellin treatment may cause appreciable stem elongation. It has been reported that the flowering response can be enhanced by gibberellin treatment in *Xanthium pennsylvanicum* under marginal day lengths which by themselves will produce some flowering, there are, nevertheless, no examples of short-day plants grown under long-day conditions, shown floral induction following gibberellin application, and in some cases gibberellin treatment has been found to reduce flowering in short-day plants under short-day conditions. Thus it is apparent that gibberellin will replace the long-day requirement for flowering but not the short-day conditions. Thus it is apparent that gibberellin will replace the long-day requirement for flowering but not the short-day requirement, and the gibberellin-treated plants show appreciable stem elongation just as do those plants growing under long-day conditions.

Studies of Lang (1956, 1957) that there exists a correlation between levels of native gibberellin-like substances and the presence or absence of bolting in biennial *Hyoscyamus niger*, have also been extended to other plants. By use of *Zea mays* bioassay, several gibberellin-like components are found in appreciably higher amounts than in non-bolting plants. Harada and Nitsch (1959) and Nitsch (1959) have also found a positive correlation between amounts of gibberellin-like substances and bolting in the cold-requiring plant *Chrysanthemum morifolium. Ram* cv. *Shuoka*, and the long-day plant *Rudbeckia speciosa* Wenderoth. They have also demonstrated increases in auxin-like substances as a result of photoinduction and cold treatment. However, since only exogenously applied gibberellins, and not auxins, induce bolting without photoperiodic or temperature induction, it may be inferred that the increase in endogenous auxin area consequence of the increased level of the native gibberellins. The accumulated evidence indicates that gibberellin may be the primary limiting factor in the bolting of at least some long-day and cold-requiring roset.ed plants.

Dwarfism

It has been found that gibberellin treatment does not always lead to stem elongation, since many cases have been recorded where the process is actually reversed by gibberellin treatment. Reduced growth results from a shortening of the internodes rather than a decrease in number of internodes. As a result, the gibberellin treated dwarfs become essentially indistinguishable from the tall or non-treated forms, by the elongation of internodes rather than by the increase in the number of internodes. In considering possible mechanisms to explain genetic dwarfism, a clear distinction should be made between the use of the general term "genetic dwarf: and the specific term "single gene dwarf". In cases of single gene dwarfism, the gene responsible for dwarfism, may in someway block a single step in a metabolic pathway. The theoritical unitary control in a single gene mutant suggests that a unitary limiting factor may be responsible for the dwarf habit of the growth. In contrast genetic dwarfs can also include cases which result from several mutant genes (multiple factors). A multiple phenomenon may have a multiple control at the physiological or biochemical level.

It has been reported by Brian and Hemming (1955) that certain dwarf cultivars of *Pisum sativum, Vicia faba* and *Phaseolus multiflorus* respond to gibberellin A_3 to become practically indistinguishable from the tall cultivars. At the same time the tall cultivars showed little or no response to exogenously applied gibberellin. It is now known that the ability of a genetic dwarf to respond to gibberellin can be under the control of a single gene. Such a unitary control of the gibberellin response has been demonstrated for the gene in *Pisum sativum* the *d* gene in *Lolium perenne* and d_1, d_2, d_3, d_5 and an_1 genes in *Zea mays.* The dwarf mutants of *Zea mays* have been usefully used for physiological studies of gibberellin action. Out of the ten mutants studied, nine were simple recessives and one a simple dominant. Five have been found to respond to gibberellin treatment by normal growth, other five showing little or no growth response. None of these mutants respond to other growth regulators or to variations in the nutrient medium. The specificity of response of the responding mutants suggests that each gene is in some way controlling a different step in a native gibberellin pathway leading to a product necessary for normal growth. The non-responsive nature of the other five mutants to all known gibberellins suggests that there are other reasons for dwarfism than merely a gibberellin treatment. There is at present no evidence, however, for inhibitor accumulation to explain the dwarf habit of growth, at least for the gibberellin-responding dwarf mutants of *Zea mays.*

Light inhibition

Many plants exhibit inhibited growth following exposure to low intensity red radiation (540-695 mμ), and this inhibition is reversed by subsequent exposure to far-red radiation (695-800 mμ). Lockhard (1956) reported that the red light induced inhibition of epicotyl elongation in etiolated seedlings of *Pisum sativum* could be specifically reversed by gibberellin treatment. In subsequent years significance has been given to the fact that the growth of irradiated and fully etiolated plants converged to a common maximum at identical saturating dosages of gibberellin, thus the ability of the plant to respond to gibberellin is dependent on its exposure to red light. These observations have led to the interpretation that the red light-induced inhibition of stem elongation is a direct consequence of a lowered availability of gibberelline-like substances but today no determinations have been reported of the amounts of gibberellin like substance from such system. This interpretation has, however, not met universal acceptance, since gibberellin and red light appear to act independently in certain other stem growth systems. In at least one plant, *Brassica alba* (*Sinapis alba*), stem elongation inhibited by red light has not been reversed by gibberellin A_3 treatment Gibberellin and red light both promote the germination of seeds of *Lactuca sativa.*

Seed and fruit growth

Parthenocarpy, induced by gibberellin has been reported in *Lycopersicum esculentum*, *Cucumis sativus* and *Solanum melongena.* In case of *Zephyranthes* the parthenocarpic fruit contained normal appearing seeds which were devoid of embryos. The growth of ovary and integuments were specific for gibberellins. Reduced grain yield has often been reported in bakanae disease as well as with exogenous gibberellin treatments. Hayashi *et al.* (1953) found that 2 mg/i GA reduced rice grain production 32 percent, although the yield of straw increased by 14 percent. Wittwer and Bukovac (1957) found gibberellin to be approximately 500 times more effective than IAA in inducing parthenocarpy of tomatoes. It appears from several examples that native gibberellins of the seed are one of the controlling factors for seed growth. Most direct information is available on the apparent level of native gibberellins in growing seed. Corcoran (1959) has shown that increases in extractable gibberellin-like substances from the seed occur during the growth of the seed and after the virtual completion of pericarp growth.

Detached plant parts

The use of detached plant parts is often advantageous in the study of gibberellin responses. Under such conditions the investigator is able to define more precisely the environment under which the material is grown. This has, however, the disadvantage in that the limiting growth requirements must be determined before the system can be studied.

Among the detached plant parts, (i) stem sections (ii) leaf sections and leaf disks, and (iii) tissue explants and tissue cultures are used. Gibberellin affects the growth of such plant parts in various ways and these studies have the obvious advantage of being able to demonstrate and analyze the interaction of the several growth factors which may be involved.

Chlorosis

Yellowing symptoms, typical of the bakanae disease are also produced by gibberellins. Fewer chloroplasts are present in diseased tissue and the chlorophyll decreases Gibberellin treatments also decrease the percent chlorophyll. It has been shown by the I.C.I group (1954) that increase nutrients markedly reduced gibberellin chlorosis. It would be really interesting to know which of the nutrients are most important.

Metabolism

It has often been reported that when gibberellins promote elongation, they do not always cause a parallel increase in dry weight. The I.C.I group has showed that the total dry weight of wheat and peas is increased by gibberellic acid treatment, provided that sufficient nutrients are available and the experiments cover a long enough period. They also emphasized that redistribution of material from the root to the shoot must be taken into account. The increased dry weight was, however, traced to more carbon fixation and the largest carbon changes involved carbohydrates. The nature of the dry weight changes as effected by gibberellins is as yet a complex one.

The Tokyo group has indicated a decrease in total sugars, dextrin, reducing sugar, and sucrose intreated rice plants. On the other hand, the British group reported marked increase in sucrose, fructose, and glucose in wheat; in peas only glucose was affected.

No starch was found and it was suggested that protein or cellulose must increase.

Recently Hayashi et al. (1956) have examined changes of enzymatic activity with time during the development of the seventh leaf of rice. Phosphatase, alkaline, pyrophosphatase, acetylesterase, maltose, β-glucosidase, α-galactosidase amylase, urease dipeptidase, ascorbidc acid oxidase, and catalase decreased on a fresh weight basis more rapidly than in non-treated controls, but peroxidase and invertase increased more rapidly than controls and gibberellin was shown to have no effect on their action. However, these enzymatic changes did not differ in any other way from the controls except for the fact that the changes were more rapid and more marked.

The effect of gibberellin on plant respiration has been little studied. Hayashi (1940) found no effect of gibberellin on yeast fermentation. Kato (1956) recorded a 15 percent increase in oxygen consumption in growing pea stem sections, the promotion was not found in non-growing tissue. He (1957) also found that cyanide, arsenite, and p-chloromercuribenzoate inhibited gibberellin-induced pea stem elongation in low concentrations whereas copper enzyme poisons did not. He thus concluded that gibberellin did not change the terminal oxidase of the tissue.

Relationship of Gibberellin to Auxin

Although gibberellins are similar to auxins in that both promote cell elongation, it is apparent that they differ from previously known auxins in several ways. They do not in short term experiments inhibit root growth, they inhibit rather than promote root initiation, they do not cause typical epinasty nor callus formation, their action on leaf growth or dwarfism is not paralleled by auxin, and they are more potent than auxin in inducing parthenocarpy. Several other properties of auxins have been compared with gibberellin. Brian et al. (1955) could find no effect of gibberellic acid on water uptake by potato tuber disks, nor did it cause any delay of leaf abscission, also it broke potato dormancy. Perhaps the most striking of all the differences is the evident indifference to light of the gibberellin effect. So there is good reason to believe that gibberellins are not auxins in the true sense of the term.

Kinetin and Cytokinins

Up to now, we have been concerned with growth hormones, synthetic and natural, whose primary function is to promote cell elongation. We have mentioned that IAA and gibberellin do promote an increase in the number of cells in certain circumstances, but this is the exception rather than the rule. The only growth regulator we have discussed so far, concerned primarily with cell division, was traumatic acid (wound harmone). However, in addition to traumatic acid, there exists in plants several compounds capable of promoting cell division. For example, coconut milk was found to be very active as a stimulation of cell division by the van Overbeek et al. (1941).

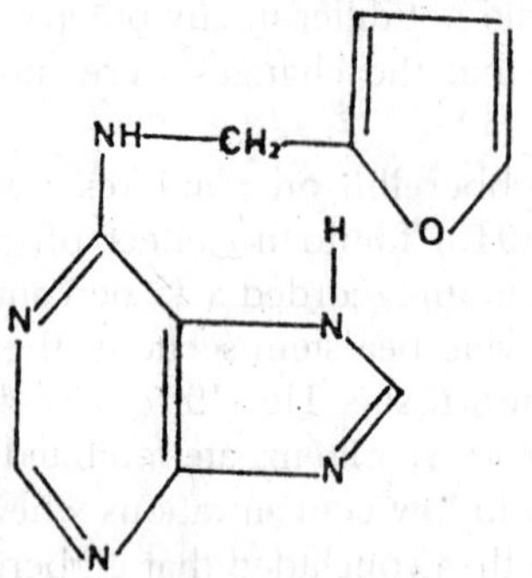

Fig. 12.12. Kinetin (6-furfuryl aminopurine).

Subsequently, this finding was supported by several investigations on a variety of plant tissues, confirming that coconut milk is, indeed an active promoter of cell division. Perhaps the most exciting discovery in the search for compounds that will induce cells to multiply is *kinetin* (6- furfury-laminopurine), a compound isolated from yeast DNA by Miller et al. (1955). Actually, kinetin is formed from deoxyadenosine, a degradation product of DNA and, as such, cannot be considered natural product of plant biosynthesis. Subsequent to its discovery, many analogs of kinetin, active in promoting cell division, were synthesized. In order to group this type of substance under one title, the generic term *cytokinin* has been given to all compounds with biological activity similar to that of kinetin. As might be expected cytokinins have widerspread occurrence in the plant world.

Substances having cytokinin activity have been extracted from about 40 species of higher plants, and in most cases, the actively dividing tissues of the plants have proven to be the best sources. Evidence is also mounting for the presence of cytokinins in micro-

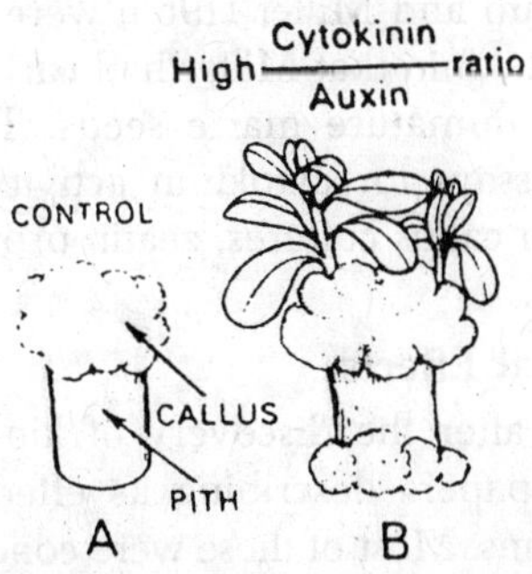

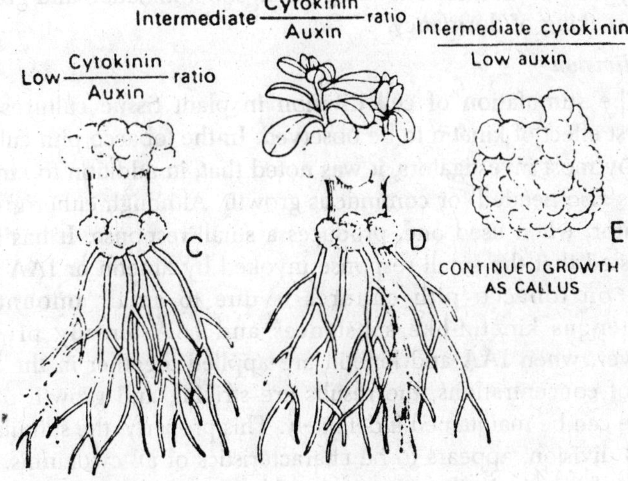

Fig. 12.13. The effect of changing ratio of cytokinin and auxin on the differentiation of buds and roots on tobacco pith cultures.

organisms. Despite the fact that many, if not all, plants contain cytokinins, it was almost 10years after discovery of kinetin before the chemical composition and properties of a natural cytokinin were described. Miller (1961) managed to extract and bring to a high state of purity a cytokinin from immature maize seeds. Attempts to crystalize and characterize the compound were unsuccessful. However, Letham (1963 : 1967) successfully extracted and purified to crystalline form a cytokinin from sweet corn. The naturally occurring cytokinin was named *zeatin.* Later, in a joint study, Letham and Miller (1965) were able to isolate in crystalline form the cytokinin that Miller had earlier isolated (in noncrystalline form) from immature maize seeds. The cytokinin proved to be zeatin. In assays for cytokinin activtty such as carrot root tissue and soybean callus cultures, zeatin proved to be much more active than kinetin.

Physiological Effects

Shortly after the discovery of kinetin, there followed a great number of papers describing its effects on many different plant growth systems. Most of these were concerned, directly, or indirectly, with kinetin's ability to pomote cell division and cell enlargement. We will discuss the effect of kinetin on cell division, cell enlargement, root initiation and growth, shoot initiation and growth, and breaking of dormancy.

Cell division

The stimulation of cell division in plant tissue cultures was the first effect of kinetin to be observed. In the tobacco pith cultures used by most investigators, it was noted that, in addition to kinetin. IAA is also needed for continuous growth. Although either growth regulator, when used one, produces a small response. It has been suggested that the small response invoked by kinetin or IAA used alone on tobacco pith cultures is due to small amounts of endogenous kinetin-like substances and IAA already present. however, when IAA and kinetin are applied together in the right ratio of concentrations, the results are striking and growth of the culture can be maintained indefinitely. This property, the stimulation of cell division, appears to be characteristics of all cytokinnis. The ability of kinetin in the presence of IAA to promote cell division.

In order that cell division may take place, an ordered sequence (DNA synthesis, mitosis, and cytokinesis) must take place. Is there a specific influence of IAA or cytokinin alone on any step in this sequence? The answer is apparently, yes Das *et al* (1956) found that both IAA and kinetin, when used alone, stimulate DNA synthesis in tobacco pith cultures. The above authors also found that both growth regulators are needed for mitosis , although IAA appears to dominate in this step. In addition, they suggested that when either kinetin or IAA is present in high concentration, the other may become limiting for at least one of the three steps needed to complete cell division. In a later paper, these authors stated that of the three steps in cell division, IAA is involved in the first two (DNA duplication and mistosis), but that the last step (cytokinesis) is controlled by kinetin. Here again, as with our discussions on gibberellins and IAA, we are presented with the importance of balance between growth hormones in plant growth and development.

How a cytokinin induces cell division is still an unsolved question. The adenine moiety of the cytokinin molecule appears essential for this process, many different substituted side chains being applicable. Strong (1958) has suggested that the side chain may influence some physical property (such as solubility) bearing on the efficiency of the growth regulator to induce cell division.

Cell enlargement

Not only do cytokinins promote cell division, but they also induce cell enlargement, an effect usually associated with IAA and gibberellin. Treatment of leaf discs cut from etiolated leaves of *Phaselous vulgaris* (bean) with kinetin causes sigificant cell enlargements. This effect of kinetin can occur in the absence of IAA. Cell enlargement after kinetin treatment has also been observed in tobacco pith cultures, tobacco roots, and in excised artichoke tissue. Stimulation of cell enlargement by cytokinins other than kinetin has also been observed. Since cytokinin-induced cell enlargement has been clearly shown, cytokinins should not be considered solely as cell division factors.

Root initiation and growth

Although relatively few studies have been made of cytokinin effects on the root system, it appears that cytokinins are able both to stimulate and to inhibit root initiation and development. Kinetin in the presence of casein hydrolysate and IAA stimulates root initiation and development in tobacco stem callus cultures. Increases in dry weight and elongation of the roots of lupin seedling were found by Frics (1960) to be promoted by kinetin. It can be seen that all concentrations of kinetin increase the dry weight of the root system even though at the higher concentration root elongation is inhibited.

In excised pea root segments, lateral root development is slightly stimulated by low concentrations of kinetin (5×10^{-5} M). At higher concentrations, however, kinetin is inhibitory. There is some evidence that interaction between cytokinins and auxin may influence the site of lateral root initiation. Bonnett and Torrey (1965), for example, demonstrated that by applying auxin and cytokinin at different concentrations to the opposite ends of excised root segments of the common bind weed (*Convolvulus*) they could alter the site of lateral root formation.

Shoot initiation and growth

In the original work with tobacco callus cultures and kinetin, it was found that callus tissue can be kept in an undifferentiated state as the proper balance of IAA and kinetin is maintained. However, if the ratio of kinetin to IAA is increased, either by the addition of more kinetin or the use of less IAA, leafy shoots are initiated. Torrey (1958) observed that kinetin initiated bud primodia on root segments of *Convolvulus arvensis* (bindweed), this effect being much more marked when the segments were grown in the dark.

Five day old bean seedlings soaked in kinetin solutions and then allowed to grow for an additional 46 hours respond with an increase in fresh weight of the epicotyl, increase in leaf expansion, and increase in the elongation of stems and petioles. In a recent study by Skoog et al (1967), auxin-cytokinin interaction in regulating growth and organ formation in tobacco callus cultres has been remarkably illustrated. The natural cytokinin, 6 (γ-γ-dimehthylallyamino) purine, is much more active than kinetin. This natural cytokinin has been found as a constituent of sRNA in yeast, corn, pea, and spinach. There have been several additional demonstrations of cytokinin induced promotions of shoot initiation and growth. However, the above studies serve to illustrate that cytokinins are active in the initiation and development of the aerial portions of the plants.

Breaking of dormancy

Earlier, we discussed apical dominance, the inhibition of lateral bud growth by auxin emanating from the apical bud. The controlling features of this phonomenon are not clearly understood and may involve not only IAA but other factors, which may interact with the auxin. This has been suggested in a study by Wickson and Thimann (1958) on the interaction of IAA and kinetin in apical dominance. They found that the growth of the lateral buds of pea stem sections in culture solutions containing IAA is inhibited, as might be expected. The growth of lateral buds on stem sections in nutrient solutions not contain in IAA, of course, is unihibited. However, the addition of kinetin along with IAA stimulates the growth of these buds. Kinetin alone has little effect. These investigators also demonstrated that the effect of kinetin on apical dominance can also be observed in entire shoots; that is, the apical bud is present. They found, as in the classical studies of apical dominance, that removal of the apical bud. stimulates the growth of the lateral buds. On the other hand, if the apical bud remains, the lateral buds are completely inhibited. However, if the instance shoot is soaked in a kinetin solution, inhibition of the lateral buds by the apical bud is overcome to a large extent. In addition to the above study, there have been several other investigations demonstrating the stimulatory influence of cytokinins on lateral bud growth. It appears that apical dominance may be controlled by a balance of concentrations between endogenous kinetin-like substance and IAA.

Other Physiological Effects

It is a well-known fact that the germination of lettuce seeds (*Lactuca sativa*) may be stimulated by red light and inhibited by infrared (far red) light. Also, the growth of bean leaf discs is sensitive to red and far red light treatment, being stimulated by red light and inhibited by far-red light treatment. In both of the above situations, the effect of kinetin treatment is similar to that of red light treatment. Only in one respect, does it differ. The stimulatory effects of kinetin are in no way inhibited by subsequent far-red treatment as in the case of red light treatment germination of seeds of white clover and carpet grass has also been observed to respond positively to kinetin treatment.

Table 12.3. Effect of kinetin and red and far-red irradiation on growth of bean leaf discs during a 48-hour growth period.

Conc.of kinetin M	*Light treatment*	*Increase in diam., mm*
0	none	1.05 ± 0.04 3
5×10^{-5}	none	2.48 ± 0.03
0	5 min red	2.48 ± 0.08
0	5 min far red	1.01 ± 0.06
0	5 min red and then 5 min far red	1.17 ± 0.07
5×10^{-5}	5 min far red	2.49 ± 0.08

Table 12.4. Effect of kinetin and red and far- red irradiation on germination of Grand Rapids lettuce seeds during a 72-hour period.

Cone of kinetin, M	*Light treatment*	*Germination %* *Expt.1*	*Expt.2*
0	none	8	7
5×10^{-5}	none	84	86
0	8 min red	96	96
0	5 min red and then		
	8 min far red	5	7
5×10^{-5}	8 min far red	86	83

Cytokinins are not only necessary factors in the growth and development of higher plants but they also markedly influence the growth of certain microorganisms. Kinetin, for example, can influence the growth of viruses, bacteria, fungi and algae all of which suggests that cytokinins are also present as natural components of lower plants. Indeed, crude extracts of natural cytokinins have been prepared from at least two micro-organisms. It is also quite evident that natural cytokinins interact with endogenous IAA to affect the growth and development of the plant. Again, we are presented with evidence supporting the concept that the growth and development of a plant is under the influence of delicate chemical balances maintained by the presence or absence of interaction between growth regulators.

INDEX